PROJECT MANAGEMENT FOR THE OIL AND GAS INDUSTRY

A WORLD SYSTEM APPROACH

Industrial Innovation Series

Series Editor

Adedeji B. Badiru

Department of Systems and Engineering Management
Air Force Institute of Technology (AFIT) – Dayton, Ohio

PUBLISHED TITLES

Carbon Footprint Analysis: Concepts, Methods, Implementation, and Case Studies,
  Matthew John Franchetti & Defne Apul

Computational Economic Analysis for Engineering and Industry, *Adedeji B. Badiru &*
  Olufemi A. Omitaomu

Conveyors: Applications, Selection, and Integration, *Patrick M. McGuire*

Global Engineering: Design, Decision Making, and Communication, *Carlos Acosta, V. Jorge Leon,*
  Charles Conrad, and Cesar O. Malave

Handbook of Industrial Engineering Equations, Formulas, and Calculations, *Adedeji B. Badiru &*
  Olufemi A. Omitaomu

Handbook of Industrial and Systems Engineering, *Adedeji B. Badiru*

Handbook of Military Industrial Engineering, *Adedeji B.Badiru & Marlin U. Thomas*

Industrial Control Systems: Mathematical and Statistical Models and Techniques, *Adedeji B. Badiru,*
  Oye Ibidapo-Obe, & Babatunde J. Ayeni

Industrial Project Management: Concepts, Tools, and Techniques, *Adedeji B. Badiru, Abidemi Badiru,*
  & Adetokunboh Badiru

Inventory Management: Non-Classical Views, *Mohamad Y. Jaber*

Kansei Engineering - 2 volume set
  • Innovations of Kansei Engineering, *Mitsuo Nagamachi & Anitawati Mohd Lokman*
  • Kansei/Affective Engineering, *Mitsuo Nagamachi*

Knowledge Discovery from Sensor Data, *Auroop R. Ganguly, João Gama, Olufemi A. Omitaomu,*
  Mohamed Medhat Gaber, & Ranga Raju Vatsavai

Learning Curves: Theory, Models, and Applications, *Mohamad Y. Jaber*

Modern Construction: Lean Project Delivery and Integrated Practices, *Lincoln Harding Forbes &*
  Syed M. Ahmed

Moving from Project Management to Project Leadership: A Practical Guide to Leading Groups,
  R. Camper Bull

Project Management: Systems, Principles, and Applications, *Adedeji B. Badiru*

Project Management for the Oil and Gas Industry: A World System Approach, *Adedeji B. Badiru &*
  Samuel O. Osisanya

Quality Management in Construction Projects, *Abdul Razzak Rumane*

Social Responsibility: Failure Mode Effects and Analysis, *Holly Alison Duckworth &*
  Rosemond Ann Moore

Statistical Techniques for Project Control, *Adedeji B. Badiru & Tina Agustiady*

STEP Project Management: Guide for Science, Technology, and Engineering Projects, *Adedeji B. Badiru*

Systems Thinking: Coping with 21st Century Problems, *John Turner Boardman & Brian J. Sauser*

Techonomics: The Theory of Industrial Evolution, *H. Lee Martin*

Triple C Model of Project Management: Communication, Cooperation, Coordination, *Adedeji B. Badiru*

PROJECT MANAGEMENT FOR THE OIL AND GAS INDUSTRY

A WORLD SYSTEM APPROACH

ADEDEJI B. BADIRU
SAMUEL O. OSISANYA

CRC Press
Taylor & Francis Group
Boca Raton London New York

CRC Press is an imprint of the
Taylor & Francis Group, an **informa** business

CRC Press
Taylor & Francis Group
6000 Broken Sound Parkway NW, Suite 300
Boca Raton, FL 33487-2742

Printed in the United States of America on acid-free paper
Version Date: 20121115

International Standard Book Number: 978-1-4200-9425-1 (Hardback)

Visit the Taylor & Francis Web site at
http://www.taylorandfrancis.com

and the CRC Press Web site at
http://www.crcpress.com

Dedicated to the loving memory of Augusta Adeola Osisanya,

who taught us all the value of life well lived

Contents

Preface

Lesson learned should be lesson practiced.

Oil and gas project management refers to the unique requirements of managing science, technology, and engineering aspects of projects in the oil and gas industry. This book presents step-by-step application of project management techniques to managing oil and gas projects. The book uses the Project Management Body of Knowledge (PMBOK®) framework from the Project Management Institute (PMI) as the platform for the topics covered. The book also strongly advocates a world systems approach to managing oil and gas projects and programs. Both quantitative and qualitative techniques are covered in the book. Thus, the book addresses technical and managerial aspects of projects in the oil and gas industry. Technical project management is the basis for sustainable national advancement, which often depends on the development of the oil and gas industry. As such, managing oil and gas projects effectively is essential for national economic vitality. Project management is the process of managing, allocating, and timing resources to achieve a given goal in an efficient and expeditious manner. The objectives that constitute the specified goal may be in terms of time, costs, or technical results. A project can range from the very simple to the very complex. Due to its expanding utility and relevance, project management has emerged as a separate body of knowledge that is embraced by various disciplines ranging from engineering and business to social services. Project management techniques are widely used in many endeavors, including construction management, banking, manufacturing, engineering management, marketing, health care delivery systems, transportation, research and development, defense, and public services. The application of project management is particularly of high value in science, technology, and engineering undertakings, such as we have in the oil and gas industry. In today's fast-changing IT-based and competitive global market, every enterprise must strive to get ahead of the competition through effective project management in all facets of its operations.

Project management represents an excellent basis for integrating various management techniques such as statistics, operations research, Six Sigma, computer simulation, and so on within the oil and gas industry operations. The purpose of this book is to present an integrated approach to project management for oil and gas projects. The integrated approach covers the concepts, tools, and techniques (both new and tested) of project management. The elements of the PMBOK provide a unifying platform for the topics covered in the book. The book is intended to serve as a reference book for planners, project operators, stakeholders, designers, project managers, business managers, consultants, project analysts, senior executives, project team members, members of project management office, project customers, functional managers, trainers, and researchers. It can also serve as a guidebook for technical consultants and as a textbook resource for students and educators. It is also useful as a supplementary reading for practicing engineers and as a handbook for field operators. It will appeal to petroleum industry professionals because of its focused treatment of oil and gas projects.

The book also contains case examples of applications of project management tools and techniques to real-life project scenarios. These could serve as lessons learned for best practices in project management.

Adedeji Badiru
Samuel Osisanya

Authors

Adedeji B. Badiru is professor and head of Systems & Engineering Management at the Air Force Institute of Technology. He was previously professor and department head of Industrial & Information Engineering at the University of Tennessee in Knoxville. Prior to that, he was professor of industrial engineering and dean of University College at the University of Oklahoma. He is a registered professional engineer (PE), a certified project management professional (PMP), a fellow of the Institute of Industrial Engineers, and a fellow of the Nigerian Academy of Engineering. He holds a BS in industrial engineering, MS in mathematics, and MS in industrial engineering from Tennessee Technological University, and PhD in industrial engineering from the University of Central Florida. His areas of interest include mathematical modeling, project modeling and analysis, economic analysis, systems engineering, and efficiency/productivity analysis and improvement. He is the author of 22 books, 34 book chapters, 65 technical journal articles, 107 conference proceedings and presentations. He also has 24 magazine articles and 20 editorials and periodicals. He is a member of several professional associations of several scholastic honor societies. Professor Badiru has won several awards for his teaching, research, and professional accomplishments. He is the recipient of the 2009 Dayton Affiliate Society Council Award for Outstanding Scientists and Engineers in the Education category with a commendation from the 128th Senate of Ohio. He also won 2010 IIE/Joint Publishers Book-of-the-Year Award for coediting *The Handbook of Military Industrial Engineering* (CRC Press, 2009). He also won 2010 ASEE John Imhoff Award for his global contributions to industrial engineering education, the 2011 Federal Employee of the Year Award in the Managerial Category from the International Public Management Association, Wright Patterson Air Force Base, the 2012 Distinguished Engineering Alum Award from the University of Central Florida, and the 2012 Medallion Award from the Institute of Industrial Engineers for his global contributions in the advancement of the profession. He has served as a consultant to several organizations around the world including Russia, Mexico, Taiwan, Nigeria, and Ghana. He has conducted customized training workshops for numerous organizations including Sony, AT&T, Seagate Technology, U.S. Air Force, Oklahoma Gas & Electric, Oklahoma Asphalt Pavement Association, Hitachi, Nigeria National Petroleum Corporation, and ExxonMobil. He has won several awards for his teaching, research, publications, administration, and professional accomplishments. He holds a leadership certificate from the University of Tennessee Leadership Institute. Professor Badiru has served as a technical project reviewer, curriculum reviewer, and proposal reviewer for several organizations including The Third-World Network of Scientific Organizations, Italy, National Science Foundation, National Research Council, and the American Council on Education. He is on the editorial and review boards of several technical journals and book publishers. Professor Badiru has also served as an industrial development consultant to the United Nations Development Program. He is also a program evaluator for ABET. In 2011, Professor Badiru led a research team to develop analytical models for Systems Engineering Efficiency Research (SEER) for the Air Force acquisitions integration office at the Pentagon. Professor Badiru has diverse areas of avocation. His professional accomplishments are coupled with his passion for writing about everyday events, interpersonal issues, and socially responsible service to the community. Outside of the academic realm, he writes self-help books, motivational poems, editorials, and newspaper commentaries, as well as engaging in paintings and crafts.

Samuel O. Osisanya is a professor in the Mewbourne School of Petroleum and Geological Engineering at the University of Oklahoma in Norman, Oklahoma, where he teaches drilling engineering, well completion and stimulation, production and reservoir engineering, horizontal well technology, and emerging technologies. The students selected him as the 1994, 2002, and 2004 outstanding professor of the year and the 2006–2007 Most Inspiring Faculty of the Student Athletes of University of Oklahoma. Formerly, Dr. Osisanya was an assistant professor at Montana Tech University, where he taught nodal analysis and surface production operations; visiting lecturer at the University of Ibadan, Nigeria from 1980 to 1983. He was a graduate teaching assistant at the University of Texas at Austin from 1986 to 1991, where he won UT College of Engineering 1989–1990 Engineering Foundation TA/AI Award for Meritorious Service in Engineering Teaching. Dr. Osisanya has 8 years of industrial experience with Dresser Magcobar, Gulf (now Chevron), Shell-BP, and ExxonMobil, where he worked as an engineer in various categories and rose to the level of a drilling foreman. Dr. Osisanya's technical interests include wellbore stability in shale formations, drilling optimization, cementing, well completion, and production engineering. National Science Foundation (NSF) Faculty Early Career Development (CAREER) award Winner in 1995 ($325,000). In 2000, Dr. Osisanya was selected by the United States National Science Foundation (NSF) as one of more than 250 NSF award winning scientists, engineers, and Nobel prize winners in the United States to visit middle school students, as part of the NSF 50th Anniversary celebration. He has authored and coauthored more than 84 technical papers in *Society of Petroleum Engineers (SPE), Journal of Canadian Petroleum Technology (JCPT), American Society for Engineering Education (ASEE),* other journals, and numerous reports. He was a member of the SPE Well Completion Technical Committee (1997–1999) and faculty advisor to the University of Oklahoma SPE Student Chapter (1995–1999). He was a member on the SPE Committee on Drilling Design and Management from 2003 to 2006. He is listed in Who's Who Among America's Teachers (1996). He is selected as SPE Distinguished

Lecturer for 2011–2012 program. He holds a BS (Summa Cum Laude) from University of Ibadan, Nigeria, MS and PhD from the University of Texas at Austin all in Petroleum Engineering. He is a registered professional engineer in Texas and a member of SPE of AIME, American Association of Drilling Engineers (AADE), and American Society of Engineering Education (ASEE). Dr. Osisanya has taught several in-house short courses for several countries and organizations, including ExxonMobil in Nigeria, Nigerian National Petroleum Corporation, Halliburton in the United States and Nigeria, and Qatar General Petroleum Corporation, Sonatrach of Algeria, PDVSA of Venezuela, Pertamina National Oil Company of Indonesia, AGIP of Italy, Sidanco & Sibneft Oil Companies of Russia, Kuwait Petroleum Corporation, TNK-BP Russia, VietSovPetro, Next-Schlumberger, Seminarium of Ecuador, Total of Nigeria, Chevron of Nigeria, Saudi Aramco of Saudi Arabia, UNIStrategic of Singapore, PetroGroup of Colombia, ZADCO of Abu-Dhabi, Repsol of Spain, Petrom of Romania, PEMEX of Mexico, and Vankoneft of Russia.

1

World Systems Framework for Oil and Gas Industry

A systems view of the world is what is required to keep industry well oiled.

Energy is critical to the success and advancement of the society. Energy supply is needed for heating, cooling, and lighting our homes, business, and industry. Modern communication facilities in the society function only when consistent energy supply is assured. Since all our modern commercial activities are highly interlinked, we must view the whole as a system of systems (SoS) of economic activities. Lorincz (2008) reminded us that the "thirst for oil and gas won't be satisfied any time soon." This is a sentiment that is shared by most people both within and outside the oil and gas industry. It is forecasted that the need to find replacement supplies of oil and gas to match the global demand for energy will continue to encourage massive capital spending in the industry around the world. Much of such investments will be directed at executing multinational projects. Based on the rapid growth of the economies of China and India, the long-term outlook for the oil and gas industry is quite positive. We just have to ensure that projects are planned, organized, controlled, and executed constructively through global alliances. The international market price of crude oil is the most crucial factor determining the consumer-level price for petroleum-based products. This means that the price of gasoline at the pump is mainly determined by the worldwide demand for and supply of crude oil. According to estimates from the International Energy Agency (IEA),* the world's primary energy needs are expected to grow by 55% between 2005 and 2030, with the demand reaching 17.7 billion tonnes of oil equivalent, compared with 11.4 billion tonnes in 2005. The current market of 2012–2013 is already experiencing being on pace with that demand curve. The challenges that oil and gas projects will continue to face fall in the following categories:

- Technical challenges
- Managerial challenges
- Human resource challenges

Rigorous and disciplined applications of project management tools and techniques can help mitigate these challenges.

---

* http://www.iea.org/stats/index.asp; accessed April 30, 2012.

A Multitiered Systems View

It is a systems world. To be successful, all industries must consider a systems view of their options. This is particularly most critical for the oil and gas industry, where many multifaceted factors are interlinked. If you do not think that everything is globally connected, consider this fact: A Boeing 747-400 has six million parts (half of which are fasteners) made in 33 different countries.

This shows that the fates of industries are interconnected throughout the entire world in terms of production and consumption activities, particularly in the oil and gas sector (Badiru, 2009). This makes it mandatory that we take a systems view of the world when dealing with oil and gas projects. With one-third of the electricity produced on Earth being used to power electric light bulbs, we must all work together to pursue more efficient practices. The first quarter of 2012 saw a significant rise in global crude oil prices. After closing 2011 at $108 per barrel, Brent spot prices exceeded $125 per barrel by early March, from whence it traded within a relatively narrow range for the remainder of the month. This price increase reflected changes in global oil supplies, as significant unplanned disruptions in production from countries that are not members of the Organization of the Petroleum Exporting Countries (OPEC) countered the recovery in Libyan production. Project management must address these types of developments in day-to-day operations in the oil and gas industry. According to an assessment by the U.S. Geological Survey (USGS),* excluding the United States, the world holds an estimated 565 billion barrels (bbo) of undiscovered, technically recoverable conventional oil; 5606 trillion cubic feet (tcf) of undiscovered, technically recoverable conventional natural gas; and 167 bbo of undiscovered, technically recoverable natural gas liquids (NGL). The report includes mean estimates of undiscovered but technically recoverable conventional oil and gas resources in 171 geologic provinces of the world. These estimates include resources beneath both onshore and offshore areas. All these numbers represent technically recoverable oil and gas resources, which are those quantities of oil and gas producible using currently available technological and industrial practices, regardless of economic or accessibility considerations. This assessment does not include reserves—accumulations of oil or gas that have been discovered, are well defined, and are considered economically viable. To tap into the proven and estimated oil and gas reserves, many countries must come together in systems-based synergistic national alliances. A systems view of the energy world is, thus, required.

This book uses the project framework view from the field of industrial and systems engineering in consonance with the fundamentals of petroleum engineering. Industrial engineering deals with the *design and development of integrated systems of people, machines, and information resources for producing*

---

* http://www.doi.gov/news/pressreleases/USGS; released April 18, 2012.

products and services. This is precisely what happens in the oil and gas industry because of the multifaceted requirements of the industry. Apart from the nuclear projects, no other projects generate as much environmental sentiments as projects in the oil and gas industry. Thus, formal project management should be an essential part of planning, organizing, scheduling, and controlling projects in the industry. Those who support or oppose oil and gas projects are often uninformed about the merits and demerits of the projects. The techniques of project management can help mitigate such awareness problems. There is a correlation between the application of project management and better business performance. For this reason, a framework based on an SoS modeling is very essential for the oil and gas industry. There are three closely linked global challenges:

1. Increasing energy demand
2. Environmental and climate change
3. National and global security

Attendant with these challenges are the related concerns of science and technology developments, management of research and development, and global economic advancement. The major activities of an oil and gas company revolve around the following elements:

- Exploration
- Drilling and production
- Transportation
- Refining
- Marketing

Each of these requires a formal and rigorous application of project management. For example, exploratory drilling does the following:

- Establishes that hydrocarbons exist
- Determines the quality; oil and gas ratio
- Establishes the extent of the reservoir
- Conducts an economic value of the resource
- Designs a development plan; how the reservoir is to be developed for maximum recovery

Drilling and production require project management for

- Designing platform
  - Substructure, top side facilities

- Fabrication
- Installation
- Drilling (early production)
  - Environmental issues
- Production (separate oil, gas, and water)
- Transportation
  - Pipelines
  - Off load directly onto tankers

Drilling economics focuses on

- Optimum drilling practices
  - Cost per foot
  - Break-even curve

Multinational projects particularly pose unique challenges pertaining to reliable power supply, efficient communication systems, credible government support, dependable procurement processes, consistent availability of technology, progressive industrial climate, trustworthy risk mitigation infrastructure, regular supply of skilled labor, uniform focus on quality of work, global consciousness, hassle-free bureaucratic processes, coherent safety and security system, steady law and order, unflinching focus on customer satisfaction, and fair labor relations. Assessing and resolving concerns about these issues in a step-by-step fashion will create a foundation of success for a large project. While no system can be perfect and satisfactory in all aspects, a tolerable trade-off on the factors is essential for project success. Figure 1.1

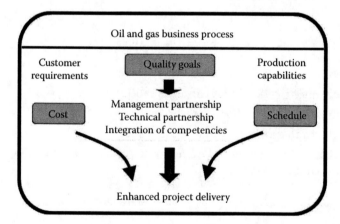

FIGURE 1.1
The oil and gas business process.

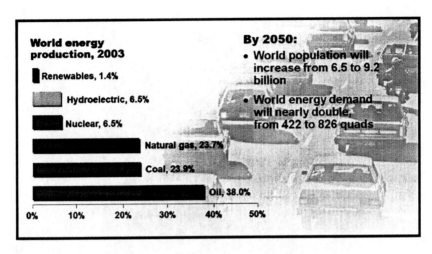

FIGURE 1.2
World energy production and projection of demand.

shows an illustration of the oil and gas business environment and the typical factors impinging upon the industry.

Figure 1.2 shows the projection of energy demands as the world population grows. The statistics show that oil and gas constitute a major portion of the world's energy production. The increasing demand for energy has led to furious searches for alternate sources as well as intensification of the exploitation of the existing sources. Figure 1.3 suggests using current technology

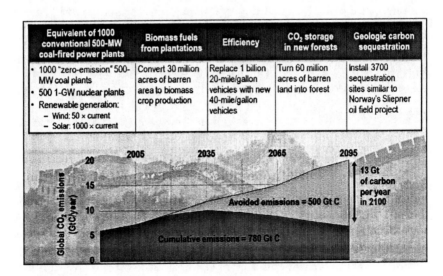

FIGURE 1.3
Using current technology to mitigate climate change due to carbon emission.

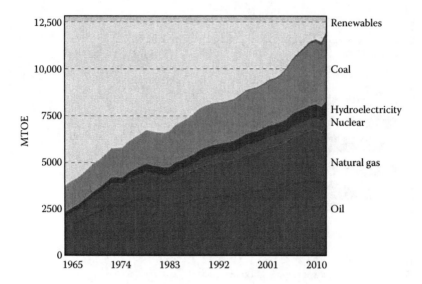

FIGURE 1.4
World energy consumption as of 2010 in millions of tons of oil equivalent. (Based on data from BP Statistical Review of World Energy, 2011.)

to mitigate the climate change problems. Fortunately, the dire energy situation often creates new innovation and creativity that can facilitate a rapid convergence of technological, political, economic, and managerial processes. Figure 1.4 shows the distribution of world energy consumption by sources as of 2010 in millions of tons of oil equivalent. As in the case of production, the data show that the oil and gas industry is a major responder to world energy demands.

Some Case Examples

Some case examples of the world system of the oil and gas industry are presented here. Figure 1.5 shows a typical small-scale "pecking" oil well on a private land in Oklahoma.

As the world's known oil and gas reserves are running out, the hunt for more is on, even in more remote places on Earth. One such area is the Barents Sea, off the northernmost tip of Norway and neighboring Russia. Both Norwegian and Russian authorities say that the potential for future fossil energy exploration in this region is quite enormous. Meanwhile, environmentalists warn of the danger to the fragile Arctic environment. Norway's StatoilHydro had won the right to join in developing the vast Shtokman gas field. This ended the long uncertainty over one of the world's

FIGURE 1.5
Private small-scale "pecking" oil well on private land in Oklahoma.

largest untapped deposits of oil and gas. Technical challenges abound in many of the exploration fields because they often exhibit diverse topographical, geographical, geological, and environmental properties with far-reaching social implications. For this reason, no oil and gas project can be undertaken in isolation. World system considerations must be brought to bear on all projects, using SoS approaches. The lessons learned from Arctic and subarctic operations can serve as starting templates for other projects around the world.

A large sinkhole that formed in south Texas's oil and gas region in 2008 is an example of the interrelationships of several factors dealing with human issues, technology availability, and economic development. The incident renewed questions about the effects of enormous volumes of barrels of saltwater injected into the ground each year as a by-product of oil and gas drilling. High energy prices have led to a surge in drilling across Texas and other states, much of it in older oil fields that tend to produce large volumes of saltwater along with crude oil. Meanwhile, new technologies for producing natural gas use millions of gallons of water to crack open gas-bearing rocks—yielding contaminated water that must then be disposed of, usually underground. This process, called fracking (hydraulic fracturing), has generated a lot of debate and concerns in many communities. Most notable in recent times is the case of fracking in Ohio. Hydraulic fracturing, known as fracking for short, involves injecting water, sand, and chemicals deep into the ground at high pressure to crack the shale and allow the oil or gas to flow out and be tapped. Some of the opponents of fracking claim that fracking is responsible for the frequent wave of earthquakes in the region. As a part of the overall project development, the full human and environmental impacts of fracking must be assessed. But at the same time, the economic development impacts must be recognized and factored into a multiattribute analysis. As energy companies snap up leasing rights in Wayne County (Ohio), some residents are optimistic, but environmentalists are worried. Only a full open-access analysis can allay the fears of all stakeholders.

In 2006, the Texas oil and gas industry injected 6.7 bbo of liquid, mostly water, beneath the ground, and experts say that the amount has been rising as new wells have multiplied and old wells are revived. Federal regulators, environmentalists, and community groups worry that lax oversight is allowing some of the water—which can be 10 times as salty as seawater and often contains oil, heavy metals, and even radioactive material—to escape from underground reservoirs. That could lead to the contamination of underground drinking water supplies, the pollution of soil and surface water, and more sinkholes as underground structures are eroded. Critics have argued that project oversights are lax or nonexistent. A comprehensive systems project management approach centered on communication, cooperation, and coordination may defuse many of the pending issues. Figure 1.6 emphasizes the interrelated challenges of energy generation, transmission, distribution, and consumption. Figure 1.7 shows an example of natural gas flaring, a practice common in some parts of gas-producing developing countries. Figure 1.8 illustrates the mega-sized structures and investments associated with energy ventures.

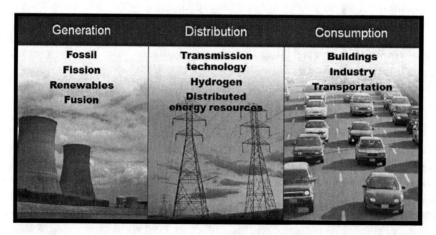

FIGURE 1.6
Interrelationships of energy generation, transmission, distribution, and consumption.

FIGURE 1.7
Natural gas flaring in a developing country.

FIGURE 1.8
Mega structures of oil and gas investments.

Energy Costs

According to a report by Pound (2010), in the United States, a state's energy portfolio can play a significant role in determining electricity prices. Figure 1.9 illustrates the three key factors that determine energy prices. These U.S. averages vary by region and market, depending on the age of the infrastructure and the fuel mix. Some of the least-expensive electricity is generated by existing coal and nuclear plants. Nonetheless, the cost of electricity from new power plants of any type is significantly higher until capital costs are recovered. When a state builds new power plants, it is likely to raise rates, regardless of the technology.

High prices encourage natural gas companies to focus on oil drilling. With oil prices above $100, even the most natural-gas-focused companies are increasing their oil drilling in the United States as a way to diversify. In recent years, U.S. energy companies have concentrated most of their domestic production efforts on natural gas as the biggest, easiest-to-get deposits of domestic crude oil are depleted. Some companies have intensified drilling for oil in Colorado and Texas, including in the Barnett Shale, which has a vast hydrocarbon reserve that had previously been known primarily for gas, not oil. The prices of both commodities have escalated drastically since the beginning of the year. However, the rise for oil has been steadier than that of gas in recent years. This is because natural gas is more difficult to transport and store. It is also more prone to seasonal fluctuations. Reports indicate that more than 80% of approximately 1,800 drilling rigs working in the United States in early 2012 were going after gas. Reports also indicate that the number of oil rigs working in the

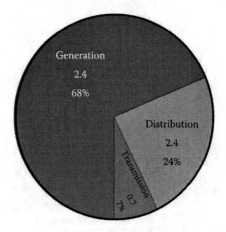

FIGURE 1.9
Major components of U.S. electricity prices (cents per kWh and share of total). (Based on data from U.S. Energy Information Administration, Annual Energy Outlook, 2010.)

United States was up 28% from 2011 to 2012 while gas-focused drilling rose by 1.4% over the same period. The Marcellus Shale, a layer of rock that stretches from upstate New York to eastern Ohio and as far south as West Virginia, may be a major source of natural gas in the future. The Marcellus Shale has been known for more than a century to contain gas, but it was generally not seen as economical to tap. Recent improvements in recovery technology, sharply higher natural gas prices, and strong drilling results in a similar shale formation in north Texas are changing the playing field of oil and gas exploration. It has been confirmed that parts of the country where energy supplies were long thought to be largely depleted are suddenly finding a new lease on exploration life cycle. The oil and gas industry is already aggressively buying mineral rights in Pennsylvania, where the Marcellus Shale appears to be thickest. It is estimated that more than 20 oil and gas companies will invest $700 million in 2012 in developing the Marcellus Shale. This bodes well for the application of systems-based project management tools and techniques. Another confirmation of the interconnectedness of markets is the 2012 report linking the sharp rise in the sale of pickup trucks to the increased activities in oil and gas explorations. Apparently, the caliber of those working or seeking employment in the oil and gas industry is in alignment with those who drive pickup trucks. The most likely type of vehicle to be spotted in the parking lot of oil and gas companies are pickup trucks, which is coveted in that industry for their rugged dependability.

Education and Workforce Development Issues

There is no other industry that is so global and aggressive in educational programs and workforce development as the oil and gas industry. The industry itself is large with operational tentacles reaching all parts of the world. The most aggressive nations pursuing oil and gas education and workforce development are the developing nations with newly discovered reserves. The resource-rich developing nations produce more oil and gas graduates than the United States and Europe. This large and steady supply of petroleum graduates from the developing nations is helping to mitigate the overall shortage of technical workforce for the industry. The nations are, thus, chipping away at the educational advantage typically held by Western countries. Traditionally, the top U.S. schools such as the University of Oklahoma and Texas A&M had provided most of the world's petroleum graduates. But in recent years, Brazil, Argentina, Nigeria, and other nations are making dramatic inroads into the production of petroleum graduates. While many of the new graduates are employed by Western oil and gas companies, many are also joining their nation-based companies. A key part of the education and skills of these foreign nationals should include training in the management fields, particularly project management, to buttress their technical assets. The educational infrastructure should be viewed and managed as a part of the overall world system of the oil and gas industry.

System of Systems Framework

This book advocates using an SoS framework so that all factors of importance can be included in the overall design and execution of projects in the oil and gas industry.

What Is a System?

A system is a collection of interrelated elements working together synergistically to achieve a set of objectives. The composite output of a system is greater than the sum of the individual outputs of its components. A systems view of a process facilitates a comprehensive inclusion of all the factors involved in the process.

What Is System of Systems?

SoS engineering is an interdisciplinary approach focusing on the synergistic integration of the elements making up a composite complex system, which is made up of other systems. Each subsystem of an SoS is a complete system in its own right. Some of the elements of an SoS include

- Self-regulation
- Self-adjustment
- Self-resolution
- Environmental sensitivity (i.e., interaction with prevailing operating conditions)
- Consolidation of capabilities
- Coordinated input–process–output relationships
- Structural interfaces between the systems within the SoS
- Composite equilibrium of the systems making up the SoS
- Value trading and transfers between the systems in the overall SoS

SoS facilitates an appreciation of what each system brings to the table in the overall assessment of the value created by the SoS. The oil and gas industry, as an SoS, will encompass such elements as the market system, the social system, the environmental system, the national development system, the national security system, the economic system, the regulatory system, the legislative system, the financial services system, and so on. Each of these examples is, itself, composed of smaller systems. Taking a world systems view of the oil and gas industry ensures that we appreciate all the components and factors that can impinge upon the success or failure of the industry. An SoS approach transforms the individual capabilities of the system entities into a collective capability to meet specific requirements as summarized below:

- Systems that are individually developed, managed, and operated function as autonomous components of one or more SoS.
- All systems within an SoS provide appropriate functional capabilities to each of the other elements in the SoS.
- The development and management of an SoS explicitly consider the political, financial, legal, technical, social, organizational, and operational characteristics of each element in the SoS.
- SoS recognizes and embraces the larger pool of stakeholders that results from bringing several systems together.
- SoS can accommodate changes to its conceptual, functional, physical, and spatial operating boundaries without negative impacts on its managerial processes.
- SoS exhibits a collective behavior representing dynamic interactions with the operating environment so that it can adapt and respond to meet or exceed requirements.

What Is a Project System?

Any project is essentially a collection of interrelated activities, people, technology, resources, work processes, and other assets brought together in the pursuit of a common goal. The goal may be in terms of generating one of the following:

1. Physical product (e.g., producing a maintenance spanner or a drill bit for oil and gas)
2. Service (e.g., providing accounting templates for oil and gas business)
3. Result (e.g., achieving a desired profit margin or meeting oil and gas production schedule)

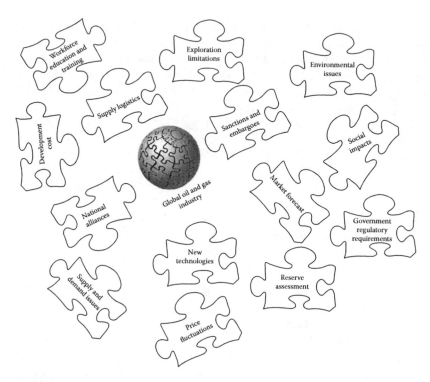

FIGURE 1.10
World systems view of the oil and gas industry.

Thus, a systems project management can be defined as follows:

Systems project management for the oil and gas industry is the process of using a systems approach to manage, allocate, and time resources to achieve enterprise-wide goals in an efficient and expeditious manner, considering all global ramifications.

Figure 1.10 illustrates a world systems view of the several factors and ramifications that affect the oil and gas industry. The most common basis for implementing an SoS view is the application of systems engineering.

What Is Systems Engineering?

Systems engineering is the application of engineering to solutions of a multifaceted problem through a systematic collection and integration of parts of the problem with respect to the life cycle of the problem. It is the branch of engineering concerned with the development, implementation, and use of large or complex systems. It focuses on specific goals of a system considering the specifications, prevailing constraints, expected services, possible behaviors, and structure of the system. It also involves a consideration of the activities required to assure that the system's performance matches the stated goals. Systems engineering addresses the integration of tools, people, and processes required to achieve a cost-effective and timely operation of the system.

Embracing Project Management

The rapid evolution of technology and environmental sensitivity in a dynamic global market creates increasing challenges for those who plan and execute oil and gas projects. Thus, project management tools and techniques are essential in the overall scheme of an oil and gas business. A coordinated application of systems engineering and project management can help increase the bottom line of the oil and gas industry. Systems engineering helps to identify an inclusive framework under which all the components would work together. Project management provides the mechanisms through which specific work and objectives within the framework would be accomplished. The following project definitions are essential:

Project: A project is traditionally defined as a unique one-of-kind endeavor with a specific goal that has a definite beginning and a definite end. The Project Management Body of Knowledge (PMBOK®) defines a project as a temporary endeavor undertaken to create a unique product, service, or result. Temporary means having a defined beginning and a definite end. The term "unique" implies that the project is different from other projects in terms of characteristics.

Project management: Badiru (2012) defines project management as the process of managing, allocating, and timing resources to achieve a given goal in an efficient and expeditious manner.

PMBOK defines project management as the application of knowledge, skills, tools, and techniques to project activities to achieve project objectives.

Identification of stakeholders: Stakeholders are individuals or organizations whose interests may be positively or negatively impacted by a project. Stakeholders must be identified by the project team for every project. A common deficiency in this requirement is that employees of the project organization are often ignored, neglected, or taken for granted during the identification of stakeholders. As the definition of stakeholders clearly suggests, if the interests of the employees can be positively or negatively affected by a project, then the employees must be viewed as stakeholders. For the oil and gas industry, in particular, the local community that can be impacted one way or the other should be included in the list of project stakeholders. A failure to do this adequately is the source of the global angst between energy-related industries and local communities around the world. For a successful overall outcome of a project, the local community should be considered as vital as the production staff of the oil and gas industry. A list of stakeholders in an oil and gas project may include all of the following:

- Customers
- Project owner
- Project sponsor
- Project operator
- Project financier
- Contractors and subcontractors
- Associated companies
- Local communities
- Project manager
- Project teams
- Industry shareholders

Figure 1.11 shows a hierarchical structure and interrelationships of some of the factors involved in the development of an oil and gas project. With an estimated $22 trillion of investment in supply infrastructure in the coming years, there will be recurring opportunities to apply the tools and techniques of project management.

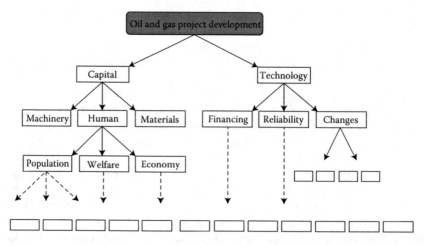

FIGURE 1.11
Hierarchical structure of factors in oil and gas project development.

The Gasoline Pump System

Since project management is found everywhere, all project participants will find the systems approach to be very beneficial. A good example of a systems design that we are all familiar with is the gas pump. Have you ever considered all the factors and requirements that go into the operation of a gas pump? The pump must exhibit a high level of reliability, dependability, and durability as it operates in a wide range of operating conditions, including extreme cold in winter and extremely high temperatures in the summer. For the pump to operate just right, all factors must interface perfectly as expected. The operation of that pump, as we have all come to expect, is similar to how a good project is supposed to operate.

Quantitative Systems Modeling

Classical control system focuses on control of the dynamics of mechanical objects, such as a pump, electrical motor, turbine, rotating wheel, and so on. The mathematical basis for such control systems can be adapted (albeit in iconical formats) for organizational management systems, including project management. This is because both technical and managerial systems are characterized by inputs, variables, processing, control, feedback, and output. This is represented graphically by input–process–output relationship block diagrams. Mathematically, it can be represented as

$$z = f(x) + \varepsilon, \tag{1.1}$$

where
  z = output
  $f(.)$ = functional relationship
  ε = error component (noise, disturbance, etc.)

For systems-based multivariable cases, the mathematical expression is represented as vector–matrix functions as shown below:

$$\mathbf{Z} = \mathbf{f}(\mathbf{X}) + \mathbf{E}, \tag{1.2}$$

where each term is a matrix. $\mathbf{Z}$ is the output vector, $\mathbf{f}(.)$ is the input vector, and $\mathbf{E}$ is the error vector. Regardless of the level or form of mathematics used, all systems exhibit the same input–process–output characteristics, either quantitatively or qualitatively. The premise of this book is that there should be a cohesive coupling of quantitative and qualitative approaches in managing a project system. In fact, it is this unique blending of approaches that makes systems application for project management more robust than what one will find in mechanical control systems, where the focus is primarily on quantitative representations.

Cost–Schedule–Quality Constraints

Systems management is the pursuit of organizational goals within the constraints of time, cost, and quality expectations. The iron triangle model depicted in Figure 1.12 shows that project accomplishments are constrained by the boundaries of quality, schedule, and cost. In this case, quality represents the composite collection of project requirements. In a situation where precise optimization is not possible, there will need to be trade-offs between these three factors of success. The concept of iron triangle is that a rigid triangle of constraints encases the project. Everything must be accomplished within the boundaries of schedule (time), cost (budget), and quality (performance requirements). If better quality is expected, a compromise along the axes of time and cost must be executed, thereby altering the shape of the triangle. The trade-off relationships are not linear and must be visualized in a multidimensional context. This is better articulated by a 3-D view of the system constraints as shown in the box. Scope requirements determine the project boundary and trade-offs must be done within that boundary. If we

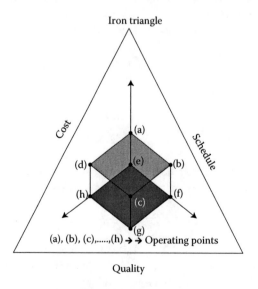

FIGURE 1.12
Project system constraints of cost, schedule, and quality.

label the eight corners of the box as (a), (b), (c), ..., (h), we can iteratively assess the best operating point for the project. For example, we can address the following two operational questions:

1. From the point of view of the project sponsor, which corner is the most desired operating point in terms of a combination of requirements, time, and cost?
2. From the point of view of the project executor, which corner is the most desired operating point in terms of a combination of requirements, time, and cost?

Note that all the corners represent extreme operating points. We notice that point (e) is the do-nothing state, where there are no requirements, no time allocation, and no cost incurrence. This cannot be the desired operating state of any organization that seeks to remain productive. Point (a) represents an extreme case of meeting all the requirements with no investment of time or cost allocation. This is an unrealistic extreme in any practical environment. It represents a case of getting something for nothing. Yet, it is the most desired operating point for the project sponsor. By comparison, point (c) provides the maximum possible for requirements, cost, and time. In other words, the highest levels of requirements can be met if the maximum possible time is allowed and the highest possible budget is allocated. This is an unrealistic expectation in any resource-conscious organization. You cannot get everything you ask for to execute a project. Yet, it is the most desired operating point for the project executor. Considering the two extreme points of (a) and (c), it is obvious that the project must be executed within some compromise region within the scope boundary.

System of Systems Value Modeling

A technique that can be used to assess the overall value-added components of a process improvement program is the systems value model (SVM), which is an adaptation of the manufacturing system value (MSV) model presented by Troxler and Blank (1989). The model provides an analytical decision aid for comparing process alternatives. Value is represented as a p-dimensional vector:

$$V = f(A_1, A_2, \ldots, A_p) \tag{1.3}$$

where $A = (A_1, \ldots, A_n)$ is a vector of quantitative measures of tangible and intangible attributes. Examples of process *attributes* are quality, throughput, capability, productivity, cost, and schedule. Attributes are considered to be a combined function of factors, x_1, expressed as

$$A_k(x_1, x_2, \ldots, x_{m_k}) = \sum_{i=1}^{m_k} f_i(x_i) \tag{1.4}$$

where $\{x_i\}$ is a set of m factors associated with attribute A_k ($k = 1, 2, \ldots, p$) and f_i is a contribution function of factor x_i to attribute A_k. Examples of factors include reliability, flexibility, user acceptance, capacity utilization, safety, and design functionality. Factors are themselves considered to be composed of indicators, v_i, expressed as

$$x_i(v_1, v_2, \ldots, v_n) = \sum_{j=1}^{n} z_j(v_j) \tag{1.5}$$

where $\{v_j\}$ is a set of n indicators associated with factor x_i ($i = 1, 2, \ldots, m$) and z_j is a scaling function for each indicator variable v_j. Examples of indicators are project responsiveness, lead time, learning curve, and work rejects. By combining the above definitions, a composite measure of the value of a process can be modeled as

$$V = f(A_1, A_2, \ldots, A_p)$$

$$= f\left\{ \left[\sum_{i=1}^{m_1} f_i \left(\sum_{j=1}^{n} z_j(v_j) \right) \right]_1, \left[\sum_{i=1}^{m_2} f_i \left(\sum_{j=1}^{n} z_j(v_j) \right) \right]_2, \ldots, \left[\sum_{i=1}^{m_k} f_i \left(\sum_{j=1}^{n} z_j(v_j) \right) \right]_p \right\} \tag{1.6}$$

where m and n may assume different values for each attribute. A subjective measure to indicate the utility of the decision maker may be included in the model by using an attribute-weighting factor, w_p to obtain a weighted *PV*:

$$PV_w = f(w_1 A_1, w_2 A_2, \ldots, w_p A_p) \tag{1.7}$$

where

$$\sum_{k=1}^{p} w_k = 1, \quad (0 \le w_k \le 1) \tag{1.8}$$

With this modeling approach, a set of process options can be compared on the basis of a set of attributes and factors. This quantitative approach fits the SoS approach perfectly.

Example of System Value Modeling

To illustrate the above modeling approach, suppose three information technology (IT) options are to be evaluated based on four attribute elements, *capability, suitability, performance,* and *productivity,* as shown in Table 1.1.

For this example, based on Equations 1.1 and 1.4, the value vector is defined as

$$V = f(\text{capability, suitability, performance, productivity}) \qquad (1.9)$$

Capability: "Capability" refers to the ability of a particular IT technology to satisfy multiple requirements. For example, a certain IT equipment may only provide computational service. A different piece of equipment may be capable of generating reports in addition to computational analysis, thus increasing the service variety that can be obtained. The levels of increase in service variety from the three competing equipment types are 38%, 40%, and 33%, respectively.

Suitability: "Suitability" refers to the appropriateness of the IT equipment for current operations. For example, the respective percentages of operating scope for which the three options are suitable are 12%, 30%, and 53%.

Performance: "Performance," in this context, refers to the ability of the IT equipment to satisfy schedule and cost requirements. In the example, the three options can, respectively, satisfy requirements on 18%, 28%, and 52% of the typical set of jobs.

Productivity: "Productivity" can be measured by an assessment of the performance of the proposed IT equipment to meet workload requirements in relation to the existing equipment. For our example, the three options, respectively, show normalized increases of 0.02, −1.0, and −1.1 on a uniform scale of productivity measurement. A plot of the histograms of the respective "values" of the three IT options is shown in Figure 1.13. Option C is the best "value" alternative in terms of suitability and performance. Option B shows the best capability measure, but its productivity is too low to justify the needed investment. Option A offers the best productivity, but its suitability measure is low. The analytical process can incorporate a lower control limit into the quantitative assessment such that any option providing a value below that point will not be acceptable.

TABLE 1.1
Comparison of Information Technology Value Options

IT Options	Suitability ($k = 1$)	Capability ($k = 2$)	Performance ($k = 3$)	Productivity ($k = 4$)
Option A	0.12	0.38	0.18	0.02
Option B	0.30	0.40	0.28	−1.00
Option C	0.53	0.33	0.52	−1.10

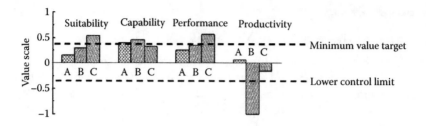

FIGURE 1.13
Relative system value weights of three IT options.

The relative weights used in many justification methodologies are based on subjective propositions of decision makers. Some of those subjective weights can be enhanced by the incorporation of utility models. For example, the weights shown in Table 1.1 could be obtained from utility functions. There is a risk of spending too much time maximizing inputs at "point-of-sale" levels with little time-defining and refining outputs at the "wholesale" systems level. Without a systems view, we cannot be sure whether we are pursuing the right outputs.

Engineering Project Management

Using a project structure allows groups to come together to pursue coordinated engineering innovation in the oil and gas industry. Many of the grand challenges of engineering, compiled by the U.S. National Academy of Engineering (NAE) in 2008, have implications for the energy industry. The extract from the NAE document on the 14 grand challenges reads: "In sum, governmental and institutional, political and economic, and personal and social barriers will repeatedly arise to impede the pursuit of solutions to problems. As they have throughout history, engineers will have to integrate their methods and solutions with the goals and desires of all society's members." The above statement emphasizes the relevance of holistic systems thinking in problem solving for sustainability solutions. The 14 grand challenges are listed below:

1. *Make solar energy economical*: Solar energy provides less than 1% of the world's total energy, but it has the potential to provide much, much more.

2. *Provide energy from fusion*: Human-engineered fusion has been demonstrated on a small scale. The challenge is to scale up the process to commercial proportions, in an efficient, economical, and environmentally benign way.

3. *Develop carbon sequestration methods*: Engineers are working on ways to capture and store excess carbon dioxide to prevent global warming.

4. *Manage the nitrogen cycle*: Engineers can help restore balance to the nitrogen cycle with better fertilization technologies and by capturing and recycling waste.

5. *Provide access to clean water*: The world's water supplies are facing new threats; affordable, advanced technologies could make a difference for millions of people around the world.

6. *Restore and improve urban infrastructure*: Good design and advanced materials can improve transportation and energy, water, and waste systems, and also create more sustainable urban environments.

7. *Advance health informatics*: Stronger health information systems not only improve everyday medical visits, but they are essential to counter pandemics and biological or chemical attacks.

8. *Engineer better medicines*: Engineers are developing new systems to use genetic information, sense small changes in the body, assess new drugs, and deliver vaccines.

9. *Reverse-engineer the brain*: The intersection of engineering and neuroscience promises great advances in health care, manufacturing, and communication.

10. *Prevent nuclear terror*: The need for technologies to prevent and respond to a nuclear attack is growing.

11. *Secure cyberspace*: It is more than preventing identity theft. The critical systems in banking, national security, and physical infrastructure may be at risk.

12. *Enhance virtual reality*: True virtual reality creates the illusion of actually being in a different space. It can be used for training, treatment, and communication.

13. *Advance personalized learning*: Instruction can be individualized based on learning styles, speeds, and interests to make learning more reliable.

14. *Engineer the tools of scientific discovery*: In the century ahead, engineers will continue to be partners with scientists in the great quest for understanding many unanswered questions of nature.

We do not know of any industry that is not touched by these global challenges. The oil and gas industry has a direct role to play in all the areas listed because energy is the foundation for accomplishing the goals stated in the grand challenges. This requires the application of project management tools and techniques to bring about new products, services, and results efficiently within cost and schedule constraints. Society will be tackling these grand challenges for the foreseeable decades, and project management is one avenue through which we can ensure that the desired products, services, and results can be achieved. With the positive outcomes of these projects achieved, we can improve the global quality of life. Some of the critical issues to address are

- Strategic implementation plans
- Strategic communication
- Knowledge management
- Evolution of virtual operating environment
- Structural analysis of projects
- Analysis of integrative functional areas
- Project concept mapping
- Prudent application of technology
- Scientific control
- Engineering research and development

Role of Emerging Nations

Ladislaw (2011) documents the roles of rapidly emerging countries in energy development trends around the world. One of the least understood, but potentially important, trends in the energy field is how the development priorities of emerging economies are shaping energy markets around the world. Emerging economies are expected to make up the bulk of growth in the demand for energy in the coming decades. It is estimated that countries outside the Organization for Economic Cooperation and Development (OECD) will account for more than 80% of the expected demand growth between 2008 and 2035. Recognized as the global focus of energy expansion, the emerging economies will increasingly influence how new energy markets develop and function. The interrelated commercial frameworks, technology development and sharing, international regulatory developments, and alternate fuel preferences will dictate how the global system will operate. This will influence how energy companies compete. Project efficiency will form a key component of the energy market dynamics. Many global energy companies are developing new exploration and marketing strategies that pay attention to the local contents of their operations. This responsive approach will help open up new energy markets in developing nations. The approach and contents of this book provide a project systems basis for increasing the success and market competitiveness for the oil and gas industry around the world.

References

Badiru, A. B. 2009, Twin fates: Partnerships will keep manufacturers' doors open, *Industrial Engineer*, March, 40–44.

Badiru, A. B. 2012, *Project Management: Systems, Principles, and Applications*, Taylor & Francis/CRC Press, Boca Raton, FL.

Ladislaw, S. O. 2011, *Energy and Development Trends: The Role of Rapidly Emerging Countries*, A Report of the Center for Strategic & International Studies, Washington, DC.

Lorincz, J. 2008, Oil powers global industry, *Manufacturing Engineering*, February, 55–57.

Troxler, J. W. and Blank, L. 1989, A comprehensive methodology for manufacturing system evaluation and comparison, *Journal of Manufacturing Systems*, 8(3), 176–183.

William, T. P. and William, T. 2010, *Meeting the Energy Challenges of the Future: A Guide for Policymakers*, Report of the National Conference of State Legislators (http://www.ncsl.org), Denver, Colorado.

2

Characteristics of Oil and Gas Projects

You have to speculate to accumulate. (An old West saying for mineral prospecting in Western United States) (Dodge, 1941)

Fundamentally, there is no business like the oil and gas business. The industry affects almost everything else in the general consumer market. The need to develop practical, efficient, and cost-effective energy infrastructure has never been more urgent. Developments in solar, wind, nuclear power, biofuels, and new oil and gas technologies have necessitated strategic project management practices. The influence sphere of oil and gas has grown dramatically over the years. A big decline in distributor costs of oil and gas directly drives a large portion of wholesale prices. Thus, it should be of interest to the general public to manage oil and gas projects so as to lower operating costs that can spread positively to consumer marketplace. As a demonstration of the influence of the industry, in 2012, Delta Airlines (Reuters, 2012) may have followed the cliché, "If you can't beat them, join them." Recognizing that its financial results and operating costs are closely tied to fuel prices, the airline decided to bid on an oil refinery. This price hedging approach is expected to be followed by other non-energy-based large corporations in the coming years. Risk is also an inherent part of the oil and gas industry. Risk management must be a core component of a company's project management portfolio in the oil and gas industry. Risks can be mitigated, but not eliminated. In fact, risk is the essence of any enterprise. In spite of government regulations designed to reduce accident risks in the energy industry, accidents will occasionally happen. Government regulators can work with oil and gas producers to monitor data and operations. This will only preempt a fraction of potential risks of incidents. For this reason, regulators must work with operators to ensure that adequate precautions are taken in all operating scenarios. Government and industry must work together in a risk-mitigation partnership, rather than in an adversarial "lording" relationship. There is no risk-free activity in the oil and gas business. For example, many of the recent petroleum industry accidents involved human elements—errors, incompetence, negligence, and so on. How do you prevent negligence? You can encourage nonnegligent operation or incentivize perfect record, but human will still be human when bad things happen. Operators and regulators must build on experiences to map out the path to risk reduction in operations.

Very few industries spark the sort of mixed sentiments that the oil and gas industry experiences. One day, ExxonMobil is hailed as the leader of Fortune 500 companies in 2012 first quarter earnings (Money, 2012). The next day, Mexican Oil Company, Petróleos, faced widespread opposition for the project to drill at extraordinary depths in the Gulf of Mexico (Businessweek, 2012). On yet another day, British Petroleum (BP) was criticized for the way it handled the initial postspill days following the Gulf of Mexico oil spill accident, which occurred in April 2010 (BBC, 2012). All the mixed sentiments, emotions, and reactions that result have direct impact on the management of oil and gas projects.

Projects in the oil and gas industry are characterized by huge investments, massive interfaces, and complex engineering endeavors. The size and complexity of these projects require special attention in the project management process. Risks are a big part of oil and gas projects. The opening quote in this chapter as well as the Latin dictum, "Spera optimum para pessimum"—Hope for the best, prepare for the worst—provide a fitting platform for approaching oil and gas projects. Because you never know what you might find (or not find) in oil and gas projects, more attention needs to be paid to project planning, organization, control, and contingencies. Many aspects of the oil and gas business can benefit from better project management practices. The span of projects of interest in the oil and gas industry includes the selected sample below:

General operations

- Economics of the market
- Financing
- Product marketing
- Government relations
- Regulations
- Inspections
- Regulatory oversight
- Quality checks
- Compliance assessment
- Corporate alliances
- Human resources
- Hardware and software infrastructure

Exploration and development

- Forecasting
- Geological research
- Discoveries
- Reserves
- Area drilling

Drilling and production

- Drilling operations
- Production operations
- Field start-ups
- Unconventional resources

Processing

- Refining
- Operations
- Construction
- Capabilities
- Petrochemicals
- Gas processing
- New plants
- Capacities
- Liquefied petroleum gas (LPG) markets

Transportation

- Tankers
- Liquefied natural gas (LNG)
- Pipelines

To set a backdrop for potential project management scenarios in the oil and gas industry, a brief collection of snip-bits in the industry is presented in the sections that follow.

Government Involvement

One of the operating constraints affecting the oil and gas industry is the typical involvement of governments of the developing countries. Under national joint venture agreements, the government of a developing country may own (or be involved in) as much as 40% of an oil company. In some cases, a government may seek to have majority ownership. In 2012, in a bold nationalistic move, Argentina sought to gain control through majority stake (i.e., >50%) in oil companies, foreign or domestic, that are operating in the country (Fox News, 2012).

Fate of Refineries

A key fact about the oil business is that the oil industry is at refining capacity, which is more of the reason why prices are so high, and not because of shortage of supply of crude. Environmental policy has stifled any new construction of refineries because of extreme regulatory requirements. Oil companies have no incentive to build new refineries because this solves a national problem (security, economy, etc.) but is not economically sound for them. They are making money with the current infrastructure and have no reason to expand. More research should be directed at reducing the overall risk to the global oil and gas system, considering spatial relationships between the several factors involved.

Oil Sands Project

It was reported in Houston in early 2012 that Koch Oil Sands Operating ULC has received conditional approval to build and operate the Gemini thermal oil sands project near Beaverdam in the Cold Lake area of Alberta (OGJ, 2012). The province's Energy Resources Conservation Board approved a two-stage recovery scheme based on steam-assisted gravity drainage to produce as much as 10,000 b/d (barrels/day) of bitumen. The first stage includes drilling of a steam assisted gravity drainage (SAGD) well pair and two observation wells and construction of a processing facility and related pipelines, leading to the production of 1200 b/d. The second stage includes drilling of as many as 23 additional well pairs from five pads and construction of pipelines and a second-stage processing facility, as well as drilling of at least 15 observation wells. The conditions set special requirements for groundwater monitoring and for a plan to mitigate potential effects on surface water near two of the drilling pads.

BP to Invest $4 Billion in Gulf of Mexico in 2012

At the 2012 Offshore Technology Conference in Houston, Texas, it was announced that BP will invest $4 billion in the Gulf of Mexico in 2012 alone (OTC, 2012). This is a massive undertaking that will require massive project management operations.

BP PLC has five offshore rigs operating in the Gulf of Mexico, which is the same number as it had before the April 2010 deepwater Macondo well blowout. By the year end, BP expects to add three more rigs focused on exploration, appraisal, and developments. BP is investing $4 billion in the gulf this year and hopes to invest at least that much every year for the next decade, creating a continuing opportunity for mega applications of project management tools and techniques. The Macondo incident prompted a fire and explosion on Transocean Ltd.'s Deepwater Horizon semisubmersible drilling rig and a massive oil spill off Louisiana in the Gulf of Mexico. In the Macondo aftermath, BP pressed ahead with its gulf exploration and development and a reinforced focus on risk management. BP currently has two rigs in operation at Thunder Horse, two at Atlantis, and one at Kaskida. Of the three rigs that BP expects to add to the gulf by year end, one of those rigs will be equipment brought back to the gulf from West Africa. The big new play is the Palaeogene or Lower Tertiary. In the Gulf of Mexico, this includes BP's discoveries at Kaskida in 2006 and Tiber in 2009. Each of these fields has accessible hydrocarbons today, but each also has resources that lie beyond our industry's current limit of 15,000 psi and 275°F. BP is working to increase its offshore exploration and production capabilities. These efforts will include sensing and monitoring systems for real-time subsea integrity management. BP is also working with others to develop subsea values, weighing 20 tons, capable of closing hydrocarbon flow in seconds. Making this vision a reality will require unprecedented collaboration across and outside of the industry—involving not only operators, vendors, and contractors but also academics and regulators, not to mention project management professionals. This will be necessary to define codes and standards for the design, operation, and reliability of the new technology. BP is also working to become more transparent so it can earn the trust and confidence of the public and the regulators. BP plans to impose higher project standards upon its operations, share what it learns, and work together to attract people with different skills to the oil and gas industry. The company plans to access the deepest areas of the Gulf of Mexico, and the other great basins of the world. These are projects that call for new techniques of management and control.

Flak of Fracking

There are cracks in the acceptance of fracking technology for oil and gas. Fracking is the industry term for hydraulic fracturing, which is the process of injecting water, sand, and chemicals deep into the ground at high pressure to crack oil shale so that oil or gas can be harvested or siphoned out. The ability to cheaply extract gas from shale rock has transformed the global energy outlook, but the method of cheap extraction is not generally accepted by everyone. Hydraulic fracturing, which was invented by Halliburton in 1947 (Halliburton, 2011), is the widening of fractures in a rock layer caused by the high-pressure injection of chemicals with water. In a hydraulic fracturing job, "fracturing fluids" or "pumping fluids" consisting primarily of water and sand are injected under high pressure into the producing formation, creating fissures that allow resources to move freely from rock pores where it is trapped. Typically, steel pipe known as surface casing is cemented into place at the uppermost portion of a well for the explicit purpose of protecting the groundwater. The depth of the surface casing is generally determined based on groundwater protection, among other factors. As the well is drilled deeper, additional casing is installed to isolate the formation(s) from which oil or natural gas is to be produced, which further protects groundwater from the producing formations in the well. Casing and cementing are critical parts of the well construction that not only protect any water zones but are also important to successful oil or natural gas production from hydrocarbon-bearing zones. Industry well design practices protect sources of drinking water from the other geologic zone of an oil and natural gas well with multiple layers of impervious rock. While 99.5% of the fluids used consist of water and sand, some chemicals are added to improve the flow. The composition of the chemical mixes varies from well to well.

Hydraulic fractures form naturally, as in the case of veins or dikes, and industrial fracturing widens or creates fractures to speed up the migration of gas and petroleum from source rocks to reservoir rocks. This process is used to release petroleum, natural gas (including shale gas, tight gas, and coal seam gas), or other substances for extraction, via a technique called induced hydraulic fracturing, often shortened to fracking or hydrofracking. This type of fracturing, known colloquially as a "frac job," creates fractures from a wellbore drilled into reservoir rock formations. A distinction can be made between low-volume hydraulic fracturing used to stimulate high-permeability reservoirs, which may consume typically 20,000–80,000 U.S. gallons (76,000–300,000 L; 17,000–67,000 imp gal) of fluid per well, with high-volume hydraulic fracturing, used in the completion of tight gas and shale gas wells; high-volume hydraulic fracturing can use as much as 2–3 million U.S. gallons (7.6–11 ML) of fluid per well. This latter practice has come under scrutiny internationally, with some countries suspending or even banning it. The first fracking job was performed in 1947, though the current fracking technique was first used in the late 1990s in the Barnett Shale in Texas. The energy from the injection of a highly pressurized fracking fluid creates new channels in the rock, which can increase the extraction rates and, ultimately, permit the recovery of fossil fuels. According to the International Energy Agency, the global use of natural gas will rise by more than 50% compared to 2010 levels, and account for over 25% of world energy demand in 2035. Proponents of fracking point to the vast amounts of formerly inaccessible hydrocarbons the process can extract. However, there remain large uncertainties in the amount of gas reserves that can be accessed in this way. Detractors point to potential environmental impacts, including contamination of ground water, risks to air quality, the migration of gases and hydraulic fracturing chemicals to the surface, surface contamination from spills and flow back, and the health effects of these.

Rules, Laws, Regulations, and Standards

Few industries are as subject to a multitude of rules, laws, regulations, and standards as the oil and gas industry. Standards provide a common basis for global commerce. Without standards, product compatibility, customer satisfaction, and production efficiency cannot be achieved. Just as quality cannot be achieved overnight, compliance with standards cannot be accomplished instantaneously. The process must be developed and incorporated into regular operating procedures over a period of time. Standards define the critical elements that must be taken into consideration to produce a high-quality product that meets customers' expectations. Each organization must then develop the best strategy to address the elements. Both regulatory and consensus standards must be taken into account when executing oil and gas projects. *Regulatory standards* refer to standards that are imposed by a governing body, such as a government agency. All firms within the jurisdiction of the agency are required to comply with the prevailing regulatory standards. *Consensus standards* refer to a general and mutual agreement among a group of companies to abide by a set of self-imposed standards. Industry alliances, such as OPEC, impose their own consensus standards and rules on themselves. There may also be *contractual standards*, which are imposed by the customer based on case-by-case or order-by-order needs. Most international standards will fall in the category of consensus. A lack of international agreements often leads to trade barriers by nations, industries, and special interest groups.

In response to the widespread expressions of concerns about fracking, the U.S. government, the Environment Protection Agency (EPA), in April 2012 issued new regulations to govern fracking (EPA, 2012).

The standards are to control air pollution from gas wells that are drilled using fracking. The government strongly supports natural gas drilling as a clean source of energy and wanted to ensure coordination to ease the production burden on the oil and gas industry. EPA maintains that the new regulations would ensure pollution is controlled without slowing natural gas production. Much of the air pollution from fracked gas wells is vented when the well transitions from drilling to actual production. This is a 3–10-day process, which is referred to as "completion." An earlier version of the rule limiting air pollution from gas wells would have required companies to install pollution-reducing equipment immediately. Under the new rules, drillers will now be given more than 2 years to employ technology to reduce emissions of smog and soot-forming pollutants during that stage. EPA would require drillers to burn off gas in the meantime. This is an alternative that can release smog-forming nitrogen oxides, but will still reduce overall emissions. This is a good example of where a systems view is instrumental because as one hole is plugged at one end, some other issue may develop at the other end. The oil and gas industry must consider these new rules as well as other production and economic considerations into their overall project management

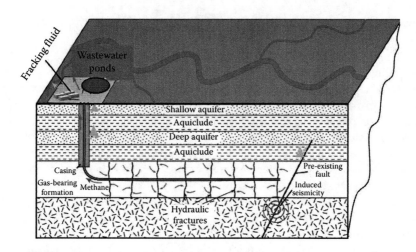

FIGURE 2.1
Schematic of a typical fracking structure.

strategies. About 25,000 wells a year undergo fracking in the United States. Figure 2.1 shows a typical schematic drawing of a fracking structure.

Keystone Oil Pipeline

The Keystone Pipeline System is a 1700-mile, $7-billion pipeline system to transport synthetic crude oil and diluted bitumen ("dilbit") from the Athabasca Oil Sands in northeastern Alberta, Canada to multiple destinations in the United States, which include refineries in Illinois, the Cushing oil distribution hub in Oklahoma, and proposed connections to refineries and export terminals along the Gulf Coast of Texas (Guardian, 2012). It consists of the operational "Keystone Pipeline" (Phase 1) and "Keystone-Cushing Extension" (Phase 2), and two proposed Keystone XL pipeline expansion segments. After the Keystone XL pipeline segments are completed, American crude oil would enter the XL pipelines at Baker, Montana and Cushing, Oklahoma. The Keystone XL has faced lawsuits from oil refineries and criticism from environmentalists and some members of the U.S. Congress. The U.S. Department of State in 2010 extended the deadline for federal agencies to decide if the pipeline is in the national interest, and in November 2011, President Obama postponed the decision until 2013. The U.S. government blocked the pipeline in early 2012, citing uncertainty over the Nebraska route, which would travel above an aquifer that provides water to eight states. In April 2012, Calgary-based TransCanada, the company planning the Keystone XL oil pipeline, proposed a new route through Nebraska that would avoid the state's environmentally sensitive Sandhills region. In addition to the U.S. EPA, the State Department is also involved because the pipeline would cross a U.S. border. The twists and turns of the diverse involvement in oil and gas projects require extraordinary systems-based project management techniques.

New Projects, New Project Management Opportunities

In April 2012, it was announced (Akers Solutions, 2012) that Statoil has issued a contract to Aker Solutions to hire a category B drilling rig, a new type of rig, which is designed to increase recovery from the company's operating fields on the Norwegian continental shelf (NCS). Statoil has developed the new type of rig and well control system in cooperation with the supplier industry. The category B rig, which falls between light intervention vessels (category A) and conventional rigs (category C), has been specially adapted to carry out well intervention and drilling operations in existing subsea wells. The new rig, with associated integrated services, is expected to reduce operating costs for well intervention by as much as 40%, Statoil reported. The rig's design provides the option for a number of different types of well interventions using wireline and coiled tubing operations. The rig type is designed for year-round well service in Statoil-operated activities. This type of rig is also designed to carry out sidetrack drilling from production tubing (through tubing drilling) in a manner that allows simultaneous production from both the new sidetrack and existing production tubing. The well services are conducted through existing subsea network. The key to maintaining the current production level on the NCS is increased recovery from existing fields, along with the development of new fields. Increasing drilling activity on mature fields is important in order to achieve the NCS's full potential. In 2011, Statoil operated about 500 subsea wells. Statoil and licensees will enter into an 8-year contract with options for three times 2 years for the category B service. The estimated value of the contract is $1.9 billion. In addition to rig rental, the contract also includes rental of the necessary equipment and services to carry out well intervention, sidetrack drilling, remotely operated vehicle operations, well testing, and cementing. If project management goes well, the rig will be in operation by 2015.

Gas Export Project Management

In February 2012, Kuwait Gulf Oil Co. let a contract to the Penspen Group (Penspen, 2012) for the project management and detailed design and construction of a natural gas and condensate export system (GCES) from Khafji, Saudi Arabia, on the Kuwait–Saudi border, to Kuwait. Penspen will manage engineering, procurement, and construction (EPC) contractor Technip. The new export system will carry 40 MMcfd of gas via 110 km of 12-in.-OD export pipeline, of which about 47 km will be offshore. Penspen will oversee Technip's detailed design work, procurement activities, and construction undertaken by Technip and its subcontractors. Penspen will also assist with commissioning the final scheme. GCES will deliver condensate and gas product to Kuwait from Saudi Arabia, reducing gas flaring in the process. Technip's operating center in Abu Dhabi will execute the project, scheduled to be completed by the second half of 2014. DLB Comanche, which entered the Technip fleet with the acquisition of Global Industries, will complete offshore operations.

Case Example of Shell's Use of PMBOK

In May 2001, Shell International Exploration and Production Inc. reported the successful use of structured project management on their Brutus project, a deepwater tension leg platform (Shell, 2012). The company used new project management tactics based on the contents of the Project Management Institute (PMI)'s PMBOK, a standardization guideline established by the PMI. Standardization is needed as various elements of the offshore energy industry need a common language and common operating procedures so that they can better communicate with each other on complex deepwater projects. The procedures developed for the Brutus project were grouped and aligned in accordance with the PMBOK. A standardized work breakdown structure and a new financial software system are just two examples of new project management tactics used in the project. The approach gave team members new insights into achieving the objectives and reinforced the basic principles of project management. Customer expectations were exceeded and efficiency improved. Future project teams can achieve benefits by similarly organizing their project management processes. Shell opined that projects must be able to stand on their own financially and be capable of paying their own bills. Cash flow is essential in project finance. People behave as rewarded. Consequently, it is suggested that organizations should tie the cash flow to the performance desired and people will respond positively. Deepwater projects are particularly costly and of high risk. Project management tactics help companies gauge the accuracy of estimates for time and cost projections, Shell concluded. Project risk is often mismanaged. If projects are approved based on anticipated return on investment (ROI) with minimal analysis of risk, a project is set up for failure. Decisions based on incomplete information will be faulty during project execution.

Hail to Project Management

In a 2003 industry study,* owners of downstream petroleum and chemical companies ranked project management capability as the main consideration in choosing engineering contractors. The owners had concerns about the viability of contracting companies and the supply of reliable engineering resources. These were some of the conclusions of a biennial survey of the global downstream EPC industry by Transmar Consult Inc., Houston in 2003. The study designated seven engineering contractors as leading firms in executing projects: Bechtel Corp., Fluor Corp., Jacobs Engineering Group Inc., JGC Corp., KBR, Snamprogetti SPA, and Technip-Coflexip.

Project Management Improve Operations

In 2000, Transocean Sedco Forex Inc.'s Discoverer Deep Seas utilized modern project management techniques to shorten cycle time and improve workflow procedures in dual-activity drillship operations. The Discoverer Deep Seas began work in the Gulf of Mexico in the fourth quarter 2000. It held a 5-year contract with Chevron Inc. at an estimated day rate of $205,000. After this vessel, only one drillship remains in the 3-year old drillship construction cycle. This ultra deep water vessel, which used a dual-activity design pioneered by the Discoverer Enterprise was delivered in the fourth quarter of 2000.

Technology Project Management

Technology project management reportedly improved multiwell shallow gas development in 1995, according to a presentation by Becker Husky Oil Operations Ltd. in Calgary, Canada (Becker, 1995). Because of time and economic constraints, a multiwell shallow gas development project in southeastern Alberta required thorough prejob planning and special drilling, cementing, and completions designs. The project took place during a period of peak industry activity, putting extra challenges on logistics and services. The Medicine Hat shallow gas project, undertaken by Husky Oil Operations Ltd. in mid-1994, was a high-volume, short–time frame, and economically tight project schedule. But using technology project management techniques, the project came through on time and within budget.

---

* http://www.ieej.or.jp/aperc/pdf/downstream_study.pdf

Project Management Improves Well Control Events

A 1995 report presented by Garold D. Oberlender (Oklahoma State University, Stillwater, Oklahoma) and L. William Abel (Wild Well Control Inc., Spring, Texas) is recounted here to illustrate how project management improved oil well control events (Oberlender and Abel, 1995). During a well control operation, the efficient use of personnel and equipment, through good project management techniques, contributes to increased safety and ensures a quality project. The key to a successful blowout control project is to use all resources in the most efficient manner. Excessive use of resources leads to unnecessary expenditures and delays in bringing the project under control.

Effective project management techniques are critical for complex, schedule-intensive, well control projects. In a well control situation, information must be gathered and organized in a systematic manner so that good decisions can be made at the right time. A project team must be organized, and all efforts must be coordinated to focus on the common goal of completing the project efficiently. Project management is a process of orderly management of numerous tasks that must be accomplished simultaneously. Project management is defined as the art and science of coordinating people, equipment, materials, money, and schedules to complete a specified project on time and within approved costs. Project management techniques have steadily improved in the last 40 years. In the 1950s, formal techniques, such as the critical path method (CPM), were developed for managing large complex engineering and construction projects for the petrochemical industry. In the 1960s, computer automation of many project management techniques became possible, with sophisticated tracking models developed for controlling the progress of projects. In the 1980s, the introduction of personal computers provided even more access to automated project management techniques to allow planning, scheduling, monitoring, and controlling of work in a real-time environment.

Lessons learned from the practice of good project management can be applied to wild well projects. In such emergency situations, the schedule of the project usually takes priority over all other considerations. Therefore, the ability to monitor progress carefully in a real-time environment is extremely important to controlling the project. Information must be gathered, organized, and given to the right person at the right time so good decisions can be made under adverse conditions. Well control projects have various levels of complexity, and as the complexity increases, the need for a project management approach becomes more important.

Several complicated and difficult well control projects occurred in the 1980s, and the cost to control these wells exceeded $200 million each. Many experienced individuals in any business will readily agree that a $200 million project justifies the use of advanced project management techniques to control and guide the efforts of all parties and ensure a successful outcome.

Several of these multimillion dollar projects, however, were run without proper planning and with little or no application of a formal system of project management. The lack of good project management principles caused inefficiencies, delays, and higher costs.

The Kuwait well control project, which involved more than 700 blowouts, was accomplished in a much shorter time (8 months) than first estimated (5 years). This improvement partly resulted from the application of sound project management techniques. These projects were prime examples of the need for a formal project management approach to handling wild well projects. There are many examples of projects that were successful in controlling wells but were economic disasters. Only through the effective application of project management can complex well control projects be completed in reasonable time frames at reasonable cost.

Team Management

To be effective, a project team must be able to make things happen, which is best accomplished by maintaining a "Can do" attitude and working together as an integrated team. Effective teamwork is mandatory for bringing any wild well under control.

The first step in organizing the team is selection of the project manager who is assigned full responsibility for all aspects of the project. This individual must have knowledge and experience in dealing with well control projects. Most importantly, he must be given the authority to make decisions and to act as the sole coordinator of all activities.

Because the project manager must focus his full attention on the wild well, he must not be involved in any other work that may distract his attention and concentration.

The first task of the project manager is to assess the condition of the well to determine viable options for bringing it under control. This task is best done if a proper contingency plan is already in place, providing for rapid deployment of predetermined action plans. The project manager must also establish proper communication channels with all appropriate authorities. He must have both a thorough knowledge of well control projects and effective management skills.

Because each wild well event is unique, the project team must be assembled, organized, and managed for the particular conditions at the well location. The project manager must assemble a team of people who have the right technical expertise to handle the job. He must be the leader of the team of individuals with diverse backgrounds and with special expertise to handle any problems that arise. Although a formal system of handling the project is needed, it is people that make things happen, and these people must have the ability to detect problems, make adjustments to the system, and make the right decisions at the right time.

Project Scope

Although each wild well control project is unique, a well-managed project generally follows this sequence:

- The team must define the scope of what must be accomplished, including giving priorities to tasks, identifying equipment and special expertise, and setting expected levels of quality, safety requirements, and reliability of operations.
- Although budget is often not a controlling factor in well control projects, eventually the allocation and responsibility of cost expenditures must be addressed, including budgeting each task in the scope definition.
- A strategy must be set for accomplishing the tasks. It is important to anticipate all events and to build contingency plans into the project for the unexpected deviations that often occur.
- A timetable must be made for the planned work to ensure an integrated sequence of all tasks.
- A tracking system must be developed to ensure the project progresses as planned, by measuring actual work done versus the schedule.
- The project should be closed out, which includes verification that all required tasks have been completed to ensure the project is in a stable condition and completed at an acceptable level of quality.

These steps represent project management in its simplest form. In practice, there is considerable overlap in the sequences because there are many parties involved, and the work of one may affect the work of others. Thus, extensive coordination is required. Most existing project management literature deals with engineering design, construction, or manufacturing work and not with wild well control projects. The issues that are discussed generally focus on expediting the work in a schedule-compressed mode, similar to a fast-track construction project. Thus, the basic principles of project management apply to schedule-driven wild well control projects. The most important task for any project is to define the scope of work. A fundamental principle of project management is that any project, regardless of its size or type, consists of three components: scope, budget, and schedule. Each of these components must be described in detail and linked to the others. For example, as the scope of work increases, the cost and time to do the work will also increase. Defining the project scope must precede defining the budget or schedule.

In a wild well project, the overall objective is to extinguish the fire and bring the well under control. Other objectives, however, may also be extremely important. For example, the objectives may include controlling the well in a way that allows drilling operations to resume so the well can eventually be produced rather than abandoned. In construction projects, the emphasis is too often incorrectly placed on budget and schedule and not on the scope or quality of work. In well control projects, there may be excessive emphasis on schedule and on controlling the well as quickly as possible without regard to the impact on budget, scope, or quality. Because of the nature of well control projects and the pressures to achieve a quick solution, it is often difficult to manage the project efficiently. Careful planning beforehand will provide a workable and rational solution for the operations group.

The well control company is best suited to be involved in the design and implementation of the control operation, provided its personnel have the skill and capabilities to perform project management duties. If the well control company has engineering expertise that is fully integrated in its field operations, they will be best suited for overall coordination and management of the project. Regardless of the type of well control project, there must be a single project manager who is responsible for coordinating all aspects of the project. The project manager is responsible for reporting to the operator or team with complete responsibility and authority for making decisions. The most successful well control projects have been completed by a competent project manager who had the technical expertise and the management skills necessary to ensure the work was performed as and when needed. Committees, with a group acting as a project manager, tend to have slow decision making and be bureaucratic.

In well control project management, four fundamental questions must be addressed: Who is going to do it? What are they going to do? When are they going to do it? And, how much is it going to cost? A lack of an answer to any of these questions during the project will usually result in additional problems. The most important skill the project manager must possess is the ability to be an effective communicator. Regardless of his abilities as an engineer or manager, he will be ineffective if he lacks communication skills. The project manager must have the technical knowledge of what needs to be done and the ability to communicate that knowledge to others. To bring a blowout under control, many different types of expertise are required. Thus, the project manager must have the ability to delegate authority and responsibility to the specialists involved so he can concentrate on the overall project objectives. The central axiom for project management is to organize the project around the work to be done rather than trying to force events to fit some predetermined organization.

Organizational Structures

Regardless of the type of management organization chosen for a well control event, the main consideration must be a bias for action. The management group must focus on five major functions: planning, organizing resources, selecting staff, coordinating work, and controlling the operation. Each well control project is unique and requires a suitable choice of organizational structure, the most common of which are discipline, functional, and matrix organizations. In a discipline-oriented organization, personnel who share the same technical expertise are placed in common departments. For example, all civil engineering expertise is located in the civil engineering department, and so on. A functional organization places individuals in departments that focus on specific processes, such as reservoir engineering, production engineering, or drilling and completion.

The matrix organization places overall management responsibilities in a management group. The management group obtains technical expertise from the various departments in the company and, if necessary, secures assistance from sources outside of the company. The most desirable organizational structure for a well control event depends on numerous factors, including the severity and complexity of the event and the availability of technical expertise. For routine projects, either the discipline or functional approach may be suitable. For a complex and technically difficult project, the matrix organization is preferable. The two structures recommended for well control events are the functional and matrix organizations. The functional organization disperses the disciplines among the departments in a company. Project teams are formed within the departmental group. Management is usually accomplished by the team leader or project manager, who also is a member of the department.

On occasion, a drilling department may be called upon to function in the role of project manager to oversee the blowout control event. This arrangement may be acceptable for smaller, more routine projects but can present some problems in dealing effectively with a major event. A common error in the management of well control projects is to attempt to direct the work from within a departmental group, which retains all of its existing responsibilities. This method dilutes the efforts of the project manager, and one or more of his projects will suffer as a result. Well control events should be managed by a single person who has no other priorities or responsibilities, a person who can focus his attention exclusively on the complex process of controlling the well.

The matrix organization increases the importance of managing cost and schedule and maintaining the general balance among all the elements of the project. A separate project management organization directs the job, with the support of technical expertise from the discipline departments. The objective is to keep the disciplines in their home environments where they function best, while making their expertise available to meet the needs of the project. The matrix organization requires two lines of communication: one to the discipline organization and the other to the project organization. A discipline manager answers any specific questions. Issues related to the project are addressed by the project manager. In the matrix organization, the project manager coordinates all aspects of the well control project. Each discipline is responsible for its portion of the project, such as quality, cost, or schedule.

Matrix organizations can be strong or weak. In a strong matrix, the project manager has the authority to decide what is best overall for the project. In a weak matrix, the discipline group has the authority to make decisions for the project. The success of the matrix organization is highly dependent on the company's philosophy, and even more so on the attitude of the employees involved. If discipline managers are more concerned about their discipline than the project, the project can have inefficiencies and delays. A key to a successful project is a balance between the technical expertise of the discipline departments and the project management group. Technical personnel generally concentrate on producing the best product possible, often without regard to cost or schedule. In a well control project, quality and schedule generally take priority, although cost is always an important consideration.

Scheduling

Planning, measuring, evaluating, forecasting, and controlling are the key responsibilities of the project management team. All aspects of the project must be tracked and managed to ensure effective control of the project. Tracking cannot take place unless a well-defined work plan, budget, and schedule have been developed. Planning should precede project scheduling. For the entire well control project, there must be an explicit operational plan that binds scope, budget, and schedule. A common error of planning is to concentrate solely on schedule and disregard the importance of scope and budget. The following are the key principles of planning:

- Begin planning before the start of work, not after.
- The people who will actually do the work should be involved in planning and scheduling from the onset.
- The plan must give balanced consideration to the aspects of scope, budget, schedule, and quality.
- The plan should be flexible, including allowances for responding to unexpected changes and with time allotted for review and approvals.
- The plan must be kept simple and without irrelevant details that prevent readability.
- The plan should be distributed to and understood by all parties involved.

Tracking Models

Planning is the first step in project scheduling, but tracking is equally essential. Tracking cannot detect deviations unless there is a master plan or schedule against which to compare actual progress. The size, complexity, duration, and importance of the well control project dictate the technique used for project scheduling. The tracking model must be functional and easy to use. Two methods meet these criteria: bar charts (sometimes called Gantt charts) and the

CPM. Because bar charts are simple and easy to interpret, they are usable by all participants in a project. Bar charts have three drawbacks: They do not show interdependency of activities; they require considerable time for updating; and they do not integrate costs or resources. A common complaint about Gantt charts is the high level of effort required to keep the charts updated. A significant amount of labor and time must be expended in keeping the charts current because of constantly changing conditions, such as man-hour requirements.

Critical Path Method

Both the CPM and the program evaluation and review technique (PERT) are often referred to as network analysis systems. CPM takes a deterministic approach, which assigns a single duration to each activity in the network, whereas PERT is a probabilistic approach that deals with three possible times for each activity (optimistic, pessimistic, and normal). Experience with fast track projects, such as blowouts, has shown CPM to be the preferred method. CPM forces the project team to direct all efforts to a single start date to kick off the project and a single completion date to bring the project under control. Both Gantt charts and CPM are suitable for well control projects. The Gantt charts are best suited for situations involving activities that do not have significant interdependencies with other activities (such as design work and public relations).

The CPM method is a preferred approach for planning and scheduling more complex well control projects, situations in which activities are highly interdependent, or where there is a need to track resources assigned to the project. CPM uses a network diagram to model the interdependencies of activities. CPM forces the project team to divide the work into definable units and to determine how work items interface with each other. A well-defined work breakdown structure (which shows the primary and secondary responsibilities for each member of the well control team) will greatly simplify the task of generating the CPM network diagram.

Many project managers use precedence diagrams for well control that only shows major events; the numerical coding system, activity duration, and resources have been omitted to simplify the illustration. These diagrams require fewer activities to describe the project and have great flexibility in sequencing activities and showing relationships between them. CPM has adapted computer-aided techniques to automate calculations for scheduling and the graphic printout of network diagrams. The use of computers in CPM allows modeling of man-hours, costs, and duration to closely monitor work progress and produce a variety of project management reports.

Decision Trees

Well control events are unique projects that require quick decisions for critical problems to be solved expeditiously. The project manager must be able to "think on his feet" and react to emergency situations in adverse working conditions, especially during kill operations. Contingency planning for likely scenarios is essential to increase the probability of success. Decision trees are useful to help the project manager run the job.

The snip-bits presented above represent just a sample of the thousands of project-related events, issues, and developments in the oil and gas industry. As can be seen, there are tremendous areas of need for project management in the industry. The chapters that follow will address specific tools, techniques, and concepts of project management as relevant for applications in the oil and gas industry. The framework used in the contents that follow center around the following knowledge areas:

- Managerial processes
- Technical systems
- Human interfaces
- Systems engineering

Both quantitative and qualitative aspects of project management are covered to the extent that they can benefit researchers, educators, students, and practitioners as reference materials.

Oil Bunkering in Developing Nations

The so-called oil bunkering occurs when thieves tap into an oil pipeline. The culprits either sell the oil or produce poorly refined petroleum products that are sold on the black market. This practice is very common in impoverished communities in developing countries and it has grave adverse impacts on the country's oil industry. The danger of explosion posed by the illegal activity creates difficulties in executing pipeline projects. The cost and time implications of dealing with the aftermaths of oil bunkering translate to schedule inconveniences as well as project execution disruptions.

References

Akers Solutions. 2012. http://www.akersolutions.com/en/Global-menu/Media/ Press-Releases/All/2012/Aker-Solutions-and-Statoil-awarded-ONS-Innovation-Award/, accessed April 25, 2012.

BBC. 2012. http://www.bbc.co.uk/news/world-us-canada-10656239, accessed May 5, 2012.

Becker, D. L. 1995. *Technology Project Management Improved Multiwell Shallow Gas Development*, http://www.ogj.com/articles/print/volume-93/issue-42/in-this-issue/drilling/technology-project-management-improved-multiwell-shallow-gas-development.html, October 16, 1995.

Businessweek. 2012. http://www.businessweek.com/stories/, accessed May 1, 2012.

Dodge, D. 1941. *Death and Taxes*, 1st Edition, MacMillan, New York, NY.

EPA. 2012. http://www.epa.gov/hydraulicfracture/, accessed April 25, 2012.

Fox News. 2012. http://www.foxnews.com/.../04/.../venezuela-rejects-threats-against-argentina..., accessed April 17, 2012.

Guardian. 2012. http://www.guardian.co.uk/environment/2012/jan/18/, accessed Febuary 1, 2012.

Halliburton. 2011. http://www.halliburton.com/public/projects/pubsdata/.../fracturing_101. htm..., accessed December 5, 2011.

Money. 2012. http://money.cnn.com/magazines/fortune/fortune500/2011/full_list/, accessed March 2, 2012.

Oberlender, G. D. and Abel, L. W. 1995. *Project Management Improves Well Control Events*, http://www.ogj.com/articles/print/volume-93/issue-28.html, July 10, 1995.

OGJ (Oil and Gas Journal). 2012. www.ogj.com/articles/.../ercb-approves-gemini-oil-sands-project.html, accessed May 6, 2012.

OTC (Offshore Technology Conference). 2012. http://www.otcnet.org/2012/, accessed March 10, 2012.

Penspen. 2012. http://www.penspen.com/News/Pages/PenspenawardedProjectManagementServicesbyKuwaitGulfOilCompanyforGasandCondensateExportSystem.aspx, accessed March 18, 2012.

Reuters. 2012. http://www.reuters.com/article/2012/04/04/delta-refinery-idUSL..., accessed April 12, 2012.

Shell. 2012. http://www.shell.us/home/content/usa/aboutshell/projects_locations/gulf_of_mexico/brutus_0308/, accessed January 7, 2012.

3

Project Management Body of Knowledge

The rapid growth of technology in the workplace has created new challenges for those who plan, organize, control, and execute complex projects. With the diversity of markets globally, project integration is of great concern. Using a consistent body of knowledge can alleviate the potential problems faced in the project environment. Projects in the oil and gas industry are particularly complex and dynamic, thus necessitating a consistent approach. The use of project management continues to grow rapidly. The need to develop effective management tools increases with the increasing complexity of new technologies and processes. The life cycle of a new product to be introduced into a competitive market is a good example of a complex process that must be managed with integrative project management approaches. The product will encounter management functions as it goes from one stage to the next. Project management will be needed throughout the design and production stages of the product. Project management will be needed in developing marketing, transportation, and delivery strategies for the product. When the product finally gets to the customer, project management will be needed to integrate its use with those of other products within the customer's organization. The need for a project management approach is established by the fact that a project will always tend to increase in size even if its scope is narrowing. The following four literary laws are applicable to any project environment:

Parkinson's law: Work expands to fill the available time or space.

Peter's principle: People rise to the level of their incompetence.

Murphy's law: Whatever can go wrong will.

Badiru's rule: The grass is always greener where you most need it to be dead.

An integrated systems project management approach can help diminish the adverse impacts of these laws through good project planning, organizing, scheduling, and control. The Project Management Institute, as a way of promoting a common language for the practice of project management, developed the Project Management Body of Knowledge (PMBOK), which has been widely adopted around the world.

Project Management Knowledge Areas

The nine knowledge areas presented in the PMBOK are listed below:

1. Integration
   - Integrative project charter
   - Project scope statement
   - Project management plan
   - Project execution management
   - Change control
2. Scope management
   - Focused scope statements
   - Cost/benefits analysis
   - Project constraints
   - Work breakdown structure
   - Responsibility breakdown structure
   - Change control
3. Time management
   - Schedule planning and control
   - PERT and Gantt charts
   - Critical path method
   - Network models
   - Resource loading
   - Reporting
4. Cost management
   - Financial analysis
   - Cost estimating
   - Forecasting
   - Cost control
   - Cost reporting
5. Quality management
   - Total quality management
   - Quality assurance
   - Quality control
   - Cost of quality
   - Quality conformance

6. Human resources management
   - Leadership skill development
   - Team building
   - Motivation
   - Conflict management
   - Compensation
   - Organizational structures
7. Communications
   - Communication matrix
   - Communication vehicles
   - Listening and presenting skills
   - Communication barriers and facilitators
8. Risk management
   - Risk identification
   - Risk analysis
   - Risk mitigation
   - Contingency planning
9. Procurement and subcontracts
   - Material selection
   - Vendor prequalification
   - Contract types
   - Contract risk assessment
   - Contract negotiation
   - Contract change orders

The above segments of the body of knowledge of project management cover the range of functions associated with any project, particularly complex ones.

Project Definitions

Project

A project is traditionally defined as a unique one-of-kind endeavor with a specific goal that has a definite beginning and a definite end. PMBOK defines a project as a temporary endeavor undertaken to create a unique product, service, or result. Temporary means having a defined beginning and a definite end. The term "unique" implies that the project is different from other projects in terms of characteristics.

Project Management

This author defines project management as the process of managing, allocating, and timing resources to achieve a given goal in an efficient and expeditious manner.

PMBOK defines project management as the application of knowledge, skills, tools, and techniques to project activities to achieve project objectives.

Other sources define project management as the collection of skills, tools, and management processes essential for executing a project successfully.

Project Management Methodology

A project management methodology defines a process that a project team uses in executing a project, from planning through phase-out. Figure 3.1 presents general framework for cross-functional application of the project management methodology. People, process, and technology assets (science and engineering) form the basis for implementing organizational goals. Human resources constitute crucial capital that must be recruited, developed, and preserved. Organizational work process must take advantage of the latest tools and techniques such as business process reengineering (BPR), continuous process improvement (CPI), Lean, Six Sigma, and systems thinking. The coordinated infrastructure represents the envelope of operations and includes physical structures, energy, leadership, operating culture, and movement of materials. The ability of an organization to leverage science and technology to move up the global value chain requires the softer side of project management in addition to the technical techniques. Another key benefit of applying integrative project management to oil and gas projects centers around systems safety. Science, technology, and engineering undertakings can be volatile and subject to safety violations through one of the following actions:

1. Systems or individuals who deliberately, knowingly, willfully, or negligently violate embedded safety requirements in science, technology, and engineering projects
2. Systems or individuals who inadvertently, accidentally, or carelessly compromise safety requirements in science, technology, and engineering projects

The above potential avenues for safety violation make safety training, education, practice, safety monitoring, and ethics very essential. An integrative approach to project management helps to cover all the possible ways for safety compromise.

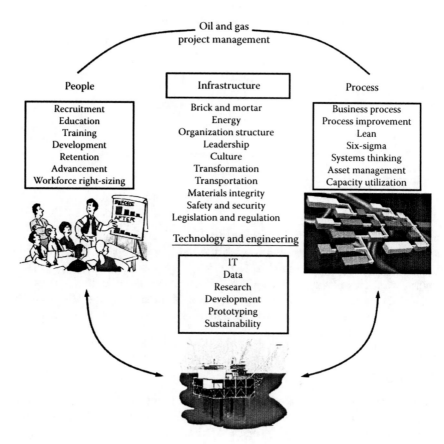

FIGURE 3.1
Framework for cross-functional application of project management.

Project Management Information System

A project management information system (PMIS) refers to an automated system or computer software used by the project management team as a tool for the execution of the activities contained in the project management plan.

Project Management System

A project management system (PMS) is the set of interrelated project elements whose collective output, through synergy, exceeds the sum of the individual outputs of the elements.

Composition of a Program

A program is defined as a recurring group of interrelated projects managed in a coordinated and synergistic manner to obtain integrated results that are

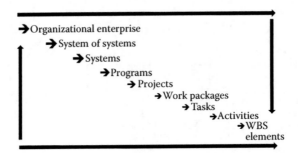

FIGURE 3.2
Hierarchy of project systems.

better than what is possible by managing the projects individually. Programs often include elements of collateral work outside the scope of the individual projects. Thus, a program is akin to having a system of systems of projects, whereby an entire enterprise might be affected. While projects have definite end points, programs often have unbounded life spans. Figure 3.2 shows the hierarchy of project systems, from organizational enterprise to work breakdown structure (WBS) elements.

Identification of Stakeholders

Stakeholders are individuals or organizations whose interests may be positively or negatively impacted by a project. Stakeholders must be identified by the project team for every project. A common deficiency in this requirement is that organization employees are often ignored, neglected, or taken for granted as stakeholders in projects going on in the organization. As the definition of stakeholders clearly suggests, if the interests of the employees can be positively or negatively affected by a project, then the employees must be viewed as stakeholders. All those who have a vested interest in the project are stakeholders and this might include the following:

- Customers
- Project sponsor
- Users
- Associated companies
- Community
- Project manager
- Owner
- Project team members
- Shareholders

Project Management Processes

The major knowledge areas of project management are administered in a structured outline covering six basic clusters as depicted in Figure 3.3. The implementation clusters represent five process groups that are followed throughout the project life cycle. Each cluster itself consists of several functions and operational steps. When the clusters are overlaid on the nine knowledge areas, we obtain a two-dimensional matrix that spans 44 major process steps.

Table 3.1 shows an overlay of the project management knowledge areas and the implementation clusters. The monitoring and controlling clusters are usually administered as one lumped process group (monitoring and controlling). In some cases, it may be helpful to separate them to highlight the essential attributes of each cluster of functions over the project life cycle. In practice, the processes and clusters do overlap. Thus, there is no crisp demarcation of when and where one process ends and where another one begins over the project life cycle. In general, project life cycle defines the following:

1. Resources that will be needed in each phase of the project life cycle
2. Specific work to be accomplished in each phase of the project life cycle

Figure 3.4 shows the major phases of project life cycle going from the conceptual phase through the close-out phase. It should be noted that project life cycle is distinguished from product life cycle. Project life cycle does not explicitly address operational issues whereas product life cycle is mostly about operational issues starting from the product's delivery to the end of its useful life. Note that for oil and gas projects, the shape of the life cycle curve may be expedited due to the rapid developments that often occur in technology. For example, for an exploration technology project, the entire life cycle may be shortened, with a very rapid initial phase, even though the

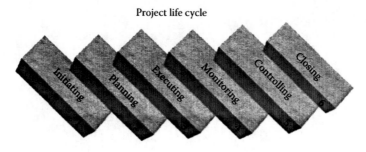

Project life cycle

FIGURE 3.3
Implementation clusters for project life cycle.

TABLE 3.1

Overlay of Project Management Areas and Implementation Clusters

Knowledge Areas	Initiating	Planning	Executing	Monitoring and Controlling	Closing
		← Project Management Process Clusters →			
Project integration	Develop project charter Develop preliminary project scope	Develop project management plan	Direct and manage project execution	Monitor and control project work Integrated change control	
Scope		Scope planning Scope definition Create WBS		Scope verification Scope control	
Time		Activity definition Activity sequencing Activity resource estimating Activity duration estimating Schedule development		Schedule control	
Cost		Cost estimating Cost budgeting		Cost control	
Quality		Quality planning	Perform quality assurance	Perform quality control	
Human resources		Human resource planning	Acquire project team Develop project team	Manage project team	
Communication		Communication planning	Information distribution	Performance reporting Manage stakeholders	
Risk	Risk management planning Risk identification Qualitative risk analysis Quantitative risk analysis Risk response planning			Risk monitoring and control	
Procurement		Plan purchases and acquisitions Plan contracting	Request seller responses Select sellers	Contract administration	Contract closure

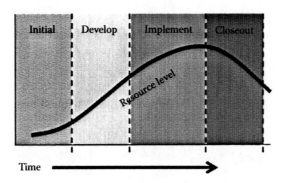

FIGURE 3.4
Phases of project life cycle.

conceptualization stage may be very long. Typical characteristics of project life cycle include the following:

1. Cost and staffing requirements are lowest at the beginning of the project and ramp up during the initial and development stages.
2. The probability of successfully completing the project is lowest at the beginning and highest at the end. This is because many unknowns (risks and uncertainties) exist at the beginning of the project. As the project nears its end, there are fewer opportunities for risks and uncertainties.
3. The risks to the project organization (project owner) are lowest at the beginning and highest at the end. This is because not much investment has gone into the project at the beginning, whereas much has been committed by the end of the project. There is a higher sunk cost manifested at the end of the project.
4. The ability of the stakeholders to influence the final project outcome (cost, quality, and schedule) is highest at the beginning and gets progressively lower toward the end of the project. This is intuitive because influence is best exerted at the beginning of an endeavor.
5. The value of scope changes decreases over time during the project life cycle while the cost of scope changes increases over time. The suggestion is to decide and finalize scope as early as possible. If there are to be scope changes, do them as early as possible.

The specific application context will determine the essential elements contained in the life cycle of the endeavor. Life cycles of business entities, products, and projects have their own nuances that must be understood and managed within the prevailing organizational strategic plan. The components of corporate, product, and project life cycles are summarized as follows:

Corporate (business) life cycle:

Planning → Needs → Business conceptualization →
Realization → Portfolio management

Product life cycle:

Feasibility studies → Development → Operations → Product obsolescence

Project life cycle:

Initiation → Planning → Execution → Monitoring and control → Closeout

This book covers the knowledge areas sequentially in Chapters 2 through 10 in the order listed above. There is no strict sequence for the application of the knowledge areas to a specific project. The areas represent a mixed bag of processes that must be followed in order to achieve a successful project. Thus, some aspects of integration may be found under the knowledge area for communications. In a similar vein, a project may start with the risk management process before proceeding into the integration process. The knowledge areas provide general guidelines. Each project must adapt and tailor the recommended techniques to the specific need and unique circumstances of the project. PMBOK seeks to standardize project management terms and definitions by presenting a common lexicon for project management activities.

Specific strategic, operational, and tactical goals and objectives are embedded within each step in the loop. For example, "initiating" may consist of project conceptualization and description. Part of "executing" may include resource allocation and scheduling. "Monitoring" may involve project tracking, data collection, and parameter measurement. "Controlling" implies taking corrective action based on the items that are monitored and evaluated. "Closing" involves phasing out or terminating a project. Closing does not necessarily mean a death sentence for a project, as the end of one project may be used as the stepping stone to the next series of endeavors.

Factors of STEP Project Success or Failure

There are several factors that impinge on the success or failure of a project. In oil and gas projects, factors that enhance project success include the following:

- Well-defined scope
- Communication among project team members

- Cooperation of project teams
- Coordination of project efforts
- Proactive management support
- Measurable metrics of project performance
- Identifiable points of accountability
- Realistic time, cost, and requirements

When projects fail, it is often due to a combination of the following factors related to project requirements:

- Requirements are incomplete
- Poor definition of project objectives
- Poor definition of scope and premature acceptance
- Requirements are unrealistic
- Requirements are ambiguous
- Requirements are inconsistent
- Changes in requirements are unbudgeted
- Poor management support
- Lack of alignment of project objectives with organizational objectives
- Poor communication
- Lack of cooperation
- Deficient coordination of project efforts

Work Breakdown Structure

WBS represents the foundation over which a project is developed and managed. WBS refers to the itemization of a project for planning, scheduling, and control purposes. WBS defines the scope of the project. In the project implementation template, WBS is developed within the scope knowledge area under the planning cluster. The WBS diagram presents the inherent components of a project in a structured block diagram or interrelationship flow chart. WBS shows the relative hierarchies of parts (phases, segments, milestone, etc.) of the project. The purpose of constructing a WBS is to analyze the elemental components of the project in detail. If a project is properly designed through the application of WBS at the project planning stage, it becomes easier to estimate cost and time requirements of the project. Project control is also enhanced by the ability to identify how components of the project link together. Tasks that are contained in the WBS collectively describe the overall project goal. Overall project planning and control can be improved by using a WBS approach. A large project may be broken down into smaller subprojects that may, in turn, be systematically broken down into task groups. Thus, WBS permits the implementation of a "divide-and-conquer" concept for project control.

Individual components in a WBS are referred to as WBS elements, and the hierarchy of each is designated by a level identifier. Elements at the same level of subdivision are said to be of the same WBS level. Descending levels provide increasingly detailed definition of project tasks. The complexity of a project and the degree of control desired determine the number of levels in the WBS. Each component is successively broken down into smaller details at lower levels. The process may continue until specific project activities (WBS elements) are reached. In effect, the structure of the WBS looks very much like an organizational chart. But it should be emphasized that WBS is not an organization chart. The basic approach for preparing a WBS is as follows:

Level 1 WBS: This contains only the final goal of the project. This item should be identifiable directly as an organizational budget item.

Level 2 WBS: This level contains the major subsections of the project. These subsections are usually identified by their contiguous location or by their related purposes.

Level 3 WBS: This level contains definable components of the level 2 subsections. In technical terms, this may be referred to as the finite element level of the project.

Subsequent levels of WBS are constructed in more specific details depending on the span of control desired. If a complete WBS becomes too crowded, separate WBS layouts may be drawn for the Level 2 components. A statement of work (SOW) or WBS summary should accompany the WBS. The SOW is a narrative of the work to be done. It should include the objectives of the work, its scope, resource requirements, tentative due date, feasibility statements, and so on. A good analysis of the WBS structure will make it easier to perform scope monitoring, scope verification, and control project work later on in the project. Figure 3.5 shows an example of a WBS structure for a hypothetical design project.

Project Organization Structures

Project organization structure provides the framework for implementing a project across functional units of an organization. Project organization structure facilitates integration of functions through cooperation and synergy. Project organizational structures are used to achieve coordinated and

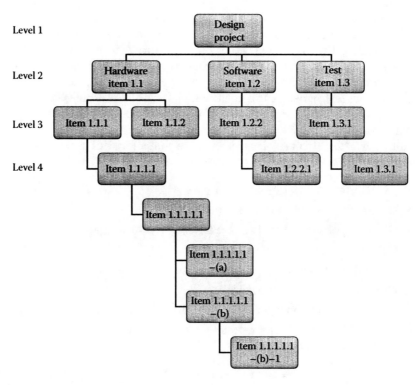

FIGURE 3.5
Example of WBS structure for a design project.

cross-functional efforts to accomplish organizational tasks. There are three basic types of organizational structures for projects:

1. Functional organization structure
2. Projectized organization structure
3. Matrix organization structure

However, some specialized or customized adaptations of the three basic structures are used in practice to meet unique project situations. Before selecting an organizational structure, the project team should assess the nature of the job to be performed and its requirements, as contained in the WBS. The structure may be defined in terms of functional specializations, departmental proximity, standard management boundaries, operational relationships, or product requirements.

Traditional Formal Organization Structures

Many organizations use the traditional formal or classical organization structures, which show hierarchical relationships between individuals or teams of individuals. Traditional formal organizational structures are effective in service enterprises because groups with similar functional responsibilities are clustered at the same level of the structure. A formal organizational structure represents the officially sanctioned structure of a functional area. An informal organizational structure, on the other hand, develops when people organize themselves in an unofficial way to accomplish a project objective. The informal organization is often very subtle in that not everyone in the organization is aware of its existence. Both formal and informal organizations exist within every project. Positive characteristics of the traditional formal organizational structure include the following:

- Availability of broad manpower base
- Identifiable technical line of control
- Grouping of specialists to share technical knowledge
- Collective line of responsibility
- Possibility of assigning personnel to several different projects
- Clear hierarchy for supervision
- Continuity and consistency of functional disciplines
- Possibility for the establishment of departmental policies, procedures, and missions

However, the traditional formal structure does have some shortcomings as summarized below:

- No one individual is directly responsible for the total project.
- Project-oriented planning may be impeded.
- There may not be a clear line of reporting up from the lower levels.
- Coordination is complex.
- A higher level of cooperation is required between adjacent levels.
- The strongest functional group may wrongfully claim project authority.

Functional Organization

The most common type of formal organization is known as the functional organization, whereby people are organized into groups dedicated to particular functions. This structure highlights the need for specialized areas of responsibilities, such as marketing, finance, accounting, engineering, production, design, and administration. In a functional organization, personnel are grouped by job function. While organizational integration is usually desired in an enterprise, there still exists a need to have service differentiation. This helps to distinguish between business units and functional responsibilities. Depending on the size and the type of auxiliary activities involved, several minor, but supporting, functional units can be developed for a project.

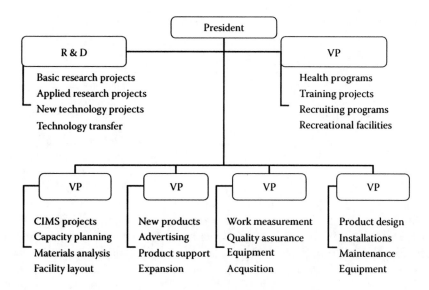

FIGURE 3.6
Functional organization structure.

Projects that are organized along functional lines normally reside in a specific department or area of specialization. The project home office or headquarters is located in the specific functional department. Figure 3.6 shows examples of projects that are organized under the functional structure. The advantages of a functional organization structure are presented below:

- Improved accountability
- Personnel within the structure have one clear chain of command (supervision)
- Discernible lines of control
- Individuals perform projects only within the boundaries of their respective functions
- Flexibility in manpower utilization
- Enhanced comradeship of technical staff
- Improved productivity of specially skilled personnel
- Potential for staff advancement along functional path
- Ability of the home office to serve as a refuge for project problems

The disadvantages of a functional organization structure include:

- Potential division of attention between project goals and regular functions
- Conflict between project objectives and regular functions

- Poor coordination similar project responsibilities
- Unreceptive attitudes on the part of the surrogate department
- Multiple layers of management
- Lack of concentrated effort

Projectized Organization

Another approach to organizing a project is to use the end product or goal of the project as the determining factor for personnel structure. This is known as the projectized structured, but often referred to as pure project organization or product organization, whereby the project is organized around a particular product (e.g., project deliverable, goal). The project is set up as a unique entity within the parent organization. It has its own dedicated technical staff and administration. It is linked to the rest of the system through progress reports, organizational policies, procedures, and funding. The interface between product-organized projects and other elements of the organization may be strict or liberal, depending on the organization. An example of a pure project organization is shown in Figure 3.7. Projects A, B, C, and D in the figure may directly represent product types A, B, C, and D. Projectized organization structure is suitable for two categories of companies:

1. Companies that use management-by-projects as a philosophy of their operations
2. Companies that derive most of their revenues from performing projects for a fee

Such organizations normally have performance systems in place to monitor, track, and control projects. For these companies, the personnel are often colocated.

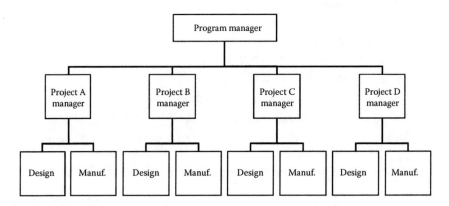

FIGURE 3.7
Projectized organization structure.

The project organization is common in industries that have multiple product lines. Unlike the functional, the project organization decentralizes functions. It creates a unit consisting of specialized skills around a given project or product. Sometimes referred to as a team, task force, or product group, the project organization is common in public, research, and manufacturing organizations where specially organized and designated groups are assigned specific functions. A major advantage of the product organization is that it gives the project members a feeling of dedication to and identification with a particular goal.

A possible shortcoming of the project organization is the requirement that the product group be sufficiently funded to be able to stand alone. The product group may be viewed as an ad hoc unit that is formed for the purpose of a specific product. The personnel involved in the project are dedicated to the particular mission at hand. At the conclusion of the mission (e.g., product phase-out), the personnel may be reassigned to other projects. Product organization can facilitate the most diverse and flexible grouping of project participants. It has the following advantages:

- Simplicity of structure
- Unity of project purpose
- Localization of project failures
- Condensed and focused communication lines
- Full authority of the project manager
- Quicker decisions due to centralized authority
- Skill development due to project specialization
- Improved motivation, commitment, and concentration
- Flexibility in determining time, cost, performance trade-offs
- Project team's reporting directly to one project manager or boss
- Ability of individuals to acquire and maintain expertise on a given project

The disadvantages of product organization are:

- Narrow view on the part of project personnel (as opposed to a global organizational view)
- The same functional expertise is replicated (or duplicated) in multiple projects
- Mutually exclusive allocation of resources (one worker to one project)
- Duplication of efforts on different but similar projects
- Monopoly of organizational resources
- Project team members may have concerns about life-after-the-project
- Reduced skill diversification

Another disadvantage of the product organization is the difficulty supervisors have in assessing the technical competence of individual team members. Since managers may supervise functional personnel in fields foreign to them, it is difficult for them to assess technical capability. For example, a project manager in a projectized structure may supervise personnel from accounting, engineering, design, manufacturing, sales, marketing, and so on. Many major organizations face this problem.

Matrix Organization Structure

The matrix organization structure is a blend of functional and projectized structures. It is a frequently used organization structure in business and industry. It is used where there is multiple managerial accountability and responsibility for a project. It combines the advantages of the traditional structure and the product organization structure. The hybrid configuration of the matrix structure facilitates maximum resource utilization and increased performance within time, cost, and performance constraints. There are usually two chains of command involving both horizontal and vertical reporting lines. The horizontal line deals with the functional line of responsibility while the vertical line deals with the project line of responsibility. An example of a matrix structure is shown in Figure 3.8. The personnel along each vertical line of reporting cross over horizontally to work on the "matrixed" project. The matrix structure is said to be *strong* if it is more closely aligned

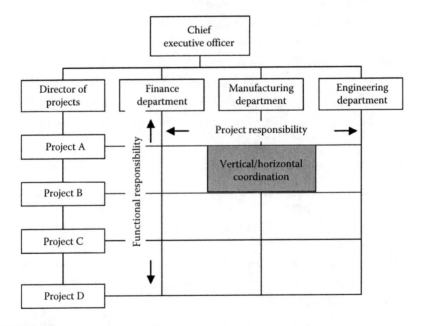

FIGURE 3.8
Matrix organization structure.

with the projectized organization structure, and it is said to be a weak matrix structure if it is more closely aligned to a functional structure. A balanced matrix structure blends projectized and functional structures equally.

Advantages of matrix organization include the following:

- Good team interaction
- Consolidation of objectives
- Multilateral flow of information
- Lateral mobility for job advancement
- Individuals have an opportunity to work on a variety of projects
- Efficient sharing and utilization of resources
- Reduced project cost due to sharing of personnel
- Continuity of functions after project completion
- Stimulating interactions with other functional teams
- Functional lines rally to support the project efforts
- Each person has a "home" office after project completion
- Company knowledge base is equally available to all projects

Some of the disadvantages of matrix organization are summarized below:

- Matrix response time may be slow for fast-paced projects
- Each project organization operates independently
- Overhead cost due to additional lines of command
- Potential conflict of project priorities
- Problems inherent in having multiple bosses
- Complexity of the structure

Traditionally, industrial projects are conducted in serial functional implementations such as R&D, engineering, manufacturing, and marketing. At each stage, unique specifications and work patterns may be used without consulting the preceding and succeeding phases. The consequence is that the end product may not possess the original intended characteristics. For example, the first project in the series might involve the production of one component while the subsequent projects might involve the production of other components. The composite product may not achieve the desired performance because the components were not designed and produced from a unified point of view. The major appeal of matrix organization is that it attempts to provide synergy within groups in an organization. Table 3.2 summarizes the levels of responsibilities and project characteristics under different organizational structures. In a projectized structure, the project manager enjoys high to almost total power for project authority and resource

TABLE 3.2

Levels of Project Characteristics under Different Organizational Structures

Project Characteristics	Organizational Structures				
	Functional	**Weak Matrix**	**Balanced Matrix**	**Strong Matrix**	**Projectized**
Project manager's authority	Low	Limited	Low to moderate	Moderate to high	High
Resource availability	Low	Limited	Low to moderate	Moderate to high	High
Control of project budget	Functional manager	Functional manager	Mixed	Project manager	Project manager
Role of project manager	Part-time	Part-time	Part-time	Full-time	Full-time
Project management staff	Part-time	Part-time	Part-time	Full-time	Full-time

availability, whereas he or she will have little power on project authority and resource availability under a functional structure.

Elements of a Project Plan

A project plan represents the roadmap for executing a project. It contains the outline of the series actions needed to accomplish the project goal. Project planning determines how to initiate a project and execute its objectives. It may be a simple statement of a project goal or it may be a detailed account of procedures to be followed during the project life cycle. In a project plan, all roles and responsibilities must be clearly defined. A project plan is not a bar chart or Gantt chart. The project manager must be versatile enough to have knowledge of most of the components of a project plan. The usual components of a detailed project plan include the following:

- Scope planning
- Scope definition
- WBS
- Activity definition
- Activity sequencing
- Activity resource estimating
- Activity duration estimating
- Schedule development
- Cost estimating
- Cost budgeting

- Quality plan
- Human resource plan
- Communications plan
- Risk management plan
- Risk identification
- Qualitative and quantitative risk analysis
- Risk response planning
- Contingencies
- Purchase plan
- Acquisition plan
- Contracting plan

Integrated Systems Approach to STEP Projects

Project management tools for STEP projects can be classified into three major categories described below:

1. *Qualitative tools*: These are the managerial tools that aid in the interpersonal and organizational processes required for project management.
2. *Quantitative tools*: These are analytical techniques that aid in the computational aspects of project management.
3. *Computer tools*: These are computer software and hardware tools that simplify the process of planning, organizing, scheduling, and controlling a project. Software tools can help in with both the qualitative and quantitative analyses needed for project management.

Managing Project Requirements

It is often said that Henry Ford offered his Model T automobile customers only one color option by saying that customers could have "any color they want, as long as it is black." But the fact is that Ford initially offered three colors: green, bright red, and green from 1908 through 1914. But when his production technology advanced to the stage of mass production on moving assembly line, the new process required a fast-drying paint, and only one particular black paint pigment met the requirements. Thus, as a result of the emergence of fast-moving mass production lines, Ford was forced to limit color options to black only. This led to the need for the famous quote. The black-only era spanned the period from 1914 through 1925, when further painting advances made it possible to have more color options. This represents a classic example of how technology limitations might dictate the

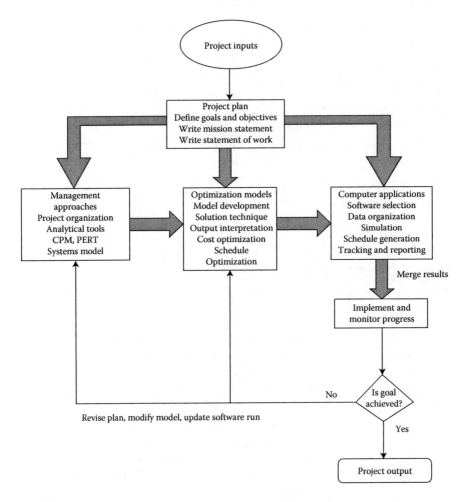

FIGURE 3.9
Flowchart of integrated STEP project management.

execution of project requirements. In oil and gas project management, an organization must remain flexible with operational choices as reflected in the flowchart in Figure 3.9.

Project Integration

Project integration management specifies how the various parts of a project come together to make up the complete project. This knowledge area recognizes the importance of linking several aspects of a project into an integrated whole. This section emphasizes the importance of "togetherness" in any project environment. Project integration management area includes the processes and activities needed to identify, define, combine, unify, and coordinate the various processes and project activities. The traditional concepts of systems analysis are applicable to project processes. The definition of a project system and its components refers to the collection of interrelated elements organized for the purpose of achieving a common goal. The elements are organized to work synergistically together to generate a unified output that is greater than the sum of the individual outputs of the components. The harmony of project integration is evident in the characteristic symbol that this book uses to denote this area of project management knowledge.

While the knowledge areas of project management, as discussed in the preceding sections, overlap and can be implemented in alternate orders, it is still apparent that project integration management is the first step of the project effort. This is particularly based on the fact that the project charter and the project scope statement are developed under the project integration process. In order to achieve a complete and unified execution of a project, both qualitative and quantitative skills must come into play.

Stepwise Project Integration

The integration component of the body of knowledge consists of the elements shown below:

Step 1: Develop project charter

Step 2: Develop preliminary project scope

Step 3: Develop project management plan

Step 4: Direct and manage project execution

Step 5: Monitor and control project work

Step 6: Perform integrated change control

Step 7: Close project

The seven elements in the block diagram are carried out across the process groups presented earlier. The overlay of the elements and the process groups are shown in Table 3.3.

In addition to the standard PMBOK inputs, tools, techniques, and outputs, the project team will add in-house items of interest to the steps presented in this section. Such in-house items are summarized below:

- *Inputs*: Other in-house (custom) factors of relevance and interest
- *Tools and techniques*: Other in-house (custom) tools and techniques
- *Outputs*: Other in-house outputs, reports, and data inferences of interest to the organization

TABLE 3.3

Implementation of Project Integration Elements across Process Groups

	Initiating	Planning	Executing	Monitoring and Controlling	Closing
Project integration	1. Develop project charter 2. Develop preliminary project scope	3. Develop project management plan	4. Direct and manage project execution	5. Monitor and control project work 6. Integrated change control	7. Close project

Table 3.4 shows the input-to-output items for developing project charter. The tabular format is useful for explicitly identifying what the project analyst needs to do or use for each step of the project management process. Tables 3.5 through 3.10 present the input-to-output entries for the other steps under integration management.

Step 1: Develop Project Charter

Project charter formally authorizes a project. It is a document that provides authority to the project manager and it is usually issued by a project initiator or sponsor external to the project organization. The purpose of a charter is to define at a high level what the project is about, what the project will deliver, what resources are needed, what resources are available, and how the project is justified. The charter also represents an organizational commitment to dedicate the time and resources to the project. The charter should be shared with all stakeholders as a part of the communication requirement for the project. Cooperating stakeholders will not only sign-off on the project,

TABLE 3.4

Tools and Techniques for Developing Project Charter within Integration Management

Step 1: Develop Project Charter		
Inputs	**Tools and Techniques**	**Output(s)**
Project contract (if applicable) Project statement of work Enterprise environmental factors Organizational process assets Other in-house (custom) factors of relevance and interest	Project selection methods Analytic hierarchy process (AHP) Project management methodology Project management information system Expert judgment Balance scorecard Process control charts Other in-house (custom) tools and techniques	Project charter Other in-house outputs, reports, and data inferences of interest to the organization

TABLE 3.5

Tools and Techniques for Developing Preliminary Project Scope Statement within Integration Management

Step 2: Develop Preliminary Project Scope Statement		
Inputs	**Tools and Techniques**	**Output(s)**
Project charter	Project management	Preliminary project scope
Project statement of work	methodology	statement
Enterprise environmental factors	Project management	Other in-house outputs,
Organizational process assets	information system	reports, and data
Other in-house (custom) factors	Expert judgment	inferences of interest to
of relevance and interest	CMMI (capability maturity	the organization
	model integration)	
	Critical chain	
	Process control charts	
	Other in-house (custom) tools	
	and techniques	

TABLE 3.6

Tools and Techniques for Developing Project Management Plan within Integration Management

Step 3: Develop Project Management Plan		
Inputs	**Tools and Techniques**	**Output(s)**
Preliminary project scope statement	Project management	Project management
Project management processes	methodology	plan
Enterprise environmental factors	Project management	
Organizational process assets	information system	
Other in-house (custom) factors of	Expert judgment	
relevance and interest		

TABLE 3.7

Tools and Techniques for Managing Project Execution within Integration Management

Step 4: Direct and Manage Project Execution		
Inputs	**Tools and Techniques**	**Output(s)**
Project management plan	Project management	Project deliverables
Approved corrective actions	methodology	Requested changes
Approved preventive actions	Project management	Implemented change requests
Approved change requests	information system	Implemented corrective actions
Approved defect repair	Process flow diagram	Implemented preventive actions
Validated defect repair	Other in-house (custom)	Implemented defect repair
Administrative closure	tools and techniques	Work performance information
procedure		Other in-house outputs, reports,
Other in-house (custom)		and data inferences of interest
factors of relevance and		to the organization
interest		

TABLE 3.8

Tools and Techniques for Monitoring and Controlling Project Work within Integration Management

Step 5: Monitor and Control Project Work		
Inputs	**Tools and Techniques**	**Output(s)**
Project management plan	Project management methodology	Recommended corrective actions
Work performance information	Project management information system	Recommended preventive actions
Rejected change requests	Earned value management	Forecasts
Other in-house (custom) factors of relevance and interest	Expert judgment	Recommended defect repair
	Other in-house (custom) tools and techniques	Requested changes
		Other in-house outputs, reports, and data inferences of interest to the organization

TABLE 3.9

Tools and Techniques for Integrated Change Control within Integration Management

Step 6: Perform Integrated Change Control		
Inputs	**Tools and Techniques**	**Output(s)**
Project management plan	Project management methodology	Approved change requests
Requested changes	Project management information system	Rejected change requests
Work performance information	Expert judgment	Update project management plan
Recommended preventive actions	Other in-house (custom) tools and techniques	Update project scope statement
Recommended corrective actions		Approved corrective actions
Deliverables		Approved preventive actions
Other in-house (custom) factors of relevance and interest		Approved defect repair
		Validated defect repair
		Deliverables
		Other in-house outputs, reports, and data inferences of interest to the organization

but also make personal pledges to support the project. Projects are usually chartered by an enterprise, a government agency, a company, a program organization, or a portfolio organization in response to one or more of the following business opportunities or organizational problems:

- Market demand
- Response to regulatory development
- Customer request
- Business need

TABLE 3.10

Tools and Techniques for Closing Project within Integration Management

Step 7: Close Project		
Inputs	**Tools and Techniques**	**Output(s)**
Project management plan	Project management	Administrative closure
Contract documentation	methodology	procedure
Enterprise environmental factors	Project management	Contract closure procedure
Organizational process assets	information system	Final product, service or result
Work performance information	Expert judgment	Updates on organizational
Deliverables	Other in-house (custom)	process assets
Other in-house (custom) factors	tools and techniques	Other in-house outputs,
of relevance and interest		reports, and data inferences
		of interest to the organization

- Exploitation of technological advance
- Legal requirement
- Social need

The driving force for a project charter is the need for an organization to make a decision about which projects to authorize to respond to operational threats or opportunities. It is desired for a charter to be brief. Depending on the size and complexity of a project, the charter should not be more than two to three pages. Where additional details are warranted, the expatiating details can be provided as addenda to the basic charter document. The longer the basic charter, the less the likelihood that everyone will read and imbibe the contents. So, brevity and conciseness are desired virtues of good project charters. The charter should succinctly establish the purpose of the project, the participants, and general vision for the project.

The project charter is used as the basis for developing project plans. While it is developed at the outset of a project, a charter should always be fluid. It should be reviewed and updated throughout the life of the project. The components of the project charter are summarized below:

- Project overview
- Assigned project manager and authority level
- Project requirements
- Business needs
- Project purpose, justification, and goals
- Impact statement
- Constraints (time, cost, performance)
- Assumptions

- Project scope
- Financial implications
- Project approach (policies, procedures)
- Project organization
- Participating organizations and their respective roles and level of participation
- Summary milestone schedule
- Stakeholder influences
- Assumptions and constraints (organizational, environmental, external)
- Business plan and expected return on investment (ROI), if applicable
- Summary budget

The project charter does not include the project plan. Planning documents, which may include project schedule, quality plan, staff plan, communication hierarchy, financial plan, and risk plan, should be prepared and disseminated separately from the charter.

- Project overview
  - The project overview provides a brief summary of the entire project charter. It may provide a brief history of the events that led to the project, an explanation of why the project was initiated, a description of project intent and the identity of the original project owner.
- Project goals
  - Project goals identify the most significant reasons for performing a project. Goals should describe improvements the project is expected to accomplish along with who will benefit from these improvements. This section should explain what various benefactors will be able to accomplish due to the project. Note that Triple C approach requires these details as a required step to securing cooperation.
- Impact statement
  - The impact statement identifies the influence the project may have on the business, operations, schedule, other projects, current technology, and existing applications. While these topics are beyond the domain of this project, each of these items should be raised for possible action.
- Constraints and assumptions
  - Constraints and assumptions identify any deliberate or implied limitations or restrictions placed on the project along with any

current or future environment the project must accommodate. These factors will influence many project decisions and strategies. The potential impact of each constraint or assumptions should be identified.

- Project scope
  - Project scope defines the operational boundaries for the project. Specific scope components are the areas or functions to be impacted by the project and the work that will be performed. The project scope should identify both what is within the scope of the project and what is outside the scope of the project.
- Project objectives
  - Project objectives identify expected deliverables from the project and the criteria that must be satisfied before the project is considered complete.
- Financial summary
  - The financial summary provides a recap of expected costs and benefits due to the project. These factors should be more fully defined in the cost–benefit analysis of the project. Project financials must be reforecast during the life of the effort.
- Project approach
  - Project approach identifies the general strategy for completing the project and explains any methods or processes, particularly policies and procedures that will be used during the project.
- Project organization
  - The project organization identifies the roles and responsibilities needed to create a meaningful and responsive structure that enables the project to be successful. Project organization must identify the people who will play each assigned role. At minimum, this section should identify who plays the roles of project owner, project manager, and core project team.
  - A project owner is required for each project.
    - This role must be filled by one or more individuals who are the fiscal trustee(s) for the project to the larger organization. This person considers the global impact of the project and deems it worthy of the required expenditure of money and time. The project owner communicates the vision for the effort and certifies the initial project charter and project plan. Should changes be required, the project owner confirms these changes and any influence on the project charter and project plan. When project decisions cannot be made at the team level, the project owner must resolve these issues. The project owner must play an active role throughout the

project, especially ensuring that needed resources have been committed to the project and remain available.

- A project manager is required for each project.
  - The project manager is responsible for initiating, planning, executing, and controlling the total project effort. Members of the project team report to the project manager for project assignments and are accountable to the project manager for the completion of their assigned work.

Definition of Inputs to Step 1:

Contract: A contract is a contractual agreement between the organization performing the project and the organizing requesting the project. It is treated as an input if the project is being done for an external customer.

Project statement of work (SOW): This is a narrative description of products or services to be supplied by the project. For internal projects, it is provided by the project initiator or sponsor. For external projects, it is provided by the customer as part of the bid document. For example, request for proposal, request for information, request for bid, or contract statements may contain specific work to be done. The SOW indicates the following:

- Business need based on required training, market demand, technological advancement, legal requirement, government regulations, industry standards, or trade consensus
- Product scope description, which documents project requirements and characteristics of the product or service that the project will deliver
- Strategic plan, which ensures that the project supports organization's strategic goals and business tactical actions

Enterprise environmental factors: These are factors that impinge upon the business environment of the organization. They include organizational structure, business culture, governmental standards, industry requirements, quality standards, trade agreements, physical infrastructure, technical assets, proprietary information, existing human resources, personnel administration, internal work authorization system, marketplace profile, competition, stakeholder requirements, stakeholder risk tolerance levels, commercial obligations, access to standardized cost estimating data, industry risk, technology variances, product life cycle, and PMIS.

Organizational process assets: These refer to the business processes used within an organization. They include standard processes, guidelines, policies, procedures, operational templates, criteria for customizing standards to specific project requirements, organization communication matrix, responsibility matrix, project closure guidelines (e.g., sunset clause), financial controls procedure, defect management procedures, change control procedures, risk control procedures, process for issuing work authorizations, processes for approving work authorizations, management of corporate knowledge base, and so on.

Definition of Tools and Techniques for Step 1:

Project selection methods: These methods are used to determine which projects an organization will select for implementation. The methods can range from basic seat-of-the-pants heuristics to highly complex analytical techniques. Some examples are benefit measurement methods, comparative measure of worth analysis, scoring models, benefit contribution, capital rationing approaches, budget allocation methods, and graphical analysis tools. Analytical techniques are mathematical models that use linear programming, nonlinear programming, dynamic programming, integer programming, multiattribute optimization, and other algorithmic tools.

Project management methodology: This defines the set of project management process groups, their collateral processes, and related control functions that are combined for implementation for a particular project. The methodology may or may not follow a project management standard. It may be an adaptation of an existing project implementation template. It can also be a formal mature process or informal technique that aids in effectively developing a project charter.

PMIS: This is a standardized set of automated tools available within the organization and integrated into a system for the purpose of supporting the generation of a project charter, facilitating feedback as the charter is refined, controlling changes to the project charter, or releasing the approved document.

Expert judgment: This is often used to assess the inputs needed to develop the project charter. Expert judgment is available from sources such as experiential database of the organization, knowledge repository, knowledge management practices, knowledge transfer protocol, business units within the organization, consultants, stakeholders, customers, sponsors, professional organizations, technical associations, and industry groups.

Definition of Output of Step 1:

Project charter: As defined earlier in this chapter, project charter is a formal document that authorizes a project. It provides authority to the project manager and it is usually issued by a project initiator or sponsor external to the project organization. It empowers the project team to carry out actions needed to accomplish the end goal of the project.

Step 2: Develop Preliminary Project Scope Statement

Project scope presents a definition of what needs to be done. It specifies the characteristics and boundaries of the project and its associated products and services, as well as the methods of acceptance and scope control. Scope is developed based on information provided by the projected initiator or sponsor. Scope statement includes the following:

- Project and product objectives
- Product characteristics
- Service requirements
- Product acceptance criteria
- Project constraints
- Project assumptions
- Initial project organization
- Initial defined risks
- Schedule milestones
- Initial WBS
- Order-of-magnitude cost estimate
- Project configuration management requirements
- Approval requirements

Definition of Inputs to Step 2:

Inputs for Step 2 are the same as defined for Step 1 covering project charter, SOW, environmental factors, and organizational process assets.

Definition of Tools and Techniques for Step 2:

The tools and techniques for Step 2 are the same as defined for Step 1 and cover project management methodology, PMIS, and expert judgment.

Definition of Output of Step 2:

The output of Step 2 is the preliminary project scope statement, which was defined and described earlier.

Step 3: Develop Project Management Plan

A project management plan includes all actions necessary to define, integrate, and coordinate all subsidiary and complementing plans into a cohesive project management plan. It defines how the project is executed, monitored and controlled, and closed. The project management plan is updated and revised through the integrated change control process. In addition, the process of developing project management plan documents the collection of outputs of planning processes and includes the following:

- Project management processes selected by the project management team
- Level of implementation of each selected process
- Descriptions of tools and techniques to be used for accomplishing those processes
- How selected processes will be used to manage the specific project
- How work will be executed to accomplish the project objectives
- How changes will be monitored and controlled
- How configuration management will be performed
- How integrity of the performance measurement baselines will be maintained and used
- The requirements and techniques for communication among stakeholders
- The selected project life cycle and, for multiphase projects, the associated project phases
- Key management reviews for content, extent, and timing

The project management plan can be a summary or integration of relevant subsidiary, auxiliary, and ancillary project plans. All efforts that are expected to contribute to the project goal can be linked into the overall project plan, each with the appropriate level of detail. Examples of subsidiary plans are the following:

- Project scope management plan
- Schedule management plan
- Cost management plan
- Quality management plan
- Process improvement plan
- Staffing management plan
- Communication management plan
- Risk management plan
- Procurement management plan
- Milestone list
- Resource calendar
- Cost baseline

- Quality baseline
- Risk register

Definition of Inputs to Step 3:

Inputs to Step 3 are the same as defined previously and include preliminary project scope statement, project management processes, enterprise environmental factors, and organizational process assets.

Definition of Tools and Techniques for Step 3:

The tools and techniques for Step 3 are project management methodology, project information system, and expert judgment. Project management methodology defines a process that aids a project management team in developing and controlling changes to the project plan. PMIS at this step covers the following segments:

- Automated system, which is used by the project team to do the following:
  - Support generation of the project management plan
  - Facilitate feedback as the document is developed
  - Control changes to the project management plan
  - Release the approved document
- Configuration management system, which is a subsystem that includes subprocesses for accomplishing the following:
  - Submitting proposed changes
  - Tracking systems for reviewing and authorizing changes
  - Providing a method to validate approved changes
  - Implementing change management system
- Configuration management system, which forms a collection of formal procedures used to apply technical and administrative oversight to do the following:
  - Identify and document functional and physical characteristics of a product or component
  - Control any changes to such characteristics
  - Record and report each change and its implementation status
  - Support audit of the products or components to verify conformance to requirements
- Change control system is the segment of PMIS that provides a collection of formal procedures that define how project deliverables and documentation are controlled.

Expert judgment, the third tool for Step 3, is applied to develop technical and management details to be included in the project management plan.

Definition of Output of Step 3:
The output of Step 3 is the project management plan.

Step 4: Direct and Manage Project Execution

Step 4 requires the project manager and project team to perform multiple actions to execute the project plan successfully. Some of the required activities for project execution are summarized below:

- Perform activities to accomplish project objectives
- Expend effort and spend funds
- Staff, train, and manage project team members
- Obtain quotation, bids, offers, or proposals as appropriate
- Implement planned methods and standards
- Create, control, verify, and validate project deliverables
- Manage risks and implement risk response activities
- Manage sellers
- Adapt approved changes into scope, plans, and environment
- Establish and manage external and internal communication channels
- Collect project data and report cost, schedule, technical and quality progress and status information to facilitate forecasting
- Collect and document lessons learned and implement approved process improvement activities

The process of directing and managing project execution also requires implementation of the following:

- Approved corrective actions that will bring anticipated project performance into compliance with the plan
- Approved preventive actions to reduce the probability of potential negative consequences
- Approved defect repair requests to correct product defects during quality process

Definition of Inputs to Step 4:
Inputs to Step 4 are summarized as follows:

- Project management plan.
- *Approved corrective actions*: These are documented, authorized directions required to bring expected future project performance into conformance with the project management plan.

- *Approved change requests*: These include documented, authorized changes to expand or contract project scope. Can also modify policies, project management plans, procedures, costs, budgets, or revise schedules. Change requests are implemented by the project team.
- *Approved defect repair*: This is documented, authorized request for product correction of defect found during the quality inspection or the audit process.
- *Validated defect repair*: This is notification that reinspected repaired items have either been accepted or rejected.
- *Administrative closure procedure*: This documents all the activities, interactions, and related roles and responsibilities needed in executing the administrative closure procedure for the project.

Definition of Tools and Techniques for Step 4:

The tools and techniques for Step 4 are project management methodology and PMIS, and they were previously defined.

Definition of Outputs of Step 4:

- Deliverables
- Requested changes
- Implemented change requests
- Implemented corrective actions
- Implemented preventive actions
- Implemented defect repair
- Work performance information

Step 5: Monitor and Control Project Work

No organization can be strategic without being quantitative. It is through quantitative measures that a project can be tracked, measured, assessed, and controlled. The need for monitoring and control can be evident in the request for quantification (RFQ) that some project funding agencies use. Some quantifiable performance measures are schedule outcome, cost effectiveness, response time, number of reworks, and lines of computer codes developed. Monitoring and controlling are performed to monitor project processes associated with initiating, planning, executing, and closing and is concerned with the following:

- Comparing actual performance against plan
- Assessing performance to determine whether corrective or preventive actions are required, and then recommending those actions as necessary

- Analyzing, tracking, and monitoring project risks to make sure risks are identified, status is reported, and response plans are being executed
- Maintaining an accurate timely information base concerning the project's products and associated documentation
- Providing information to support status reporting, progress measurement, and forecasting
- Providing forecasts to update current cost and schedule information
- Monitoring implementation of approved changes

Definition of Inputs to Step 5:

Inputs to Step 5 include the following:

- Project management plan
- Work performance plan
- Rejected change requests
    - Change requests
    - Supporting documentation
    - Change review status showing disposition of rejected change requests

Definition of Tools and Techniques for Step 5:

- Project management methodology.
- PMIS.
- Earned value technique: This measures performance as project moves from initiation through closure. It provides means to forecast future performance based on past performance.
- Expert judgment.

Definition of Outputs of Step 5:

- *Recommended corrective actions*: Documented recommendations required to bring expected future project performance into conformance with the project management plan
- *Recommended preventive actions*: Documented recommendations that reduce the probability of negative consequences associated with project risks
- *Forecasts*: Estimates or predictions of conditions and events in the project's future based on information available at the time of the forecast

- *Recommended defect repair*: Some defects found during quality inspection and audit process recommended for correction
- Requested changes

Step 6: Integrated Change Control

Integrated change control is performed from project inception through completion. It is required because projects rarely run according to plan. Major components of integrated change control include the following:

- Identifying when a change needs to occur or when a change has occurred
- Amending factors that circumvent change control procedures
- Reviewing and approving requested changes
- Managing and regulating flow of approved changes
- Maintaining and approving recommended corrective and preventive actions
- Controlling and updating scope, cost, budget, schedule, and quality requirements based upon approved changes
- Documenting the complete impact of requested changes
- Validating defect repair
- Controlling project quality to standards based on quality reports

Combining configuration management system with integrated change control includes identifying, documenting, and controlling changes to the baseline. Project-wide application of the configuration management system, including change control processes, accomplishes three major objectives:

- Establishes evolutionary method to consistently identify and request changes to established baselines and to assess the value and effectiveness of those changes
- Provides opportunities to continuously validate and improve the project by considering the impact of each change
- Provides the mechanism for the project management team to consistently communicate all changes to the stakeholders

Integrated change control process includes some specific activities of the configuration management as summarized below:

- *Configuration identification*: This provides the basis from which the configuration of products is defined and verified, products and documents are labeled, changes are managed, and accountability is maintained.

- *Configuration status accounting*: This involves capturing, storing, and accessing configuration information needed to manage products and product information effectively.
- *Configuration verification and auditing*: This involves confirming that performance and functional requirements defined in the configuration documentation have been satisfied.

Under integrated change control, every documented requested change must be either accepted or rejected by some authority within the project management team or an external organization representing the initiator, sponsor, or customer. Integrated change control can, possibly, be controlled by a change control board.

Definition of Inputs to Step 6:

The inputs to Step 6 include the following items, which were all described earlier:

- Project management plan
- Requested changes
- Work performance information
- Recommended preventive actions
- Deliverables

Definition of Tools and Techniques for Step 6:

- *Project management methodology*: This defines a process that helps a project management team in implementing integrated change control for the project.
- *PMIS*: This is an automated system used by the team as an aid for the implementation of an integrated change control process for the project. It also facilitates feedback for the project and controls changes across the project.
- *Expert judgment*: This refers to the process whereby the project team uses stakeholders with expert judgment on the change control board to control and approve all requested changes to any aspect of the project.

Definition of Outputs of Step 6:

The outputs of Step 6 include the following:

- Approved change requested
- Rejected change requests
- Project management plan (updates)
- Project scope statement (updates)

- Approved corrective actions
- Approved preventive actions
- Approved defect repair
- Validated defect repair
- Deliverables

Step 7: Close Project

At its completion, a project must be formally closed. This involves performing the project closure portion of the project management plan or closure of a phase of a multiphase project. There are two main procedures developed to establish interactions necessary to perform the closure function:

- *Administrative closure procedure*: This provides details of all activities, interactions, and related roles and responsibilities involved in executing the administrative closure of the project. It also covers activities needed to collect project records, analyze project success or failure, gather lessons learned, and archive project information.
- *Contract closure procedure*: This involves both product verification and administrative closure for any existing contract agreements. Contract closure procedure is an input to the close contract process.

Definition of Inputs to Step 7:
The inputs to Step 7 are the following:

- Project management plan.
- Contract documentation: This is an input used to perform the contract closure process and includes the contract itself as well as changes to the contract and other documentation, such as technical approach, product description, or deliverable acceptance criteria and procedures.
- Enterprise environmental factors.
- Organizational process assets.
- Work performance information.
- Deliverables, as previously described, and also as approved by the integrated change control process.

Definition of Tools and Techniques of Step 7:

- Project management methodology
- PMIS
- Expert judgment

Definition of Outputs of Step 7:

- Administrative closure procedure
  - Procedures to transfer the project products or services to production and/or operations are developed and established at this stage
  - This stage covers a step-by-step methodology for administrative closure that addresses the following:
    - Actions and activities to define the stakeholder approval requirements for changes and all levels of deliverables
    - Actions and activities confirm project has met all sponsor, customer, and other stakeholders' requirements
    - Actions and activities to verify that all deliverables have been provided and accepted
    - Actions and activities to validate completion and exit criteria for the project
- Contract closure procedure
  - This stage provides a step-by-step methodology that addresses the terms and conditions of the contracts and any required completion or exit criteria for contract closure
  - Actions performed at this stage formally close all contracts associated with the completed project
- Final product, service, or result
  - Formal acceptance and handover of the final product, service, or result that the project was authorized to provide
  - Formal statement confirming that the terms of the contract have been met
- Organizational process assets (updates)
  - Development of the index and location of project documentation using the configuration management system
  - Formal acceptance documentation, which formally indicates the customer or sponsor has officially accepted the deliverables
  - Project files, which contain all documentation resulting from the project activities
  - Project closure documents, which consist of a formal documentation indicating the completion of the project and transfer of deliverables
  - Historical information, which is transferred to knowledge base of lessons learned for use by future projects
  - Traceability of process steps

Project Sustainability

Project efforts must be sustained for a project to achieve the intended end results in the long run. Project sustainability is not often addressed in project management, but it is very essential particularly for projects in the oil and gas industry.

Sustainability, in ordinary usage, refers to the capacity to maintain a certain process or state indefinitely. In day-to-day parlance, the concept of sustainability is applied more specifically to living organisms and systems, particularly environmental systems. As applied to the human community, sustainability has been expressed as meeting the needs of the present without compromising the ability of future generations to meet their own needs. The term has its roots in ecology as the ability of an ecosystem to maintain ecological processes, functions, biodiversity, and productivity into the future. When applied to systems, sustainability brings out the conventional attributes of a system in terms of having the following capabilities:

- Self-regulation
- Self-adjustment
- Self-correction
- Self-recreation

To be sustainable, nature's resources must only be used at a rate at which they can be replenished naturally. Within the environmental science community, there is a strong belief that the world is progressing on an unsustainable path because the Earth's limited natural resources are being consumed more rapidly than they are being replaced by nature. Consequently, a collective human effort to keep human use of natural resources within the sustainable development aspect of the Earth's finite resource limits has become an issue of urgent importance. Unsustainable management of natural resources puts the Earth's future in jeopardy.

Sustainability has become a widespread, controversial, and complex issue that is applied in many different ways, including the following:

- Sustainability of ecological systems or biological organization (e.g., wetlands, prairies, forests)
- Sustainability of human organization (e.g., ecovillages, eco-municipalities, sustainable cities)
- Sustainability of human activities and disciplines (e.g., sustainable agriculture, sustainable architecture, sustainable energy)
- Sustainability of projects (e.g., operations, resource allocation, cost control)

For project integration, the concept of sustainability can be applied to facilitate collaboration across project entities. The process of achieving continued improvement in operations, in a sustainable way, requires that engineers create new technologies that facilitate interdisciplinary thought exchanges. Under the project methodology of this book sustainability means asking questions that relate to the consistency and long-term execution of the project plan. Essential questions that should be addressed include the following:

- Is the project plan supportable under current operating conditions?
- Will the estimated cost remain stable within some tolerance bounds?
- Are human resources skills able to keep up with the ever-changing requirements of a complex project?
- Will the project team persevere toward the project goal, through both rough and smooth times?
- Will interest and enthusiasm for the project be sustained beyond the initial euphoria?

Figure 3.10 illustrates a potential distribution of how organizations embrace the tools and techniques of project management. The innovators are those who are always on the cutting edge of the applications of project management. They find creative ways to use existing tools and invest in creating new and enhanced tools. The early adopters are those who capitalize on using project management whenever an opportunity develops. Early champions are those who provide support for and encourage the application of project management. Most managers will fall in this category. Late champions are those who say "show me the money" and I will believe. They eventually come around to the side of project management once they see and experience the benefits directly. The laggards are those who remain obstinate no matter what. They deprive themselves of the structured benefits of project management. This is where organizations should focus more efforts to encourage the laggards to move in the direction of embracing and applying project management.

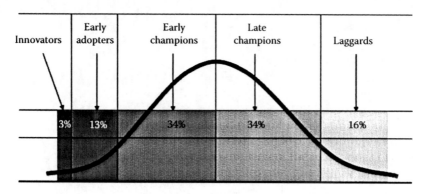

FIGURE 3.10
Percentage categories of how organizations embrace project management.

4

Oil and Gas Project Communication

Communication is the foundation of project success.

Communication is vital to everything in a project (Mooz and Howard, 2003). Any successful project manager would spend 90% of his or her time on communication activities. This is a vital function that is even more crucial in the oil and gas business because of the multitude of stakeholders. Communications management refers to the functional interface between individuals and groups within the project environment. This involves proper organization, routing, and control of information needed to facilitate work. Good communication is in effect when there is a common understanding of information between the communicator and the target. Communications management facilitates unity of purpose in the project environment. The success of a project is directly related to the effectiveness of project communication.

The project team should employ all possible avenues to get the project information across to everyone.

Communications Management: Step-by-Step Implementation

The communications management component of the PMBOK consists of the four elements shown below:

Step 1: Communications planning
Step 2: Information distribution
Step 3: Performance reporting
Step 4: Manage stakeholders

Table 4.1 shows an implementation of communications management across the process groups of project management. Tables 4.2 through 4.5 present the inputs, tools, techniques, and outputs of each step of communications management. Communications planning involves determining the information and communication needs of the stakeholders regarding who needs what information, when, where, and how. Information distribution involves making the needed information available to project stakeholders in a timely manner and in appropriate dosage. Performance reporting involves collecting

TABLE 4.1

Implementation of Communications Management across Process Groups

Initiating	Planning	Executing	Monitoring and Controlling	Closing
Project communications management	1. Communications planning	2. Information distribution	3. Performance reporting	
			4. Manage stakeholders	

Source: PMI. 2004. *A Guide to the Project Management Body of Knowledge (PMBOK Guide)*, 3rd Edition, Project Management Institute, Newtown Square, PA.

and disseminating performance information, which includes status reporting, progress measurement, and forecasting. Managing stakeholders involves managing communications to satisfy the requirements of the stakeholders so as to resolve issues that develop.

Complexity of Multiperson Communication

Communication complexity increases with an increase in the number of communication channels. It is one thing to wish to communicate freely, but it is another thing to contend with the increased complexity when more people are involved. The statistical formula of combination can be used to estimate the complexity of communication as a function of the number of communication channels or number of participants. The combination formula is used

TABLE 4.2

Tools and Techniques for Communications Planning within Project Communications Management

Step 1: Communications Planning		
Inputs	**Tools and Techniques**	**Output(s)**
Enterprise environmental factors	Communications requirement analysis	Communications management plan
Organizational process assets	Communications technology	Other in-house outputs, reports, and data inferences of interest to the organization
Project scope statement	Communications responsibility matrix	
Project constraints and assumptions	Collaborative alliance	
Other in-house (custom) factors of relevance and interest	Other in-house (custom) tools and techniques	

TABLE 4.3

Tools and Techniques for Information Distribution within Project Communications Management

Step 2: Information Distribution		
Inputs	**Tools and Techniques**	**Output(s)**
Communications management plan	Communication modes and skills	Organizational process assets (updates)
Personnel distribution list	Social networking	Other in-house outputs, reports, and data inferences of interest to the organization
Other in-house (custom) factors of relevance and interest	Influence networking	
	Meetings and dialogues	
	Communication relationships	
	Information gathering and retrieval systems	
	Information distribution methods	
	Lessons learned	
	Best practices	
	Information Eexchange	
	Other in-house (custom) tools and techniques	

TABLE 4.4

Tools and Techniques for Performance Reporting within Project Communications Management

Step 3: Performance Reporting		
Inputs	**Tools and Techniques**	**Output(s)**
Work performance information	Information presentation tools	Performance reports
Performance measurements	Performance information gathering and compilation	Forecasts
Forecasted completion	Status review meetings	Requested changes
Quality control measurements	Time reporting systems	Recommended corrective actions
Project performance measurement baseline	Cost reporting systems	Organizational process assets
Approved change requests		Other in-house outputs, reports, and data inferences of interest to the organization
List of deliverables	Other in-house (custom) tools and techniques	
Other in-house (custom) factors of relevance and interest		

TABLE 4.5

Tools and Techniques for Managing Stakeholders within Project Communications Management

Step 4: Manage Stakeholders		
Inputs	**Tools and Techniques**	**Output(s)**
Communications management plan	Communications methods	Resolved issues
Organizational process assets	Issue logs	Conflict resolution report
Other in-house (custom) factors of relevance and interest	Other in-house (custom) tools and techniques	Approved change requests
		Approved corrective actions
		Organizational process assets (updates)
		Other in-house outputs, reports, and data inferences of interest to the organization

to calculate the number of possible combinations of r objects from a set of n objects. This is written as

$$_nC_r = \frac{n!}{r![n-r]!} \tag{4.1}$$

In the case of communication, for illustration purposes, we assume that communication is between two members of a team at a time. That is, combination of two from n team members. That is, the number of possible combinations of two members out of a team of n people. Thus, the formula for communication complexity reduces to the expression below, after some of the computation factors cancel out:

$$_nC_2 = \frac{n(n-1)}{2} \tag{4.2}$$

In a similar vein, Badiru (2008) introduced a formula for cooperation complexity based on the statistical concept of permutation. Permutation is the number of possible arrangements of k objects taken from a set of n objects. The permutation formula is written as

$$_nP_k = \frac{n!}{(n-k)!} \tag{4.3}$$

Thus, for the number of possible permutations of two members out of a team of n members is estimated as

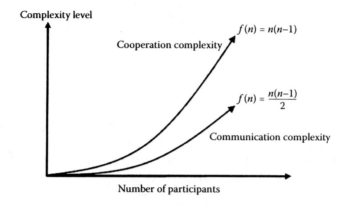

FIGURE 4.1
Plots of communication and cooperation complexities.

$$_nP_2 = n(n-1) \tag{4.4}$$

The permutation formula is used for cooperation because cooperation is bidirectional. Full cooperation requires that if A cooperates with B, then B must cooperate with A. But A cooperating with B does not necessarily imply B cooperating with A. In notational form, that is

$$A \rightarrow B \text{ does not necessarily imply } B \rightarrow A.$$

Figure 4.1 shows the relative plots of communication complexity and cooperation complexity as function of project team size, n. It is seen that complexity increases rapidly as the number of communication participants increases. Coordination complexity is even more exponential as the number of team members increases. Interested readers can derive their own coordination complexity formula based on the standard combination and permutation formulas or other statistical measures. The complexity formulas indicate a need for a more structured approach to implement the techniques of project management. The communications templates and guidelines presented in this chapter are useful for the general management of STEP projects. Each specific project implementation must adapt the guidelines to the prevailing scenario and constraints of a project.

Using the Triple C Model

The Triple C model of communication, cooperation, and coordination presented by Badiru (2008) is a viable tool for ensuring effective communication in the project environment. The Triple C model states that project management can be enhanced by implementing it within the following integrated and hierarchical processes:

- Communication
- Cooperation
- Coordination

The model facilitates a systematic approach to project planning, organizing, scheduling, and control. The Triple C model requires communication to be the first and foremost function in the project endeavor. The model explicitly provides an avenue to address questions such as the following:

When will the project be accomplished?

Which tools are available for the project?

What training is needed for the project execution?

What resources are available for the project?

Who will participate on the project team?

Figure 4.2 illustrates the three elements of the Triple C model with respect to cost, schedule, and performance improvement goals.

Figure 4.3 presents a framework for the application of the Triple C model within PMBOK.

Triple C highlights what must be done and when. It can also help to identify the resources (personnel, equipment, facilities, etc.) required for each

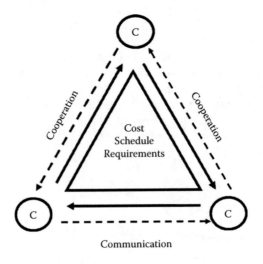

FIGURE 4.2
Triple C for planning, scheduling, and control.

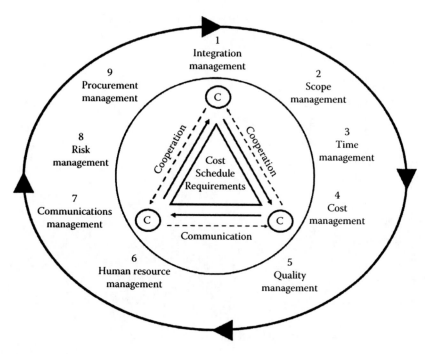

FIGURE 4.3
Implementing Triple C within PMBOK.

effort in the project. It points out important questions such as

- Does each project participant know what the objective is?
- Does each participant know his or her role in achieving the objective?
- What obstacles may prevent a participant from playing his or her role effectively?

Triple C can mitigate disparity between idea and practice because it explicitly solicits information about the critical aspects of a project. The different types of communication, cooperation, and coordination are outlined below.

Types of communication:

- Verbal
- Written
- Body language
- Visual tools (e.g., graphical tools)
- Sensual (use of all five senses: sight, smell, touch, taste, and hearing—olfactory, tactile, and auditory)
- Simplex (unidirectional)

- Half-duplex (bidirectional with time lag)
- Full-duplex (real-time dialogue)
- One-on-one
- One-to-many
- Many-to-one

Types of cooperation:

- Proximity
- Functional
- Professional
- Social
- Romantic
- Power influence
- Authority influence
- Hierarchical
- Lateral
- Cooperation by intimidation
- Cooperation by enticement

Types of coordination:

- Teaming
- Delegation
- Supervision
- Partnership
- Token-passing
- Baton hand-off

Typical Triple C Questions

Questioning is the best approach to getting information for effective project management. Everything should be questioned. By upfront questions, we can preempt and avert project problems later on. The typical questions to ask under Triple C approach are:

- What is the purpose of the project?
- Who is in charge of the project?

- Why is the project needed?
- Where is the project located?
- When will the project be carried out?
- How will the project contribute to increased opportunities for the organization?
- What is the project designed to achieve?
- How will the project affect different groups of people within the organization?
- What will be the project approach or methodology?
- What other groups or organizations will be involved (if any)?
- What will happen at the end of the project?
- How will the project be tracked, monitored, evaluated, and reported?
- What resources are required?
- What are the associated costs of the required resources?
- How do the project objectives fit the goal of the organization?
- What respective contribution is expected from each participant?
- What level of cooperation is expected from each group?
- Where is the coordinating point for the project?

How to Accomplish Triple C Communication

Communication makes working together possible. The communication function of project management involves making all those concerned become aware of project requirements and progress. Those who will be affected by the project directly or indirectly, as direct participants or as beneficiaries, should be informed as appropriate regarding the following:

- Scope of the project
- Personnel contribution required
- Expected cost and merits of the project
- Project organization and implementation plan
- Potential adverse effects if the project should fail
- Alternatives, if any, for achieving the project goal
- Potential direct and indirect benefits of the project

The communication channel must be kept open throughout the project life cycle. In addition to internal communication, appropriate external sources should also be consulted. The project manager must:

- Exude commitment to the project
- Utilize the communication responsibility matrix
- Facilitate multichannel communication interfaces
- Identify internal and external communication needs
- Resolve organizational and communication hierarchies
- Encourage both formal and informal communication links

When clear communication is maintained between management and employees and among peers, many project problems can be averted. Project communication may be carried out in one or more of the following formats:

- One-to-many
- One-to-one
- Many-to-one
- Written and formal
- Written and informal
- Oral and formal
- Oral and informal
- Nonverbal gestures

Good communication is affected when what is implied is perceived as intended. Effective communications are vital to the success of any project. Despite the awareness that proper communications form the blueprint for project success, many organizations still fail in their communications functions. The study of communication is complex. Factors that influence the effectiveness of communication within a project organization structure include the following:

1. *Personal perception.* Each person perceives events on the basis of personal psychological, social, cultural, and experimental background. As a result, no two people can interpret a given event the same way. The nature of events is not always the critical aspect of a problem situation. Rather, the problem is often the different perceptions of the different people involved.

2. *Psychological profile.* The psychological makeup of each person determines personal reactions to events or words. Thus, individual needs and level of thinking will dictate how a message is interpreted.

3. *Social environment.* Communication problems sometimes arise because people have been conditioned by their prevailing social environment to interpret certain things in unique ways. Vocabulary, idioms, organizational status, social stereotypes, and economic situation are among the social factors that can thwart effective communication.

4. *Cultural background.* Cultural differences are among the most pervasive barriers to project communications, especially in today's multinational organizations. Language and cultural idiosyncrasies often determine how communication is approached and interpreted.

5. *Semantic and syntactic factors.* Semantic and syntactic barriers to communications usually occur in written documents. Semantic factors are those that relate to the intrinsic knowledge of the subject of communication. Syntactic factors are those that relate to the form in which the communication is presented. The problems created by these factors become acute in situations where response, feedback, or reaction to the communication cannot be observed.

6. *Organizational structure.* Frequently, the organization structure in which a project is conducted has a direct influence on the flow of information and, consequently, on the effectiveness of communication. Organization hierarchy may determine how different personnel levels perceive a given communication.

7. *Communication media.* The method of transmitting a message may also affect the value ascribed to the message and consequently, how it is interpreted or used. The common barriers to project communications are:

   - Inattentiveness
   - Lack of organization
   - Outstanding grudges
   - Preconceived notions
   - Ambiguous presentation
   - Emotions and sentiments
   - Lack of communication feedback
   - Sloppy and unprofessional presentation
   - Lack of confidence in the communicator
   - Lack of confidence by the communicator
   - Low credibility of the communicator
   - Unnecessary technical jargon
   - Too many people involved
   - Untimely communication

- Arrogance or imposition
- Lack of focus

Some suggestions on improving the effectiveness of communication are presented next. The recommendations may be implemented as appropriate for any of the forms of communications listed earlier. The recommendations are for both the communicator and the audience.

1. Never assume that the integrity of the information sent will be preserved as the information passes through several communication channels. Information is generally filtered, condensed, or expanded by the receivers before relaying it to the next destination. While preparing a communication that needs to pass through several organization structures, one safeguard is to compose the original information in a concise form to minimize the need for recomposition of the project structure.

2. Give the audience a central role in the discussion. A leading role can help make a person feel a part of the project effort and responsible for the projects' success. One can then have a more constructive view of project communication.

3. Do homework and think through the intended accomplishment of the communication. This helps eliminate trivial and inconsequential communication efforts.

4. Carefully plan the organization of the ideas embodied in the communication. Use indexing or points of reference whenever possible. Grouping ideas into related chunks of information can be particularly effective. Present the short messages first. Short messages help create focus, maintain interest, and prepare the mind for the longer messages to follow.

5. Highlight why the communication is of interest and how it is intended to be used. Full attention should be given to the content of the message with regard to the prevailing project situation.

6. Elicit the support of those around you by integrating their ideas into communication. The more people feel that they have contributed to the issue, the more expeditious they are in soliciting the cooperation of others. The effect of the multiplicative rule can quickly garner support for the communication purpose.

7. Be responsive to the feelings of others. It takes two to communicate. Anticipate and appreciate the reactions of members of the audience. Recognize their operational circumstances and present your message in a form they can relate to.

8. Accept constructive criticism. Nobody is infallible. Use criticism as a springboard to higher communication performance.

9. Exhibit interest in the issue to arouse the interest of your audience. Avoid delivering your messages as a matter of a routine organizational requirement.

10. Obtain and furnish feedback promptly. Clarify vague points with examples.

11. Communicate at the appropriate time, at the right place, and to the right people.

12. Reinforce words with positive action. Never promise what cannot be delivered. Value your credibility.

13. Maintain eye contact in oral communication and read the facial expressions of your audience to obtain real-time feedback.

14. Concentrate on listening as much as speaking. Evaluate both the implicit and explicit meanings of statements.

15. Document communication transactions for future references.

16. Avoid asking questions that can be answered yes or no. Use relevant questions to focus the attention of the audience. Use questions that make people reflect upon their words, such as, "How do you think this will work?" compared to "Do you think this will work?"

17. Avoid patronizing the audience. Respect their judgment and knowledge.

18. Speak and write in a controlled tempo. Avoid emotionally charged voice inflections.

19. Create an atmosphere for formal and informal exchange of ideas.

20. Summarize the objectives of the communication and how they will be achieved.

SMART Communication

The key to getting everyone on board with a project is to ensure that task objectives are clear and comply with the principle of SMART as outlined below:

Specific: The task objective must be specific.

Measurable: The task objective must be measurable.

Aligned: The task objective must be achievable and aligned with an overall project goal.

Realistic: The task objective must be realistic and relevant to the organization.

Timed: The task objective must have a time basis.

If a task has the above intrinsic characteristics, then the function of communicating the task will more likely lead to personnel cooperation. A communication responsibility matrix shows the linking of sources of communication and targets of communication. The cells within the matrix indicate the subject of the desired communication. There should be at least one filled cell in each row and each column of the matrix. This assures that each individual of a department has at least one communication source or a target associated with him or her. With a communication responsibility matrix, a clear understanding of what needs to be communicated to whom can be developed. Communication in a project environment can take any of several forms. The specific needs of a project may dictate the most appropriate mode. Three popular computer communication modes are discussed next in the context of communicating data and information for project management.

Simplex communication. This is a unidirectional communication arrangement in which one project entity initiates communication with another entity or individual within the project environment. The entity addressed in the communication does not have a mechanism or capability for responding to the communication. An extreme example of this is a one-way, top–down communication from the top management to the project personnel. In this case, the personnel have no communication access or input to the top management. A budget-related example is a case where the top management allocates budget to a project without requesting and reviewing the actual needs of the project. Simplex communication is common in authoritarian organizations.

Half-duplex communication. This is a bidirectional communication arrangement whereby one project entity can communicate with another entity and receive a response within a certain time lag. Both entities can communicate with each other but not at the same time. An example of half-duplex communication is a project organization that permits communication with the top management without a direct meeting. Each communicator must wait for a response from the target of the communication. Request and allocation without a budget meeting is another example of half-duplex data communication in project management.

Full-duplex communication. This involves a communication arrangement that permits a dialogue between the communicating entities. Both individuals and entities can communicate with each other at the same time or face-to-face. As long as there is no clash of words, this appears to be the most receptive communication mode. It allows participative project planning in which each project personnel has an opportunity to contribute to the planning process.

Each member of a project team needs to recognize the nature of the prevailing communication mode in the project. The management must evaluate the prevailing communication structure and attempt to modify it if necessary to enhance the project functions. An evaluation of who is to communicate with whom about what may help improve the project data/information communication process. A communication matrix may include notations about the desired modes of communication between individuals and groups in the project environment.

How to Achieve Cooperation

The cooperation of team members should never be taken for granted. It must be explicitly elicited and facilitated. Good project communication is what is needed to facilitate cooperation. Merely voicing consent for a project is not enough assurance of full cooperation. Most off-the-cuff expression of cooperation is just "faking it." A project buy-in or sign-off that does not have a solid foundation of communication is groundless. Project participants and beneficiaries must be convinced of the merits of the project. That is how intrinsic and sustainable cooperation can be achieved. Some of the factors that influence cooperation in a project environment include personnel requirements, resource requirements, budget limitations, past experiences, conflicting priorities, and lack of uniform organizational support. As the lead author often says:

Our level of willing involvement and participation on project is shaped by past experiences.

We should, thus, leverage successful cooperation on one project to build cooperative groundwork to future project alliances. A structured approach to seeking cooperation should clarify the following:

- Cooperative efforts required
- Precedents for future projects
- Implication of lack of cooperation
- Criticality of cooperation to project success
- Organizational impact of cooperation
- Time frame involved in the project
- Rewards of good cooperation

Cooperation is a basic virtue of human interaction. More projects fail due to a lack of cooperation and commitment than any other project factors. To secure and retain the cooperation of the project participants, you must elicit a positive first reaction to the project. The most positive aspects of a project should be the first items of project communication. For project management, there are different types of cooperation that should be understood.

Functional cooperation. This is cooperation induced by the nature of the functional relationship between two groups. The two groups may be required to perform related functions that can only be accomplished through mutual cooperation.

Social cooperation. This is the type of cooperation effected by the social relationship between two groups. The prevailing social relationship motivates cooperation that may be useful in getting the project work done.

Legal cooperation. Legal cooperation is the type of cooperation that is imposed through some authoritative requirement. In this case, the participants may have no choice other than to cooperate.

Administrative cooperation. This is cooperation brought on by administrative requirements that make it imperative that two groups work together on a common goal.

Associative cooperation. This type of cooperation may also be referred to as collegiality. The level of cooperation is determined by the association that exists between two groups.

Proximity cooperation. Cooperation due to the fact that two groups are geographically close is referred to as proximity cooperation. Being close makes it imperative that the two groups work together.

Dependency cooperation. This is cooperation caused by the fact that one group depends on another group for some important aspect. Such dependency is usually of a mutual two-way nature. One group depends on the other for one thing while the latter group depends on the former for some other thing.

Imposed cooperation. In this type of cooperation, external agents must be employed to induce cooperation between two groups. This is applicable for cases where the two groups have no natural reason to cooperate. This is where the approaches presented earlier for seeking cooperation can become very useful.

Lateral cooperation. Lateral cooperation involves cooperation with peers and immediate associates. Lateral cooperation is often easy to achieve because existing lateral relationships create an environment that is conducive for project cooperation.

Vertical cooperation. Vertical or hierarchical cooperation refers to cooperation that is implied by the hierarchical structure of the project. For example, subordinates are expected to cooperate with their vertical superiors.

Whichever type of cooperation is available in a project environment, the cooperative forces should be channeled toward achieving project goals. Documentation of the prevailing level of cooperation is useful for winning further support for a project. Clarification of project priorities will facilitate personnel cooperation. Relative priorities of multiple projects should be specified so that any prioritized project will hold the same level of priority for all groups within the organization. Some guidelines for securing cooperation for most projects are:

- Establish achievable goals for the project.
- Clearly outline the individual commitments required.
- Integrate project priorities with existing priorities.
- Eliminate the fear of job loss due to industrialization.
- Anticipate and eliminate potential sources of conflict.

- Use an open-door policy to address project grievances.
- Remove skepticism by documenting the merits of the project.

Commitment. Cooperation must be supported with commitment. To cooperate is to support the ideas of a project. To commit is to willingly and actively participate in project efforts again and again through the thick and thin of the project. Provision of resources is one way by which the management can express commitment to a project.

Implementing Coordination

After the communication and cooperation functions have been successfully initiated, the efforts of the project personnel must be coordinated. Coordination facilitates a harmonious organization of project efforts. The construction of a responsibility chart can be very helpful at this stage. A responsibility chart is a matrix consisting of columns of individual or functional departments and rows of required actions. Cells within the matrix are filled with relationship codes that indicate who is responsible for what. Table 4.6 illustrates an example of a responsibility matrix for the planning for a seminar program. The matrix helps avoid neglecting crucial communication requirements and obligations. It can help resolve questions such as:

- Who is to do what?
- How long will it take?
- Who is to inform whom of what?
- Whose approval is needed for what?
- Who is responsible for which results?
- What personnel interfaces are required?
- What support is needed from whom and when?

Conflict Resolution Using Triple C Approach

Conflicts can and do develop in any work environment. Conflicts, whether intended or inadvertent, prevent an organization from getting the most out of the work force. When implemented as an integrated process, the Triple C model can help avoid conflicts in a project. When conflicts do develop, it can help in resolving the conflicts. The key to conflict resolution is open and direct communication, mutual cooperation, and sustainable coordination. Several sources of conflicts can exist in projects. Some of these are discussed below.

TABLE 4.6

Example of Responsibility Matrix for Project Coordination

Tasks	Person Responsible				Status of Task			
	Staff A	Staff B	Staff C	Mgr	31 Jan	15 Feb	28 Mar	21 Apr
Brainstorming meeting	R	R	R	R	D			
Identify speakers				R		O		
Select seminar location	I	R	R			O		
Select banquet location	R	R				D		
Prepare publicity materials		C	R	I	O	O	D	
Draft brochures		C	R					D
Develop a schedule			R			L	L	
Arrange for visual aids			R		L	L	L	
Coordinate activities			R				L	
Periodic review of tasks	R	R	R	S				D
Monitor progress of the program	C	R	R			O	L	
Review the program progress	R				O	O	L	L
Closing arrangements	R							L
Post-program review and evaluation	R	R	R	R			D	

Responsibility codes: R, responsible; I, inform; S, support; C, consult.
Task codes: D, done; O, on track; L, late.

Schedule conflict. Conflicts can develop because of improper timing or sequencing of project tasks. This is particularly common in large multiple projects. Procrastination can lead to having too much to do at once, thereby creating a clash of project functions and discord among project team members. Inaccurate estimates of time requirements may lead to infeasible activity schedules. Project coordination can help avoid schedule conflicts.

Cost conflict. Project cost may not be generally acceptable to the clients of a project. This will lead to project conflict. Even if the initial cost of the project is acceptable, a lack of cost control during implementation can lead to conflicts. Poor budget allocation approaches and the lack of a financial feasibility study will cause cost conflicts later on in a project. Communication and coordination can help prevent most of the adverse effects of cost conflicts.

Performance conflict. If clear performance requirements are not established, performance conflicts will develop. Lack of clearly defined performance standards can lead each person to evaluate his or her own performance based on personal value judgments. To uniformly evaluate the quality of work and monitor the project progress, performance standards should be established by using the Triple C approach.

Management conflict. There must be a two-way alliance between the management and the project team. The views of the management should be understood by the team. The views of the team should be appreciated by the management. If this does not happen, management conflicts will develop. A lack of a two-way interaction can lead to strikes and industrial actions, which can be detrimental to project objectives. The Triple C approach can help create a conducive dialogue environment between the management and the project team.

Technical conflict. If the technical basis of a project is not sound, technical conflict will develop. New industrial projects are particularly prone to technical conflicts because of their significant dependence on technology. Lack of a comprehensive technical feasibility study will lead to technical conflicts. Performance requirements and systems specifications can be integrated through the Triple C approach to avoid technical conflicts.

Priority conflict. Priority conflicts can develop if project objectives are not defined properly and applied uniformly across a project. Lack of a direct project definition can lead each project member to define his or her own goals that may be in conflict with the intended goal of a project. Lack of consistency of the project mission is another potential source of priority conflicts. Overassignment of responsibilities with no guidelines for relative significance levels can also lead to priority conflicts. Communication can help defuse priority conflict.

Resource conflict. Resource allocation problems are a major source of conflict in project management. Competition for resources, including personnel, tools, hardware, software, and so on, can lead to disruptive clashes among project members. The Triple C approach can help secure resource cooperation.

Power conflict. Project politics lead to a power play that can adversely affect the progress of a project. Project authority and project power should be clearly delineated. Project authority is the control that a person has by virtue of his or her functional post. Project power relates to the clout and influence that a person can exercise due to connections within the administrative structure. People with popular personalities can often wield a lot of project power in spite of low or nonexistent project authority. The Triple C model can facilitate a positive marriage of project authority and power to the benefit of project goals. This will help define clear leadership for a project.

Personality conflict. Personality conflict is a common problem in projects involving a large group of people. The larger the project, the larger the size of the management team needed to keep things running. Unfortunately, the larger management team creates an opportunity for personality conflicts. Communication and cooperation can help defuse personality conflicts. In summary, conflict resolution through Triple C can be achieved by observing the following guidelines:

1. Confront the conflict and identify the underlying causes
2. Be cooperative and receptive to negotiation as a mechanism for resolving conflicts
3. Distinguish between proactive, inactive, and reactive behaviors in a conflict situation
4. Use communication to defuse internal strife and competition
5. Recognize that short-term compromise can lead to long-term gains
6. Use coordination to work toward a unified goal
7. Use communication and cooperation to turn a competitor into a collaborator

It is the little and often neglected aspects of a project that lead to project failures. Several factors may constrain the project implementation. All the relevant factors can be evaluated under the Triple C model right from the project-initiation stage.

A summary of lessons to be inferred from a Triple C approach are

- Use proactive planning to initiate project functions.
- Use preemptive planning to avoid project pitfalls.
- Use meetings strategically. Meeting is not *work*. Meeting should be done to facilitate work.
- Use project assessment to properly frame the problem, adequately define the requirements, continually ask the right questions, cautiously analyze risks, and effectively scope the project.
- Be bold to terminate a project when termination is the right course of action. Every project needs an exit plan. In some cases, there is victory in capitulation.

The sustainability of the Triple C approach is summarized below:

1. For effective communication, create good communication channels.
2. For enduring cooperation, establish partnership arrangements.
3. For steady coordination, use a workable organization structure for communications team building.

References

Badiru, A. B. 2008, *Triple C Model of Project Management*, Taylor & Francis/CRC Press, Boca Raton, FL.

Mooz, H., Kevin, F., and Howard C. 2003, *Communicating Project Management*, John Wiley & Sons, Hoboken, New Jersey.

PMI. 2004, *A Guide to the Project Management Body of Knowledge (PMBOK Guide)*, 3rd Edition, Project Management Institute, Newtown Square, PA.

5

Critical Path Method for Oil and Gas Projects

On the critical path, the shortest distance between two poor points is a curve.

Like a pipeline network, activities that make up a project form a network of interrelationships. Consider the network in Figure 5.1. The complexity of the activity network in a large project increases rapidly with increase in the number of activities. Network analysis is essential for making sense out of the jumble of activities. Project scheduling is the time-phased sequencing of network activities subject to precedence relationships, time constraints, and resource limitations to accomplish specific objectives. The computational approaches to project network analysis using PERT, CPM, and PDM (precedence diagramming method) are presented. Several graphical variations of Gantt charts are presented. CPM network charts and Gantt charts are excellent visual communication tools for conveying project scope, requirements, and lines of responsibility.

Because of the long-run nature of large projects in the oil and gas industry, activity scheduling and long-term coordination are very important. There are five main categories of scheduling as listed below:

1. Stochastic project scheduling
2. Fuzzy project scheduling
3. Proactive project scheduling
4. Reactive project scheduling
5. Hybrid predictive project scheduling

Stochastic scheduling recognizes the fact that variability exists in the attributes of the schedule elements. These could be in terms of time, cost, requirements, and human resources. Fuzzy scheduling considers the imprecision associated with the parameters of the activities in the project schedule. Instead of precise parameter end points, we have shades of overlap of parameter values. In proactive scheduling, advance contingencies are built into the project schedule. This can be very useful in preempting problems in the project schedule down the line. In reactive scheduling, the project team is poised to react and make adjustments when adverse events develop in the project

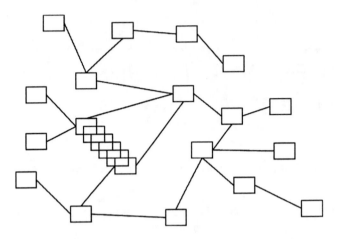

FIGURE 5.1
Network of project activities.

schedule. In predictive scheduling, a projection is developed about what the future holds for the project as things unfold in the project environment.

Activity Networks

Activity network analysis is distinguished from job shop, flow shop, and other production sequencing problems because of the unique nature of many of the activities that make up a project. In production scheduling, the scheduling problem follows a standard procedure that determines the characteristics of production operations. A scheduling technique that works for one production run may be expected to work equally effectively for succeeding and identical production runs. By contrast, projects usually involve one-time endeavors that may not be duplicated in identical circumstances. In some cases, it may be possible to duplicate the concepts of the whole project or a portion of it. Several techniques have been developed for the purpose of planning, scheduling, and controlling projects. Project schedules may be complex, unpredictable, and dynamic. Complexity may be due to interdependencies of activities, multiple resource requirements, multiple concurrent events, conflicting objectives, technical constraints, and schedule conflicts. Unpredictability may be due to equipment breakdowns, raw material inconsistency (delivery and quality), operator performance, labor absenteeism, and unexpected events. Dynamism may be due to resource variability, changes in work orders, and resource substitutions. We define *predictive scheduling* as a scheduling approach that attempts to anticipate the potential causes of schedule problems. These problems are corrected by contingency plans. We define *reactive scheduling* as a scheduling approach that reacts to problems that develop in the scheduling environment.

Critical Path Method

The most widely used scheduling aids involve network techniques, two of which are the CPM and the PERT. The network of activities contained in a project provides the basis for scheduling the project. The PDM is also commonly used particularly in scheduling concurrent projects. A project network is the graphical representation of the contents and objectives of the project. The basic project network analysis is typically implemented in three phases: network planning phase, network scheduling phase, and network control phase. *Network planning* is sometimes referred to as activity planning. This involves the identification of the relevant activities for the project. The required activities and their precedence relationships are determined. Precedence requirements may be determined on the basis of technological, procedural, or imposed constraints. The activities are then represented in the form of a network diagram. The two popular models for network drawing are the *activity-on-arrow (AOA)* and the *activity-on-node (AON)* conventions. In the AOA approach, arrows are used to represent activities, while nodes represent starting and ending points of activities. In the AON approach, nodes represent activities, while arrows represent precedence relationships. Time, cost, and resource requirement estimates are developed for each activity during the network planning phase. The estimates may be based on historical records, time standards, forecasting, regression functions, or other quantitative models. *Network scheduling* is performed by using forward pass and backward pass computational procedures. These computations give the earliest and latest starting and finishing times for each activity. The amount of slack or float associated with each activity is determined. The activity path with the minimum slack in the network is used to determine the critical activities. This path also determines the duration of the project. Resource allocation, and time–cost trade-offs are other functions performed during network scheduling. *Network control* involves tracking the progress of a project on the basis of the network schedule and taking corrective actions when needed. An evaluation of actual performance versus expected performance determines deficiencies in the project progress. The advantages of project network analysis are as follows:

- Advantages for communication

    Clarify project objectives

    Establish the specifications for project performance

    Provide a starting point for more detailed task analysis

    Present a documentation of the project plan

    Serve as a visual communication tool

- Advantages for control

  Present a measure for evaluating project performance

  Help determine what corrective actions are needed

  Give a clear message of what is expected

  Encourage team interactions
- Advantages for team interaction

  Offer a mechanism for a quick introduction to the project

  Specify functional interfaces on the project

  Facilitate ease of application

Figure 5.2 shows the graphical representation for an AON network. The components of the network are explained next.

1. *Node*: A node is a circular representation of an activity.
2. *Arrow*: An arrow is a line connecting two nodes and having an arrowhead at one end. The arrow implies that the activity at the tail of the arrow precedes the one at the head of the arrow.
3. *Activity*: An activity is a time-consuming effort required to perform a part of the overall project. An activity is represented by a node in the AON system or by an arrow in the AOA system. The job the activity represents may be indicated by a short phrase or symbol inside the node or along the arrow.
4. *Restriction*: A restriction is a precedence relationship that establishes the sequence of activities. When one activity must be completed

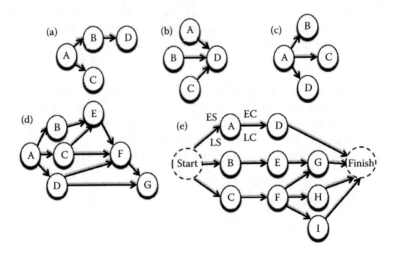

FIGURE 5.2
CPM activity network (a)–(e).

before another activity can begin, the first is said to be a predecessor of the second.

5. *Dummy*: A dummy is used to indicate one event of a significant nature (e.g., milestone). It is denoted by a dashed circle and treated as an activity with zero time duration. A dummy is not required in the AON method. However, it may be included for convenience, network clarification, or to represent a milestone in the progress of the project.

6. *Predecessor activity*: A predecessor activity is one that immediately precedes the one being considered. In Figure 5.2a, A is a predecessor of B and C.

7. *Successor activity*: A successor activity is one that immediately follows the one being considered. In Figure 5.2a, activities B and C are successors to A.

8. *Descendent activity*: A descendent activity is any activity restricted by the one under consideration. In Figure 5.2a, activities B, C, and D are all descendants of activity A.

9. *Antecedent activity*: An antecedent activity is any activity that must precede the one being considered. Activities A and B are antecedents of D. Activity A is antecedent of B and A has no antecedent.

10. *Merge point*: A merge point (see Figure 5.2b) exists when two or more activities are predecessors to a single activity. All activities preceding the merge point must be completed before the merge activity can commence.

11. *Burst point*: A burst point (see Figure 5.2c) exists when two or more activities have a common predecessor. None of the activities emanating from the same predecessor activity can be started until the burst point activity is completed.

12. *Precedence diagram*: A precedence diagram (see Figure 5.2d) is a graphical representation of the activities making up a project and the precedence requirements needed to complete the project. Time is conventionally shown to be from left to right, but no attempt is made to make the size of the nodes or arrows proportional to time.

Precedence relationships in CPM fall into three major categories:

1. Technical precedence
2. Procedural precedence
3. Imposed precedence

Technical precedence requirements are caused by the technical relationships among activities in a project. For example, in conventional construction, walls must be erected before the roof can be installed. Procedural precedence requirements are determined by policies and procedures. Such policies and procedures are often subjective, with no concrete justification. Imposed precedence requirements can be classified as resource-imposed, state-imposed, or environment-imposed. For example, resource shortages may require that one task be before another. The current status of a project (e.g., percent completion) may determine that one activity be performed before another. The environment of a project, for example, weather changes or the effects of concurrent projects, may determine the precedence relationships of the activities in a project. The primary goal of a CPM analysis of a project is the determination of the *critical path*. The critical path determines the minimum completion time for a project. The computational analysis involves *forward pass* and *backward pass* procedures. The forward pass determines the earliest start time and the earliest completion time for each activity in the network. The backward pass determines the latest start time and the latest completion time for each activity. Conventional network logic is always drawn from left to right. If this convention is followed, there is no need to use arrows to indicate the directional flow in the activity network. The notations used for activity A in the network are explained as follows:

A: Activity identification
ES: Earliest starting time
EC: Earliest completion time
LS: Latest starting time
LC: Latest completion time
t: Activity duration

During the forward pass analysis of the network, it is assumed that each activity will begin at its earliest starting time. An activity can begin as soon as the last of its predecessors is finished. The completion of the forward pass determines the earliest completion time of the project. The backward pass analysis is the reverse of the forward pass analysis. The project begins at its latest completion time and ends at the latest starting time of the first activity in the project network. The rules for implementing the forward pass and backward pass analyses in CPM are presented below. These rules are implemented iteratively until the ES, EC, LS, and LC have been calculated for all nodes in the activity network.

Rule 1

Unless otherwise stated, the starting time of a project is set equal to time 0. That is, the first node, *node* 1, in the network diagram has an earliest start time of 0. Thus

$$ES\ (1) = 0$$

If a desired starting time, t_0, is specified, then

$$ES\ (1) = t_0$$

Rule 2

The ES for any node (activity j) is equal to the maximum of the EC of the immediate predecessors of the node. That is

$$ES(i) = Max\{EC(j)\}$$
$$j \in P\{i\}$$

$P\{i\}$ = {set of immediate predecessors of activity i}.

Rule 3

The EC of activity i is the activity's earliest start time plus its estimated time, t_i. That is

$$EC(i) = ES(i) + t_i$$

Rule 4

The earliest completion time of a project is equal to the earliest completion time of the very last node, *node n*, in the project network. That is

$$EC(Project) = EC(n)$$

Rule 5

Unless the LC of a project is explicitly specified, it is set equal to the earliest completion time of the project. This is called the *zero project slack convention*. That is

$$LC(Project) = EC(Project)$$

Rule 6

If a desired deadline, T_p, is specified for the project, then

$$LC(Project) = T_p$$

It should be noted that a latest completion time or deadline may sometimes be specified for a project on the basis of contractual agreements.

Rule 7

The LC for activity j is the smallest of the latest start times of the activity's immediate successors. That is

$$LC(i) = Min\{LS(j)\}$$
$$j \in S\{i\}$$

where $S\{i\}$ = {immediate successors of activity i}.

Rule 8

The latest start time for activity j is the latest completion time minus the activity time. That is

$$LS(i) = LC(i) - t_i$$

CPM Example

Table 5.1 presents the data for a simple project network. This network and extensions of it will be used for computational examples in this chapter and subsequent chapters. The AON network for the example is given in Figure 5.3. Dummy activities are included in the network to designate single starting and ending points for the network.

Forward Pass

The forward pass calculations are shown in Figure 5.4. Zero is entered as the ES for the initial node. Since the initial node for the example is a dummy

TABLE 5.1

Data for Sample Project for CPM Analysis

Activity	Predecessor	Duration (Days)
A	—	2
B	—	6
C	—	4
D	A	3
E	C	5
F	A	4
G	B, D, E	2

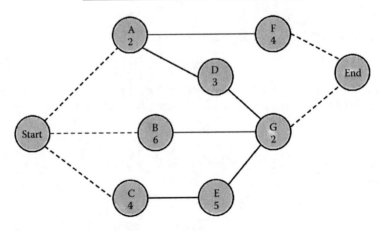

FIGURE 5.3
Example of activity network.

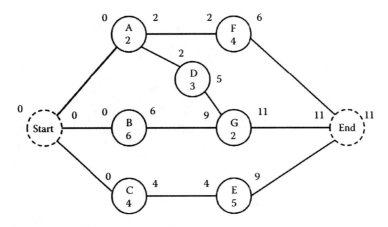

FIGURE 5.4
Forward pass analysis for CPM example.

node, its duration is 0. Thus, the EC for the starting node is equal to its ES. The ES values for the immediate successors of the starting node are set equal to the EC of the start node and the resulting EC values are computed. Each node is treated as the start node for its successors. However, if an activity has more than one predecessor, the maximum of the ECS of the preceding activities is used as the activity's starting time. This happens in the case of activity G, whose ES is determined as Max{6, 5, 9} = 9. The earliest project completion time for the example is 11 days. Note that this is the maximum of the immediately preceding earliest completion time: Max{6, 11} = 11. Since the dummy ending node has no duration, its earliest completion time is set equal to its earliest start time of 11 days.

Backward Pass

The backward pass computations establish the LS and LC for each node in the network. The results of the backward pass computations are shown in Figure 5.5. Since no deadline is specified, the latest completion time of the project is set equal to the earliest completion time. By backtracking and using the network analysis rules presented earlier, the latest completion and latest start times are determined for each node. Note that in the case of activity A with two immediate successors, the latest completion time is determined as the minimum of the immediately succeeding latest start times. That is, Min{6, 7} = 6. A similar situation occurs for the dummy starting node. In that case, the latest completion time of the dummy start node is Min{0, 3, 4} = 0. Since this dummy node has no duration, the latest starting time of the project is set equal to the node's latest completion time. Thus, the project starts at time 0 and is expected to be completed by time 11.

Within a project network, there are usually several possible paths and a number of activities that must be performed sequentially and some activities

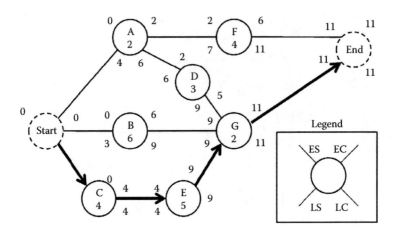

FIGURE 5.5
Backward pass analysis for CPM example.

that may be performed concurrently. If an activity has ES and EC times that are not equal, then the actual start and completion times of that activity may be flexible. The amount of flexibility an activity possesses is called a slack time. The slack time is used to determine the critical activities in the network as discussed next.

The Critical Path

The critical path is defined as the path with the least slack in the network diagram. All the activities on the critical path are said to be critical activities. These activities can create bottlenecks in the network if they are delayed. The critical path is also the longest path in the network diagram. In some networks, particularly large ones, it is possible to have multiple critical paths or a critical path subnetwork. If there are a large number of paths in the network, it may be very difficult to visually identify all the critical paths. The slack time of an activity is also referred to as its *float*. There are four basic types of activity slack:

1. *Total slack (TS)*. TS is defined as the amount of time an activity may be delayed from its earliest starting time without delaying the latest completion time of the project.

   The TS of activity *i* is the difference between the LC and the EC of the activity, or the difference between the LS and the ES of the activity.

$$TS_i = LC_i - EC_i$$

or

$$TS_i = LS_i - ES_i$$

TS is the measure that is used to determine the critical activities in a project network. The critical activities are identified as those having the minimum TS in the network diagram. If there is only one critical path in the network, then all the critical activities will be on that one path.

2. *Free slack (FS)*. FS is the amount of time an activity may be delayed from its earliest starting time without delaying the starting time of any of its immediate successors. Activity FS is calculated as the difference between the minimum earliest starting time of the activity's successors and the earliest completion time of the activity.

$$FS_i = Min\{ES_j\} - EC_i$$

$$j \in S(i)$$

FS_i = free slack for activity i

ES_j = earliest starting time of a succeeding activity, j, from the set of successors of activity i

$S(i)$ = set of successor of activity i

EC_i = earliest completion of activity i

3. *Interfering slack (IS)*. IS or interfering float is the amount of time by which an activity interferes with (or obstructs) its successors when its TS is fully used. This is rarely used in practice. The interfering float is computed as the difference between the TS and the FS.

$$IS_i = TS_i - FS_i$$

4. *Independent float (IF)*. IF or independent slack is the amount of float that an activity will always have regardless of the completion time of its predecessors or the starting times of its successors. IF is computed as

$$IF_i = Max\{0, (Min\ ES_j - Max\ LC_k - t_i)\}$$

$$j \in S(i)\ \ k \in P(i)$$

ES_j = earliest starting time of succeeding activity j

LC_k = latest completion time of preceding activity k

t_i = duration of the activity i, whose IF is being calculated

IF takes a pessimistic view of the situation of an activity. It evaluates the situation whereby the activity is pressured from either side, that is, when its predecessors are delayed as late as possible while its successors are to be started as early as possible. IF is useful for conservative planning purposes, but it is not used much in practice. Despite its low level of use, IF does have practical implications for better project management. Activities can be buffered with IFs as a way to handle contingencies. For Figure 5.5, the TS and the FS for activity A are calculated, respectively, as

$$TS = 6 - 2 = 4 \text{ days}$$

$$FS = Min\{2, 2\} - 2 = 2 - 2 = 0$$

Similarly, the TS and the FS for activity F are

$$TS = 11 - 6 = 5 \text{ days}$$

$$FS = Min\{11\} - 6 = 11 - 6 = 5 \text{ days}$$

Table 5.2 presents a tabulation of the results of the CPM example. The table contains the earliest and latest times for each activity as well as the TS and FS. The results indicate that the minimum TS in the network is 0. Thus, activities C, E, and G are identified as the critical activities. The critical path is highlighted in Figure 5.4 and consists of the following sequence of activities: START–C–E–G–END.

The TS for the overall project itself is equal to the TS observed on the critical path. The minimum slack in most networks will be zero since the ending LC is set equal to the ending EC. If a deadline is specified for a project, then we would set the project's latest completion time to the specified deadline. In that case, the minimum TS in the network would be given by

$$TS_{min} = \text{project deadline} - \text{EC of the last node in the network}$$

TABLE 5.2

Result of CPM Analysis for Sample Project

Activity	Duration	ES	EC	LS	LC	TS	FS	Criticality
A	2	0	2	4	6	4	0	—
B	6	0	6	3	9	3	3	—
C	4	0	4	0	4	0	0	Critical
D	3	2	5	6	9	4	4	—
E	5	4	9	4	9	0	0	Critical
F	4	2	6	7	11	5	5	—
G	2	9	11	9	11	0	0	Critical

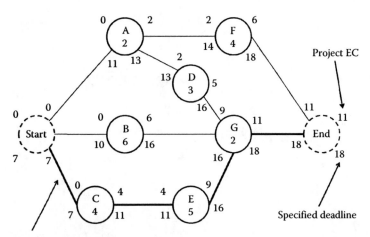

Critical path with float of 7 days

FIGURE 5.6
CPM network with deadline.

This minimum TS will then appear as the TS for each activity on the critical path. If a specified deadline is lower than the EC at the finish node, then the project will start out with a negative slack. That means that it will be behind schedule before it even starts. It may then become necessary to expedite some activities (i.e., crashing) in order to overcome the negative slack. Figure 5.6 shows an example with a specified project deadline. In this case, the deadline of 18 days comes after the earliest completion time of the last node in the network.

Gantt Charts

When the results of a CPM analysis are fitted to a calendar time, the project plan becomes a schedule. The Gantt chart is one of the most widely used tools for presenting a project schedule. A Gantt chart can show the planned and actual progress of activities. The time scale is indicated along the horizontal axis, while horizontal bars or lines representing activities are ordered along the vertical axis. As a project progresses, markers are made on the activity bars to indicate actual work accomplished. Gantt charts must be updated periodically to indicate project status. Figure 5.7 presents the Gantt chart for our illustrative example using the ES from Table 5.2. Figure 5.8 presents the Gantt chart for the example based on the LS. Critical activities are indicated by the shaded bars.

In Figure 5.7, the starting time of activity F can be delayed from day 2 until day 7 (i.e., TS = 5) without delaying the overall project. Likewise, A, D, or both

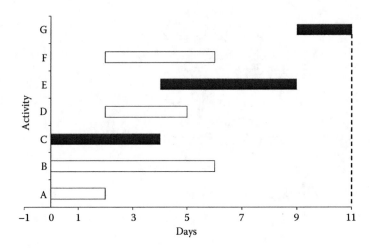

FIGURE 5.7
Gantt chart based on earliest starting times.

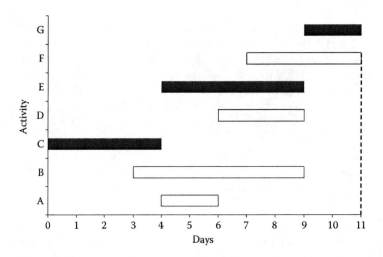

FIGURE 5.8
Gantt chart based on latest starting times.

may be delayed by a combined total of four days (TS = 4) without delaying the overall project. If all the 4 days of slack are used up by A, then D cannot be delayed. If A is delayed by 1 day, then D can be delayed by up to 3 days without causing a delay of G, which determines the project completion. The Gantt chart also indicates that activity B may be delayed by up to 3 days without affecting the project completion time. In Figure 5.8, the activities are scheduled by their latest completion times. This represents the extreme case where activity slack times are fully used. No activity in this schedule can be delayed without delaying the project. In Figure 5.8, only one activity is scheduled over the first three days. This may be compared to the schedule in Figure 5.7, which has three starting activities. The schedule in Figure 5.8 may be useful if there is a situation that permits only a few activities to be scheduled in the early stages of the project. Such situations may involve shortage of project personnel, lack of initial budget, time for project initiation, time for personnel training, allowance for learning period, or general resource constraints. Scheduling of activities based on ES times indicates an optimistic view. Scheduling on the basis of LS times represents a pessimistic approach.

Schedule Compression

Schedule compression refers to reducing the length of a project network. This is often accomplished by crashing activities. *Crashing*, sometimes referred to as expediting, reduces activity durations, thereby reducing project duration. Crashing is done as a trade-off between shorter task duration and higher task cost. It must be determined whether the total cost savings realized from reducing the project duration is enough to justify the higher costs associated with reducing individual task durations. If there is a delay penalty associated with a project, it may be possible to reduce the total project cost even though individual task costs are increased by crashing. If the cost savings on a delay penalty are higher than the incremental cost of reducing the project duration, then crashing is justified. Under conventional crashing, the further the duration of a project is compressed, the higher the total cost of the project. The objective is to determine at what point to terminate further crashing in a network. *Normal task duration* refers to the time required to perform a task under normal circumstances. *Crash task duration* refers to the reduced time required to perform a task when additional resources are allocated to it.

If each activity is assigned a range of time and cost estimates, then several combinations of time and cost values will be associated with the overall project. Iterative procedures are used to determine the best time or cost combination for a project. Time–cost trade-off analysis may be conducted, for example, to determine the marginal cost of reducing the duration of the project by one time unit. Table 5.3 presents an extension of the data for the example problem to include normal and crash times as well as normal and crash costs for each activity. The normal duration of the project is 11 days, as seen earlier, and the normal cost is $2775.

If all the activities are reduced to their respective crash durations, the total crash cost of the project will be $3545. In that case, the crash time is found by CPM analysis to be 7 days. The CPM network for the fully crashed project is shown in Figure 5.9. Note that activities C, E, and G remain critical. Sometimes, the crashing of activities may result in additional critical paths.

TABLE 5.3

Normal and Crash Time and Cost Data

Activity	Normal Duration	Normal Cost ($)	Crash Duration	Crash Cost ($)	Crashing Ratio
A	2	210	2	210	0
B	6	400	4	600	100
C	4	500	3	750	250
D	3	540	2	600	60
E	5	750	3	950	100
F	4	275	3	310	35
G	2	100	1	125	25
		2775		3545	

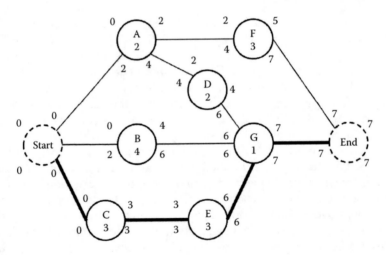

FIGURE 5.9

Example of fully crashed CPM network.

The Gantt chart in Figure 5.10 shows a schedule of the crashed project using the ES times. In practice, one would not crash all activities in a network. Rather, some heuristic would be used to determine which activity should be crashed and by how much. One approach is to crash only the critical activities or those activities with the best ratios of incremental cost versus time reduction. The last column in Table 5.3 presents the respective ratios for the activities in our example. The crashing ratio is computed as

$$r = \frac{\text{crash cost} - \text{normal cost}}{\text{normal duration} - \text{crash duration}}$$

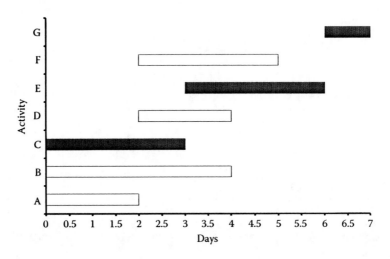

FIGURE 5.10
Gantt chart of fully crashed CPM network.

This method of computing the crashing ratio gives crashing priority to the activity with the lowest cost slope. It is a commonly used approach to expediting in CPM networks. Activity G offers the lowest cost per unit time reduction of $25. If our approach is to crash only one activity at a time, we may decide to crash activity G first and evaluate the increase in project cost versus the reduction in project duration. The process can then be repeated for the next best candidate for crashing, which in this case is activity F. The project completion time is not reduced any further since activity F is not a critical activity. After F has been crashed, activity D can then be crashed. This approach is repeated iteratively in order of activity preference until no further reduction in project duration can be achieved or until the total project cost exceeds a specified limit. A more comprehensive analysis is to evaluate all possible combinations of the activities that can be crashed. However, such a complete enumeration would be prohibitive, since there would be a total of 2^c crashed networks to evaluate, where c is the number of activities that can be crashed out of the n activities in the network ($c \leq n$). For our example, only 6 out of the 7 activities in the network can be crashed. Thus, a complete enumeration will involve $2^6 = 64$ alternate networks. Table 5.4 shows 7 of the 64 crashing options. Activity G, which offers the best crashing ratio, reduces the project duration by only 1 day. Even though activities F, D, and B are crashed by a total of 4 days at an incremental cost of $295, they do not generate any reduction in project duration. Activity E is crashed by 2 days and it generates a reduction of 2 days in project duration. Activity C, which is crashed by 1 day, generates a further reduction of 1 day in the project duration. It should be noted that the activities that generate

TABLE 5.4

Selected Crashing Options for CPM Example

Option Number	Activities Crashed	Network Duration	Time Reduction	Incremental Cost ($)	Total Cost ($)
1	None	210	2	210	0
2	G	400	4	600	100
3	G, F	500	3	750	250
4	G, F, D	540	2	600	60
5	G, F, D, B	750	3	950	100
6	G, F, D, B, E	275	3	310	35
7	G, F, D, B, E, C	100	1	125	25

reductions in project duration are the ones that were identified earlier as the critical activities.

Figure 5.11 shows the crashed project duration versus the crashing options and a plot of the total project cost after crashing. As more activities are crashed, the project duration decreases while the total project cost increases. If full enumeration were performed, the plot would contain additional points between the minimum possible project duration of 7 days (fully crashed) and the normal project duration of 11 days (no crashing). Similarly, the plot for total project cost would contain additional points between the normal cost

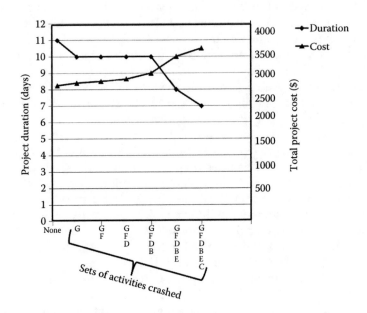

FIGURE 5.11
Plot of duration and cost as a function of crashing options.

of \$2775 and the crash cost of \$3545. In general, there may be more than one critical path, so one needs to check for the set of critical activities with the least total crashing ratio in order to minimize the total crashing cost. Also, one needs to update the critical paths every time a set of activities is crashed because new activities may become critical in the meantime. For the network shown earlier in Figure 5.9, the path C–E–G is the only critical path throughout $7 \le T \le 11$. Therefore, one need not consider crashing other jobs since the incurred cost will not affect the project completion time. There are 12 possible ways one can crash activities C, G, and E in order to reduce the project time.

Table 5.5 defines possible strategies and crashing costs for durations of $7 \le T \le 11$. Again, the strategies involve only critical arcs (activities), since crashing a noncritical arc is clearly fruitless. Figure 5.12 is a plot of the strategies with respect to cost and project duration values. The optimal strategy for each T value is the strategy with the minimum cost. Optimal strategies are connected in the figure. This piecewise linear and convex curve is referred to as the time–cost trade-off curve. Several other approaches exist for determining which activities to crash in a project network. Two alternate approaches are presented below for computing the crashing ratio, r. The first one directly uses the criticality of an activity to determine its crashing ratio while the second one uses a computational expression as shown below:

$$r = \text{criticality index}$$

$$r = \frac{\text{crash cost} - \text{normal cost}}{(\text{normal duration} - \text{crash duration})(\text{criticality index})}$$

TABLE 5.5

Project Compression Strategies

Project Duration	Crashing Strategy	Description of Crashing	Total Cost (\$)
$T = 11$	S_1	Activities at normal duration	\$2775
$T = 10$	S_2	Crash G by 1 unit	2800
	S_3	Crash C by 1 unit	3025
	S_4	Crash E by 1 unit	2875
$T = 9$	S_5	Crash G and C by 1 unit	3050
	S_6	Crash G and E by 1 unit	2900
	S_7	Crash C and E by 1 unit	3125
	S_8	Crash E by 2 units	2975
$T = 8$	S_9	Crash G, C, and E by 1 unit	3150
	S_{10}	Crash G by 1 unit, E by 2 units	3000
	S_{11}	Crash C by 1 unit, E by 2 units	3225
$T = 7$	S_{12}	Crash G and C by 1 unit, and E by 2 units	3250

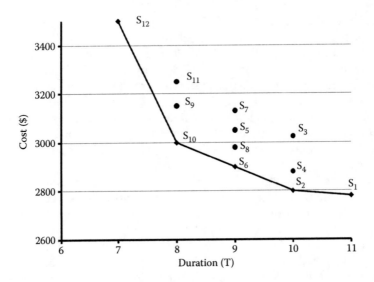

FIGURE 5.12
Time–cost plots for the strategies in Table 5.5.

The first approach gives crashing priority to the activity with the highest probability of being on the critical path. In deterministic networks, this refers to the critical activities. In stochastic networks, an activity is expected to fall on the critical path only a percentage of the time. The second approach is a combination of the approach used for the illustrative example and the criticality index approach. It reflects the process of selecting the least-cost expected value. The denominator of the expression represents the expected number of days by which the critical path can be shortened. For different project networks, different crashing approaches should be considered, and the one that best fits the nature of the network should be selected.

Program Evaluation and Review Technique

The PERT is an extension of CPM that incorporates variability in activity durations into project network analysis. PERT has been used extensively and successfully in practice. In real life, activities are often prone to uncertainties that determine the actual durations of the activities. In CPM, activity durations are assumed to be deterministic. In PERT, the potential uncertainties in activity durations are accounted for by using three time estimates for each activity. The three time estimates represent the spread of the estimated activity duration. The greater the uncertainty of an activity, the wider the range of the estimates.

PERT Formulas

PERT uses the three time estimates and simple equations to compute the expected duration and variance for each activity. The PERT formulas are based on a simplification of the expressions for the mean and variance of a beta distribution. The approximation formula for the mean is a simple weighted average of the three time estimates, with the end points assumed to be equally likely and the mode four times as likely. The approximation formula for PERT is based on the recognition that most of the observations from a distribution will lie within plus or minus three standard deviations, or a spread of six standard deviations. This leads to the simple method of setting the PERT formula for standard deviation equal to one-sixth of the estimated duration range. While there is no theoretical validation for these approximation approaches, the PERT formulas do facilitate ease of use. The formulas are presented below:

$$t_e = \frac{a + 4m + b}{6}$$

$$s^2 = \frac{(b - a)^2}{36}$$

where
a = optimistic time estimate
m = most likely time estimate
b = pessimistic time estimate ($a < m < b$)
t_e = expected time for the activity
s^2 = variance of the duration of the activity

After obtaining the estimate of the duration for each activity, the network analysis is carried out in the same manner as previously illustrated for the CPM approach. The major steps in PERT analysis are as follows:

1. Obtain three time estimates a, m, and b for each activity.

2. Compute the expected duration for each activity by using the formula for t_e.

3. Compute the variance of the duration of each activity from the formula for s^2. It should be noted that CPM analysis cannot calculate variance of activity duration, since it uses a single time estimate for each activity.

4. Compute the expected project duration, T_e. As in the case of CPM, the duration of a project in PERT analysis is the sum of the durations of the activities on the critical path.

5. Compute the variance of the project duration as the sum of the variances of the activities on the critical path. The variance of the project duration is denoted by s^2. It should be recalled that CPM cannot compute the variance of the project duration, since variances of activity durations are not computed.

6. If there are two or more critical paths in the network, choose the one with the largest variance to determine the project duration and the variance of the project duration. Thus, PERT is pessimistic with respect to the variance of project duration when there are multiple critical paths in the project network. For some networks, it may be necessary to perform a mean–variance analysis to determine the relative importance of the multiple paths by plotting the expected project duration versus the path duration variance.

7. If desired, compute the probability of completing the project within a specified time period. This is not possible under CPM.

In practice, a question often arises as to how to obtain good estimates of a, m, and b. Several approaches can be used in obtaining the required time estimates for PERT. Some of the approaches are:

- Estimates furnished by an experienced person
- Estimates extracted from standard time data
- Estimates obtained from historical data
- Estimates obtained from simple regression and/or forecasting
- Estimates generated by simulation
- Estimates derived from heuristic assumptions
- Estimates dictated by customer requirements

The pitfall of using estimates furnished by an individual is that they may be inconsistent since they are limited by the experience and personal bias of the person providing them. Individuals responsible for furnishing time estimates are usually not experts in estimation, and they generally have difficulty in providing accurate PERT time estimates. There is often a tendency to select values of a, m, and b that are optimistically skewed. This is because a conservatively large value is typically assigned to b by inexperienced individuals. The use of time standards, on the other hand, may not reflect the changes occurring in the current operating environment due to new technology, work simplification, new personnel, and so on. The use of historical data and forecasting is very popular because estimates can be verified and validated by actual records. In the case of regression and forecasting, there is the danger of extrapolation beyond the data range used for fitting the regression and forecasting models.

PERT Example

Suppose we have the project data presented in Table 5.6. The expected activity durations and variances as calculated by the PERT formulas are shown in the two right-hand columns of the table. Figure 5.13 shows the PERT network. Activities C, E, and G are shown to be critical, and the project completion time is 11 time units.

The probability of completing the project on or before a deadline of 10 time units (i.e., $T_d = 10$) is calculated as

$$T_e = 11$$

$$s^2 = V[C] + V[E] + V[G]$$

$$= 0.25 + 0.25 + 0.1111 = 0.6111$$

$$s = \sqrt{0.6111} = 0.7817$$

$$P(T \le T_d) = P(T \le 10)$$

$$= P\left(z \le \frac{10 - T_e}{s}\right)$$

$$= P\left(z \le \frac{10 - 11}{0.7817}\right)$$

$$= P(z \le -1.2793)$$

$$= 1 - P(z \le 1.2793)$$

$$= 1 - 0.8997 = 0.1003$$

TABLE 5.6

PERT Project Data

Activity	Predecessors	a	m	b	t_e	s^2
A	—	1	2	4	2.17	0.2500
B	—	5	6	7	6.00	0.1111
C	—	2	4	5	3.83	0.2500
D	A	1	3	4	2.83	0.2500
E	C	4	5	7	5.17	0.2500
F	A	3	4	5	4.00	0.1111
G	B, D, E	1	2	3	2.00	0.1111

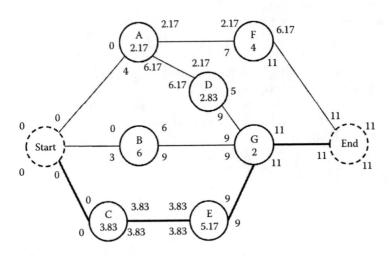

FIGURE 5.13
PERT network example.

Thus, there is just over 10% probability of finishing the project within 10 days. By contrast, the probability of finishing the project in 13 days is calculated as

$$P(T \leq 13) = P\left(z \leq \frac{13 - 11}{0.7817}\right)$$

$$= P(z \leq 2.5585) = 0.9948$$

This implies that there is over 99% probability of finishing the project within 13 days. Note that the probability of finishing the project in exactly 13 days will be 0. An exercise at the end of this chapter requires the reader to show that $P(T = T_d) = 0$. If we desire the probability that the project can be completed within a certain lower limit (T_L) and a certain upper limit (T_U), the computation will proceed as follows: Let $T_L = 9$ and $T_U = 11.5$. Then

$$P(T_L \leq T \leq T_U) = P(9 \leq T \leq 11.5)$$

$$= P(T \leq 11.5) - P(T \leq 9)$$

$$= P\left(z \leq \frac{11.5 - 11}{0.7817}\right) - P\left(z \leq \frac{9 - 11}{0.7817}\right)$$

$$= P(z \leq 0.6396) - P(z \leq -2.5585)$$

$$= P(z \leq 0.6396) - [1 - P(z \leq 2.5585)]$$

$$= 0.7389 - [1 - 0.9948] = 0.7337$$

Precedence Diagramming Method

The PDM was developed in the early 1960s as an extension of the basic PERT/CPM network analysis. PDM permits mutually dependent activities to be performed partially in parallel instead of serially. The usual finish-to-start dependencies between activities are relaxed to allow activities to be overlapped. This facilitates schedule compression. An example is the requirement that concrete should be allowed to dry for a number of days before drilling holes for handrails. That is, drilling cannot start until so many days after the completion of concrete work. This is a finish-to-start constraint. The time between the finishing time of the first activity and the starting time of the second activity is called the *lead–lag* requirement between the two activities. Figure 5.14 shows the graphical representation of the basic lead–lag relationships between activity A and activity B. The terminology is explained as follows:

SS_{AB} (start-to-start) lead: This specifies that activity B cannot start until activity A has been in progress for at least SS time units.

FF_{AB} (finish-to-finish) lead: This specifies that activity B cannot finish until at least FF time units after the completion of activity A.

FS_{AB} (finish-to-start) lead: This specifies that activity B cannot start until at least FS time units after the completion of activity A. Note that PERT/CPM approaches use $FS_{AB} = 0$ for network analysis.

SF_{AB} (start-to-finish) lead: This specifies that there must be at least SF time units between the start of activity A and the completion of activity B.

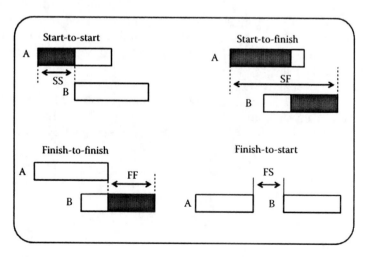

FIGURE 5.14
Lead–lag relationships in PDM.

The leads or lags may, alternately, be expressed in percentages rather than time units. For example, we may specify that 25% of the work content of activity A must be completed before activity B can start. If percentage of work completed is used for determining lead–lag constraints, then a reliable procedure must be used for estimating the percent completion. If the project work is broken up properly using WBS, it will be much easier to estimate percent completion by evaluating the work completed at the elementary task levels. The lead–lag relationships may also be specified in terms of *at most* relationships instead of *at least* relationships. For example, we may have at most an FF lag requirement between the finishing time of one activity and the finishing time of another activity. Splitting activities often simplify the implementation of PDM, as will be shown later with some examples. Some of the factors that will determine whether or not an activity can be split are technical limitations affecting splitting of a task, morale of the person working on the split task, setup times required to restart split tasks, difficulty involved in managing resources for split tasks, loss of consistency of work, and management policy about splitting jobs.

Figure 5.15 presents a simple CPM network consisting of three activities. The activities are to be performed serially and each has an expected duration of 10 days. The conventional CPM network analysis indicates that the duration of the network is 30 days. The earliest times and the latest times are as shown in the figure.

The Gantt chart for the example is shown in Figure 5.16. For comparison, Figure 5.17 shows the same network but with some lead–lag constraints. For example, there is an SS constraint of 2 days and an FF constraint of 2 days between activities A and B. Thus, activity B can start as early as 2 days after activity A starts, but it cannot finish until 2 days after the completion of A. In other words, *at least* 2 days must be between the starting times of A and B. Likewise, *at least* 2 days must separate the finishing time of A and the finishing time of B. A similar precedence relationship exists between activity B and activity C. The earliest and latest times obtained by considering the lag constraints are indicated in Figure 5.17.

The calculations show that if B is started just 2 days after A is started, it can be completed as soon as 12 days as opposed to the 20 days obtained in the case of conventional CPM. Similarly, activity C is completed at time 14, which is considerably less than the 30 days calculated by conventional CPM. The lead–lag constraints allow us to compress or overlap activities.

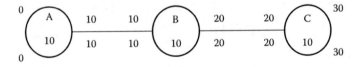

FIGURE 5.15
Serial activities in CPM network.

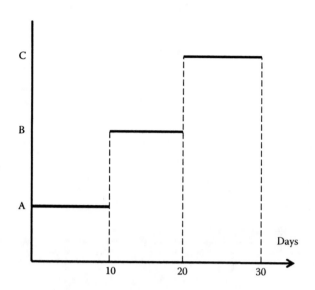

FIGURE 5.16
Gantt chart of serial activities in CPM example.

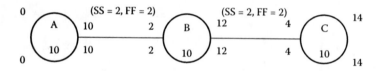

FIGURE 5.17
PDM network example.

Depending on the nature of the tasks involved, an activity does not have to wait until its predecessor finishes before it can start. Figure 5.18 shows the Gantt chart for the example incorporating the lead–lag constraints. It should be noted that a portion of a succeeding activity can be performed simultaneously with a portion of the preceding activity.

A portion of an activity that overlaps with a portion of another activity may be viewed as a distinct portion of the required work. Thus, partial completion of an activity may be elevated. Figure 5.19 shows how each of the three activities is partitioned into contiguous parts. Even though there is no physical break or termination of work in any activity, the distinct parts (beginning and ending) can still be identified. This means that there is no physical splitting of the work content of any activity. The distinct parts are determined on the basis of the amount of work that must be completed before or after another activity, as dictated by the lead–lag relationships. Note that activity A is partitioned into parts A_1 and A_2. The duration of A_1 is 2 days because there is an $SS = 2$ relationship between activity A and activity B. Since the original duration of A is 10 days, the duration of A_2 is then calculated to be $10 - 2 = 8$ days.

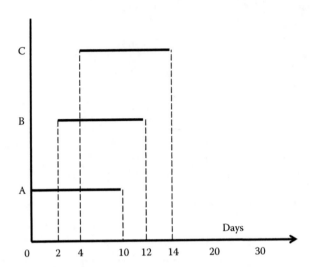

FIGURE 5.18
Gantt chart for PDM example.

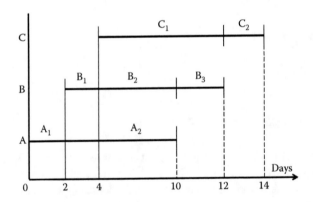

FIGURE 5.19
Partitioning of activities in PDM example.

Likewise, activity B is partitioned into parts B_1, B_2, and B_3. The duration of B_1 is 2 days because there is an SS = 2 relationship between activity B and activity C. The duration of B_3 is also 2 days because there is an FF = 2 relationship between activity A and activity B. Since the original duration of B is 10 days, the duration of B_2 is calculated to be $10 - (2 + 2) = 6$ days. In a similar fashion, activity C is partitioned into C_1 and C_2. The duration of C_2 is 2 days because there is an FF = 2 relationship between activity B and activity C. Since the original duration of C is 10 days, the duration of C_1 is then calculated to be $10 - 2 = 8$ days. Figure 5.20 shows a conventional CPM network drawn for the three activities after they are partitioned into distinct parts.

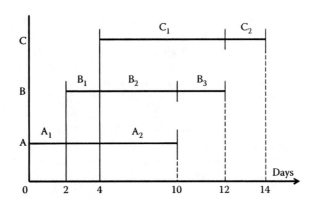

FIGURE 5.20
CPM network of partitioned activities.

The conventional forward and backward passes reveal that all the activity parts are on the critical path. This makes sense, since the original three activities are performed serially and no physical splitting of activities has been performed. Note that there are three critical paths, each with a length of 14 days. It should also be noted that the distinct parts of each activity are performed contiguously.

Figure 5.21 shows an alternate example of three serial activities. The conventional CPM analysis shows that the duration of the network is 30 days. When lead–lag constraints are introduced into the network as shown in Figure 5.22, the network duration is compressed to 18 days.

In the forward pass computations in Figure 5.22, note that the earliest completion time of B is time 11, because there is an $FF = 1$ restriction between

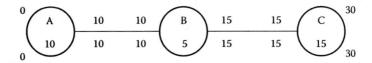

FIGURE 5.21
Another CPM example of serial activities.

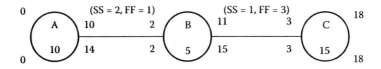

FIGURE 5.22
Compressed PDM network.

activity A and activity B. Since A finishes at time 10, B cannot finish until at least time 11. Even though the earliest starting time of B is time 2 and its duration is 5 days, its earliest completion time cannot be earlier than time 11. Also note that C can start as early as time 3 because there is an SS = 1 relationship between B and C. Thus, given a duration of 15 days for C, the earliest completion time of the network is 3 + 15 = 18 days. The difference between the earliest completion time of C and the earliest completion time of B is 18 – 11 = 7 days, which satisfies the FF = 3 relationship between B and C. In the backward pass, the latest completion time of B is 15 (i.e., 18 – 3 = 15), since there is an FF = 3 relationship between activity B and activity C. The latest start time for B is time 2 (i.e., 3 – 1 = 2), since there is an SS = 1 relationship between activity B and activity C. If we are not careful, we may errone- ously set the latest start time of B to 10 (i.e., 15 – 5 = 10). But that would violate the SS = 1 restriction between B and C. The latest completion time of A is found to be 14 (i.e., 15 – 1 = 14), since there is an FF = 1 relationship between A and B. All the earliest times and latest times at each node must be evalu- ated to ensure that they conform to all the lead–lag constraints. When com- puting earliest start or earliest completion times, the smallest possible value that satisfies the lead–lag constraints should be used. By the same reason- ing, when computing the latest start or latest completion times, the largest possible value that satisfies the lead–lag constraints should be used. Manual evaluations of the lead–lag precedence network analysis can become very tedious for large networks. A computer program may be used to simplify the implementation of PDM. If manual analysis must be done for PDM com- putations, it is suggested that the network be partitioned into more manage- able segments. The segments may then be linked after the computations are completed. The expanded CPM network in Figure 5.23 was developed on

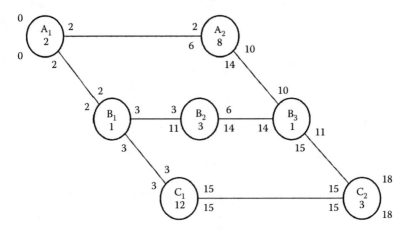

FIGURE 5.23
CPM expansion of second PDM example.

the basis of the precedence network in Figure 5.22. It is seen that activity A is partitioned into two parts, activity B is partitioned into three parts, and activity C is partitioned into two parts. The forward and backward passes show that only the first parts of activities A and B are on the critical path. Both parts of activity C are on the critical path.

Figure 5.24 shows the corresponding earliest-start Gantt chart for the expanded network. Looking at the earliest start times, one can see that activity B is physically split at the boundary of B_2 and B_3 in such a way that B_3 is separated from B_2 by 4 days. This implies that work on activity B is temporarily stopped at time 6 after B_2 is finished and is not started again until time 10. Note that despite the 4-day delay in starting B_3, the entire project is not delayed. This is because B_3, the last part of activity B, is not on the critical path. In fact, B_3 has a TS of 4 days. In a situation like this, the duration of activity B can actually be increased from 5 to 9 days without any adverse effect on the project duration. It should be recognized, however, that increasing the duration of an activity may have negative implications for project cost and personnel productivity.

If the physical splitting of activities is not permitted, then the best option available in Figure 5.24 is to stretch the duration of B_2 so as to fill up the gap from time 6 to time 10. An alternative is to delay the starting time of B_1 until time 4 so as to use up the 4-day delay slack right at the beginning of activity B. Unfortunately, delaying the starting time of B_1 by 4 days will delay the overall project by 4 days, since B_1 is on the critical path as shown in Figure 5.23. The project analyst will need to evaluate the appropriate trade-offs among splitting activities, delaying activities, increasing activity durations, and incurring higher project costs. The prevailing project scenario should be considered when making such trade-off decisions. Figure 5.25 shows the Gantt chart for the compressed PDM schedule based on latest start times. In this case, it will be necessary to split both activities A

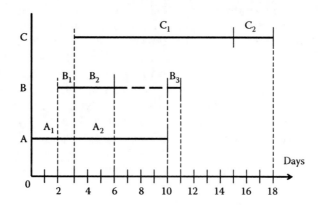

FIGURE 5.24
Compressed PDM schedule based on ES times.

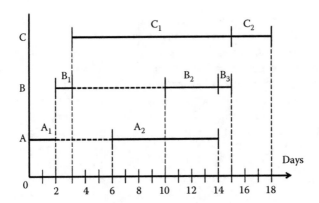

FIGURE 5.25
Compressed PDM schedule based on LS times.

and B even though the total project duration remains the same at 18 days. If activity splitting is to be avoided, then we can increase the duration of activity A from 10 to 14 days and the duration of B from 5 to 13 days without adversely affecting the entire project duration. The important benefit of precedence diagramming is that the ability to overlap activities facilitates flexibility in manipulating individual activity times and compressing the project duration.

Work Rate Analysis

Work rate and work time are essential components of estimating the cost of specific tasks in project management. Given a certain amount of work that must be done at a given work rate, the required time can be computed. Once the required time is known, the cost of the task can be computed on the basis of a specified cost per unit time. Work rate analysis is important for resource substitution decisions. The analysis can help identify where and when the same amount of work can be done with the same level of quality and within a reasonable time span by a less expensive resource. The results of learning curve analysis can yield valuable information about expected work rate. The general relationship among work, work rate, and time is given by

$$\text{work done} = (\text{work rate})(\text{time})$$

This is expressed mathematically as

$$w = rt,$$

where

w = the amount of actual work done expressed in appropriate units. Example of work units are miles of road completed, lines of computer code typed, gallons of oil spill cleaned, units of widgets produced, and surface area painted.

r = the rate at which the work is accomplished (i.e., work accomplished per unit time).

t = the total time required to perform the work excluding any embedded idle times.

It should be noted that work is defined as a physical measure of accomplishment with uniform density. That means, for example, that one line of computer code is as complex and desirable as any other line of computer code. Similarly, cleaning 1 gallon of oil spill is as good as cleaning any other gallon of oil spill within the same work environment. The production of one unit of a product is identical to the production of any other unit of the product. If uniform work density cannot be assumed for the particular work being analyzed, then the relationship presented above may lead to erroneous conclusions. Uniformity can be enhanced if the scope of the analysis is limited to a manageable size. The larger the scope of the analysis, the more the variability from one work unit to another, and the less uniform the overall work measurement will be. For example, in a project involving the construction of 50 miles of surface road, the work analysis may be done in increments of 10 miles at a time rather than the total 50 miles. If the total amount of work to be analyzed is defined as one whole unit, then the relationship below can be developed for the case of a single resource performing the work, with the parameters below:

Resource: Machine A
Work rate: r
Time: t
Work done: 100% (1.0)

The work rate, r, is the amount of work accomplished per unit time. For a single resource to perform the whole unit (100%) of the work, we must have the following:

$$rt = 1.0$$

For example, if machine A is to complete one work unit in 30 min, it must work at the rate of 1/30 of the work content per unit time. If the work rate is too low, then only a fraction of the required work will be performed. The information about the proportion of work completed may be useful for

TABLE 5.7

Work Rate Tabulation for Multiple Resources

Resource, i	Work Rate, r_i	Time, t_i	Work Done, w
RES 1	r_1	t_1	$(r_1)(t_1)$
RES 2	r_2	t_2	$(r_2)(t_2)$
...	...	...	...
RES n	r_n	t_n	$(r_n)(t_n)$
		Total	1.0

productivity measurement purposes. In the case of multiple resources performing the work simultaneously, the work relationship is as presented in Table 5.7.

Even though the multiple resources may work at different rates, the sum of the work they all performed must equal the required whole unit. In general, for multiple resources, we have the following relationship:

$$\sum_{i=1}^{n} r_i t_i = 1.0$$

where

n = number of different resource types
r_i = work rate of resource type i
t_i = work time of resource type i

For partial completion of work, the relationship is

$$\sum_{i=1}^{n} r_i t_i = p$$

where p is the proportion of the required work actually completed.

Work Rate Examples

Machine A, working alone, can complete a given job in 50 min. After machine A has been working on the job for 10 min, machine B was brought in to work with machine A in completing the job. Both machines working together finished the remaining work in 15 min. What is the work rate for machine B?

SOLUTION

The amount of work to be done is 1.0 whole unit.
The work rate of machine A is 1/50.
The amount of work completed by machine A in the 10 min it worked alone is $(1/50)(10) = 1/5$ of the required total work.
Therefore, the remaining amount of work to be done is 4/5 of the required total work.

Table 5.8 shows the two machines working together for 15 min. The computation yields

$$\frac{15}{50} + 15(r_2) = \frac{4}{5}$$

which yields $r_2 = 1/30$. Thus, the work rate for machine B is 1/30. That means machine B, working alone, could perform the same job in 30 min.

In this example, it is assumed that both machines produce an identical quality of work. If quality levels are not identical, then the project analyst must consider the potentials for quality/time trade-offs in performing the required work. The relative costs of the different resource types needed to perform the required work may be incorporated into the analysis as shown in Table 5.9.

TABLE 5.8

Work Rate Tabulation for Machines A and B

Resource, i	Work Rate, r_i	Time, t_i	Work Done, w
Machine A	1/50	15	15/50
Machine B	r_2	15	$15(r_2)$
		Total	4/5

TABLE 5.9

Incorporation of Resource Cost into Work Rate Analysis

Resource, i	Work Rate, r_i	Time, t_i	Work Done, w	Pay Rate, p_i	Pay, P_i
Machine A	r_1	t_1	$(r_1)(t_1)$	p_1	P_1
Machine B	r_2	t_2	$(r_2)(t_2)$	p_2	P_2
...	...	...	...	...	...
Machine n	r_n	t_n	$(r_n)(t_n)$	p_n	P_n
		Total	1.0		Budget

Using the above relationship for work rate and cost, the work crew can be analyzed to determine the best strategy for accomplishing the required work, within the required time, and within a specified budget. For another simple example of possible application scenarios, consider a case where an IT technician can install new IT software in three computers every 4 h. At this rate, it is desired to compute how long it would take the technician to install the same software in five computers. We know, from the information given that we can write the proportion three computers is to 4 h as the proportion that five computers is to x hours, where x represents the number of hours the technician would take to install the software in five computers. This gives the following ratio relationship:

$$\frac{3\,\text{computers}}{4\,\text{h}} = \frac{5\,\text{computers}}{x\,\text{h}},$$

which simplifies to yield $x = 6$ h, 40 min. Now consider a situation where the technician's competence with the software installation degrades over time for whatever reason. We will see that the time requirements for the IT software installation will vary depending on the current competency level of the technician. Half-life analysis can help to capture such situations so that an accurate work time estimate can be developed. Consider another example where a worker can assemble parts at the rate of 120 parts per minute. Another worker can inspect the parts at the rate of three per second. How many inspectors are needed to keep up with 18 assemblers? At the assembling rate given, one assembler can complete the task at the rate of two per second (i.e., 120 parts/60 s). So, 18 assemblers would complete 36 assembles per second. Now, let x be the number of inspector needed to keep up with the 18 assemblers. Since one inspector completes three inspections per second, x inspectors would inspect $3x$ assembles per second. That is, $3x = 36$, which yields $x = 12$.

A Fuel Consumption Rate Example

Similar to the personnel work rate example is a fuel consumption rate computation. Suppose the 36 cars in a car service use a total of 5000 gallons of gasoline per week. If each of the cars uses the same amount of gasoline, then, at this rate, what is the number of gallons used by 5 of the cars in 2 weeks? The solution proceeds as follows:

The car service uses a total of 5000 gallons of gasoline per week, and each of the cars uses the same amount of gasoline, so each of the 36 cars uses $5000/36 = 138.89$ gallons of gasoline per week. Therefore, 5 of the cars use a total of $5(5000)/36 = 694.44$ gallons per week. It follows that 5 of the cars use $(2)[5(5000)/36] = 1388.89$ gallons of gasoline in 2 weeks.

Team Work Rate Analysis

When resources work concurrently at different work rates, the amount of work accomplished by each may be computed by the procedure for work rate analysis. The critical resource diagram and the resource schedule chart provide information to identify when, where, and which resources work concurrently.

Example

Suppose the work rate of RES 1 is such that it can perform a certain task in 30 days. It is desired to add RES 2 to the task so that the completion time of the task can be reduced. The work rate of RES 2 is such that it can perform the same task alone in 22 days. If RES 1 has already worked 12 days on the task before RES 2 comes in, find the completion time of the task. Assume that RES 1 starts the task at time 0.

SOLUTION

The amount of work to be done is 1.0 whole unit (i.e., the full task).

The work rate of RES 1 is 1/30 of the task per unit time.

The work rate of RES 2 is 1/22 of the task per unit time.

The amount of work completed by RES 1 in the 12 days it worked alone is $(1/30)(12) = 2/5$ (or 40%) of the required work.

Therefore, the remaining work to be done is 3/5 (or 60%) of the full task.

Let T be the time for which both resources work together.

The two resources working together to complete the task yield Table 5.10. Thus, we have

$$T/30 + T/22 = 3/5$$

which yields $T = 7.62$ days. Thus, the completion time of the task is $(12 + T) = 19.62$ days from time zero. The results of this example are summarized graphically in Figure 5.26. It is assumed that both resources produce identical quality of work and that the respective work rates remain consistent. The respective costs of the different resource types may be incorporated into the work rate analysis. The CRD and RS charts are

TABLE 5.10

Tabulation of Resource Work Rates for RES 1 and RES 2

Resource Type, i	Work Rate, r_i	Time, t_i	Work Done, w_i
RES 1	1/30	T	$T/30$
RES 2	1/22	T	$T/22$
		Total	3/5

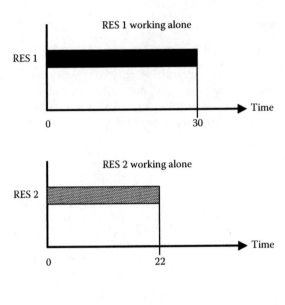

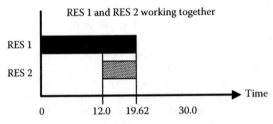

FIGURE 5.26
Resource schedule charts for RES 1 and RES 2.

simple extensions of very familiar tools. They are simple to use and they convey resource information quickly. They can be used to complement existing resource management tools. Users can find innovative ways to modify or implement them for specific resource planning, scheduling, and control purposes. For example, resource-dependent task durations and resource cost can be incorporated into the CRS and RS procedures to enhance their utility for resource management decisions.

Learning Curve Analysis

Learning curves present the relationship between cost (or time) and level of activity on the basis of the effect of learning. An early study disclosed the "80 percent learning" effect, which indicates that a given operation is subject to a 20% productivity improvement each time the activity level or production volume doubles. A learning curve can serve as a predictive tool for obtaining time estimates for tasks in a project environment. Typical learning rates that have been encountered in practice range from 70 to 95%. A learning curve is also referred to as a *progress function*, a *cost–quantity relationship*, a *cost curve*, a *product acceleration curve*, an *improvement curve*, a *performance curve*, an *experience curve*, and an *efficiency curve*.

Several alternate models of learning curves have been presented in the literature. Some of the most notable models are the *log–linear model*, the *S-curve*, the *Stanford-B model*, *DeJong's learning formula*, *Levy's adaptation function*, *Glover's learning formula*, *Pegels' exponential function*, *Knecht's upturn model*, and *Yelle's product model*. The univariate learning curve expresses a dependent variable (e.g., production cost) in terms of some independent variable (e.g., cumulative production). The log–linear model is by far the most popular and most used of all the learning curve models. The model states that the improvement in productivity is constant (i.e., it has a constant slope) as output increases. There are two basic forms of the log–linear model:

1. Average cost model
2. Unit cost model

The average cost model is more common than the unit cost model. It specifies the relationship between the cumulative average cost per unit and cumulative production. The relationship indicates that cumulative cost per unit will decrease by a constant percentage as the cumulative production volume doubles. The model is expressed as

$$A_x = C_1 x^b$$

$$\log A_x = \log C_1 = b \log x$$

where
A_x = cumulative average cost of producing x units
C_1 = cost of the first unit
x = cumulative production count
b = the learning curve exponent (i.e., constant slope of on log–log paper)

The relationship between the learning curve exponent, b, and the learning rate percentage, p, is as shown below:

$$b = \frac{\log p}{\log 2}$$

$$p = 2^b$$

Learning Curve Example 1

Assume that 50 units of an item are produced at a cumulative average cost of \$20 per unit. Suppose we want to compute the learning percentage when 100 units are produced at a cumulative average cost of \$15 per unit. The learning curve analysis would proceed as follows:

Initial production level = 50 units; average cost = \$20
Double production level = 100 units; cumulative average cost = \$15
Using the log relationship, we obtain the following equations:

$$\log 20 = \log C_1 + b \log 50$$

$$\log 15 = \log C_1 + b \log 100$$

Solving the equations simultaneously yields

$$b = \frac{\log 20 - \log 15}{\log 50 - \log 100} = -0.415$$

Thus

$$p = (2)^{-0.415} = 0.75$$

That is a 75% learning rate. In general, the learning curve exponent, b, may be calculated directly from actual data or computed analytically. That is

$$b = \frac{\log A_{x1} - \log A_{x2}}{\log x_1 - \log x_2}$$

$$b = \frac{\ln(p)}{\ln(2)}$$

where
x_1 = first production level
x_2 = second production level
A_{x1} = cumulative average cost per unit at the first production level
A_{x2} = cumulative average cost per unit at the second production level
p = learning rate percentage

Figure 5.27 shows the profile of a typical learning curve.
Using the basic cumulative average cost function, the total cost of producing x units is computed as

$$TC_x = (x)\, A_x = (x)\, C_1\, x^b = C_1\, x^{(b+1)}$$

The unit cost of producing the xth unit is given by

$$U_x = C_1 x^{(b+1)} - C_1(x-1)^{(b+1)}$$

$$= C_1 x \left[x^{(b+1)} - (x-1)^{(b+1)} \right]$$

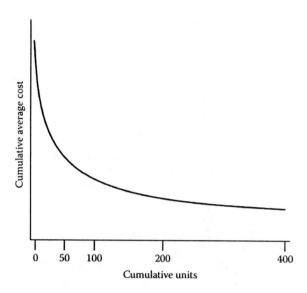

FIGURE 5.27
The log–linear learning curve.

The marginal cost of producing the xth unit is given by

$$MC_x = \frac{d[TC_x]}{dx} = (b+1)C_1 x^b$$

Learning Curve Example 2

Suppose in a production run of a certain product it is observed that the cumulative hours required to produce 100 units is 100,000 h with a learning curve effect of 85%. For project planning purposes, an analyst needs to calculate the number of hours spent in producing the fiftieth unit. Following the notation used previously, we have the following information:

$$p = 0.85$$
$$X = 100 \text{ units}$$
$$A_x = 100,000 \text{ h}/100 \text{ units} = 1000 \text{ h/unit}$$

Now

$$0.85 = 2^b$$

Therefore, $b = -0.2345$
Also

$$100,000 = C_1(100)^b$$

Therefore, $C_1 = 2944.42$ h. Thus

$$C_{50} = C_1(50)^b = 1176.50 \text{ h}$$

That is, the cumulative average hours for 50 units is 1176.50 h. Therefore, cumulative total hours for 50 units = 58,824.91 h. Similarly

$$C_{49} = C_1(49)^b = 1182.09 \text{ h}$$

That is, the cumulative average hours for 49 units is 1182.09 h. Therefore, cumulative total hours for 49 units = 57,922.17 h. Consequently, the number of hours for the fiftieth unit is given by

$$58,824.91 \text{ h} - 57,922.17 \text{ h} = 902.74 \text{ h}$$

6

The Drag Efficient: The Missing Quantification of Time on the Critical Path*

Introduction

As discussed in the preceding chapters, the oil and gas industry can benefit from the proven tools and techniques of project management that have been used in other industries before the oil and gas industry came of age. With appropriate adaptation, customization, and extension of the tools and techniques, the unique aspects of projects in the oil and gas industry can be attended to. This chapter, reprinted from the *Defense AT&L* magazine, illustrates a specific extension that is relevant for the oil and gas project environment. The chapter introduces the methodology of incorporating drag and drag cost into the conventional critical path method (CPM). It can serve as a template for scheduling, planning, and control in all types of projects.

Drag and Drag Cost

Critical path analysis has been around for more than half a century. An argument can be made that no project management technique is more important. Yet, in project management theory and in scheduling software, there is the significant omission of two vital critical path metrics: *drag* and *drag cost*. Critical path drag is a key metric in the planning and scheduling of a project. Its greatest value is to the contractor who must manage the schedule. But it is also crucial for the customer to know that the project team is using this metric both to generate an efficient schedule and to target the most appropriate work packages when slippage occurs.

---

* Reprinted with permission from Devaux, S. A. *Defense AT&L*, January–February 2012, 18–24.

The drag cost of an activity has even greater implications for the customer; it is the amount of value that the project is losing due to delivery being delayed by that activity's critical path drag. Unfortunately, financial analysis of project work tends to focus almost exclusively on budget. Benjamin Franklin wrote that time is money. Every customer knows that the time required for a project comes at great cost. Those funding projects often would willingly pay significantly more to accelerate deployment of a mission-critical system. Since it is exclusively critical path activities that are delaying project completion, the cost of delay is an invisible and expensive cost of critical path work.

The problem is the inability to identify which critical path activities are costing time and money—that is, their drag and drag cost. This chapter will show that the use of these concepts is vital to on-time delivery, schedule recovery, and the generation of maximum customer value.

Impact of Critical Path on Project Investment

All projects, without exception, are investments, undertaken to create greater value than the cost of the required resources. No customer or sponsor would ever *knowingly* invest $5 million worth of resources if the total value from the final product, from all sources, was only expected to be $4.9 million. The difference between the value of the final product and the cost of producing it, what we might call *project profit*, should be a key metric for project performance (as it is for all other investments!). The cost of a project investment is always carefully tracked—but the return, or the *expected monetary value* (EMV) of the scope is little analyzed and often ignored.

One of the main factors that can affect the EMV of a project is changes in delivery date. It is usually the case that the earlier the delivery date, the greater the value of the project investment. Delivery date is always determined by the project's longest, or critical, path. This may start as a planned critical path, but will finish as the *actual* longest path, or what the construction industry terms the "as-built critical path" (ABCP). The project manager should recognize the overwhelming importance of this path, and manage it. During project postmortem, the ABCP and the changes from plan that may have generated it should be a vital artifact and a generator of lessons learned.

Gaps in Traditional Critical Path Data

Whether dealing with the planned critical path or the ABCP, it is important to recognize that both the gods and the devils are in the details. Good schedule management requires knowing the contribution of each activity (as well as technical difficulties, scope changes, resource insufficiencies, schedule constraints, etc.) that contributes time to the length of the path. And here, unfortunately, we enter an area in which critical path theory, as beneficial and vital as it is, is silent. What does critical path analysis tell us about each activity in our project? If an activity is *not* on the critical path, both critical path theory and traditional PM software quantify something called either *total float* or *total slack* (depending on the software): the maximum amount of time that an activity can be delayed without making its path the longest in the project. Figure 6.1 shows a simple network logic diagram of a project with the earliest and latest dates for each activity filled in on top and at bottom, respectively. Let us assume that this is the schedule of a project with a 45-day deadline, with each additional day reducing investment value by $10,000.

As the network shows, the critical path is A, C, E, H, I, and the project duration would be 60 days. The total floats of the noncritical activities would be

F = 10
G = 10
H = 8
I = 3

But since total float quantification is all *off* the critical path, this gives us little help in knowing where to compress the schedule. And unfortunately, no similar quantification is performed for activities that are *on* the critical path! For all critical path activities, the software (and all traditional PM theory, including the PMBOK® Guide) simply says zero—that its total float is zero. Of course, project schedules are much more complex than the simple

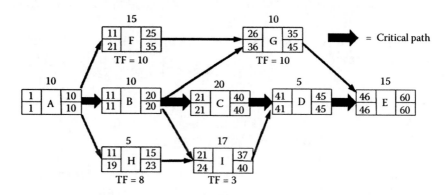

FIGURE 6.1
A simple network logic diagram showing forward and backward passes and total float.

example shown in Figure 6.1. But no matter how large or complex the schedule, the project manager's approach should always be to make the project schedule as efficient as possible, providing the customer with the greatest value for the least cost. The trouble is that most traditional project management metrics are silent about what we all know is really important: the critical path. What we need to know is

1. Of all the activities on the critical path, which are adding the most time to project duration and offer the greatest "bang for the buck" if shortened?
2. How much money is each activity's added time costing, and how much would it cost to compress it?

The first metric that addresses this issue is not float—it is the much more important metric, critical path drag as introduced by Devaux (1999). Just as drag is what slows down a submarine or an airplane, critical path drag is the amount of time by which a critical path activity is slowing down the project. *And it is vital information for any project manager to know about the activities in his/her project!*

- Float is always off the critical path, whereas drag is always on critical activities.
- Float usually does not cost the project time and money, whereas drag almost invariably does!

There is an old saying: "What is measured is what is emphasized." As a result of the standard CPM metric of total float, the emphasis winds up being on precisely the wrong things—the work that is not on the critical path! What the project manager needs to know is: how much time is each critical path activity adding to my project duration so that I can target the best tasks for compression. This is critical path drag. In Figure 6.2, we show the drag totals on the critical path activities.

Although "manual" drag computation in a large network with complex dependencies (Six Sigma, lag, etc.) can be intimidating and time consuming, it is relatively easy in a simple network such as the one in Figure 6.2:

- Step 1: Only critical path activities have drag.
- Step 2: If an activity has nothing else in parallel (e.g., A and in Figure 6.2), its drag equals its duration.
- Step 3: If a critical path activity has other activities in parallel, its drag is *whichever is less*: the total float of the parallel activity with the *least* total float (B and C in Figure 6.2), *or* its own duration (D, whose duration of 5 days is *less* than the 10 days of total float in each of the parallel activities F and G).

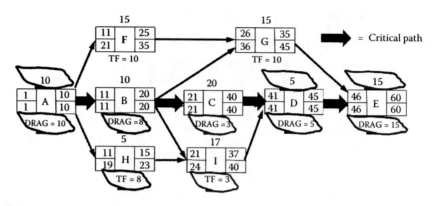

FIGURE 6.2
A simple network logic diagram with drag computed.

Today, three software packages compute drag:

1. Project Optimizer from Sumatra.com (an MSProject 2007 add-on)
2. PlanontheNet.com
3. Spider Project

Of course, there is more to schedule optimization than drag computation. Just because activity E has drag of 15 and activity B's drag is only 8 does not necessarily mean that you can shorten E more than B.

- Some activities are less "resource-elastic" than others, that is, adding resources may do little to shorten their durations.
- Shortening some activities may increase risk unacceptably, decrease quality, or otherwise reduce project value and profit.
- The resources needed to reduce one activity by each unit of time may be much more costly than those needed for an equal or greater reduction on a different activity.

However, when trying to shorten the project duration (either up front during planning, or during execution when schedule slippage may leave the project manager seeking alternatives), we may be searching through a network of not five activities but 500 or 5000! Then there needs to be a way of focusing the process of schedule reduction onto those candidates that will provide the greatest reward. These are almost always the activities with the greatest drag.

In Figure 6.2, even though activity C has a duration of 20 days, it is only adding 3 days to the project schedule. By contrast, even though activity D has a duration of just 5 days, it is adding 2 more days to the critical path than is activity C. And, all else being equal, activity E may offer the greatest opportunity with 15 days of drag.

Computing the Drag Cost of an Activity

Benjamin Franklin's statement that "time is money!" is never more accurate than when applied to projects. The key is to tie the cost of project delay to each individual activity generating the delay. The cost of this delay is caused by the activity's critical path drag, and is the activity's drag cost.

Drag cost represents the synthesis of the concept of *project profit* with a truly scope/cost/schedule-integrated plan. It is the reduction in the net value of the project because of the delay in project completion due to the time impact of each activity's drag. It may be caused either because the delay reduces the project's EMV or because the delay increases the indirect costs (overhead and opportunity costs). Figure 6.3 computes the drag cost of each activity if the cost of delay beyond 45 days is $10,000 per day.

Drag cost assigns the cost of project time to the individual critical path activities that are adding that time to the schedule. Suddenly, not only does Benjamin Franklin's dictum apply to projects—it now applies to individual work items in the project, and to the resources performing that work. This allows the project manager to assess the relative cost of each work item, and to target additional resources to reduce the drag cost.

Computing the True Cost of an Activity

Although finance departments have taught us to identify the cost of work with the price of the resources doing that work, this is simply not true of work performed on the critical path of a project! A week's work by a minimum-wage

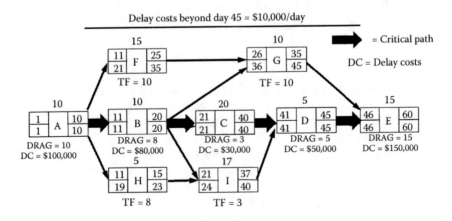

FIGURE 6.3
A simple network logic diagram with drag cost computed at $10,000/day.

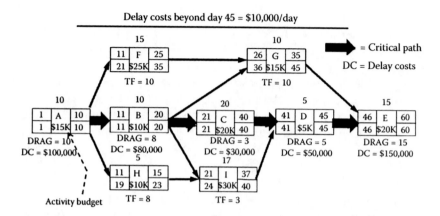

FIGURE 6.4
A simple network logic diagram with both drag cost and activity budgets.

laborer can be much more costly than a week's work by a Nobel laureate physicist—if the physicist's work has float while the laborer's work is on the critical path with lots of drag cost! The *true cost* of project work is the sum of the resource cost and the drag cost (which of course is zero if the work is not on the critical path). In Figure 6.4, we have provided the budget for each activity's resources.

Even though most financial analysis would determine that activity I is the most costly work, with a budget of $30,000, since it has no drag cost, it is actually not even close. Since activity I is not on the critical path, its true cost is only its resources. Conversely, activity E's true cost is the sum of its $20,000 budget and its $150,000 of drag cost, or $170,000. The true cost of each activity is as follows:

$$A = \$15,000 + \$100,000 = \$115,000$$
$$B = \$10,000 + \$80,000 = \$90,000$$
$$C = \$20,000 + \$30,000 = \$50,000$$
$$D = \$5,000 + \$50,000 = \$55,000$$
$$E = \$20,000 + \$150,000 = \$170,000$$
$$F = \$55,000$$
$$G = \$15,000$$
$$H = \$10,000$$
$$I = \$30,000$$

Computing the true cost of an activity can provide huge benefit to customer, the project manager, and to the organization performing the project:

- Additional resources can be targeted to the activities with large true cost. For example, if doubling the resources on Activity E reduced its duration and drag from 15 days to 10 days, its budget would increase by $20,000, but its drag cost would be reduced by $50,000 and its new true cost would be $140,0000 ($40,000 + $100,000), or $30,000 less.

- Some optional activities ("nice-to-haves" rather than "must-haves") often wind up delaying a project by more than they are worth. Drag cost computation would allow both the customer and the project manager to recognize the true cost of optional work when it migrates to the critical path and determine if it is of sufficient value or whether it should be jettisoned. (This analysis should be performed any time that the critical path changes, loading a new set of activities with drag cost) during project performance.

- Any organization in the business of performing multiple simultaneous projects should conduct quarterly assessments of the true cost of specific resource types (mechanical engineer, programmer, etc.) and create Pareto charts highlighting those that have the greatest true cost. Increases in such resources will usually result in decreases in the drag cost component of their summed true costs.

A Concluding Anecdote

A few years ago, while teaching the concept of drag in a seminar, an engineer who worked with a large defense contractor told an illuminating story. The customer had requested that a specific deliverable that was not part of the project's critical path be pulled in by 5 weeks. The transcontinental team all flew to a central site and spent a full day suggesting the changes they thought would meet the new scheduling needs. When they were finished, they incorporated the changes into the master schedule—and the deliverable came in by 1 day! The team then spent the rest of the week engaged in pure trial-and-error: "What if we could do this in 8 days instead of 12? Nope, no change." "What if we made this 5 days instead of 14? Okay, we gained 3 days!" The engineer told me: "If we'd understood the concept of drag, we'd never have even left our offices. We could have accomplished our goal in a half-hour conference call."

Drag Cost in Human Lives

Benjamin Franklin's dictum that time is money sometimes understates the case: on some projects, time can be measured in human suffering and death. Examples can be found in pharmaceutical development, hospital systems, emergency response—any endeavor in which projects are undertaken to save lives. Deployment of homeland security and defense systems are prime examples of efforts where human lives are often on the line. To identify just one example, earlier deployment of a counter-measure to defend against MANPADS (man-portable air-defense systems) could protect aircraft in a combat zone and save many American lives. If the annual loss of life in a combat zone due to MANPADS is determined to be 50, and a planned countermeasure deployment would reduce that number by half, then decreasing the drag of *any critical path activity* by 2 weeks would eliminate an estimated drag cost, over and above the dollars, of the death of an American soldier.

A Historical Example of Drag Cost in Human Lives

In 1991, during the first Gulf War, it was discovered that a software bug in the radar of the Patriot antimissile system was causing the timing system to lose a small fraction of a second for every hour that a battery had been operational. Quoting from the February 4, 1992 report of the Information Management and Technology Division of the United States General Accounting Office (GAO):

> On February 21, 1991, the Patriot Project Office sent a message to Patriot users stating that very long run times could cause a shift in the range gate, resulting in the target being offset. The message also said a software change was being sent that would improve the system's targeting. However, the message did not specify what constitutes very long run times...
>
> ...Alpha Battery, the battery in question, was to protect the Dhahran Air Base. On February 25, Alpha Battery had been in operation for over 100 consecutive hours. Because the system had been on so long, the resulting inaccuracy in the time calculation caused the range gate to shift so much that the system could not track the incoming Scud. Consequently, Alpha Battery did not engage the Scud, which then struck an Army barracks and killed 28 American soldiers.
>
> On February 26, the next day, the modified software, which compensated for the inaccurate time calculation, arrived in Dhahran. According to Army officials, the delay in distributing the software from the United States to all Patriot locations was due to the time it took to arrange for air and ground transportation in a wartime environment. (GAO, 1992)

Although there is always a strong tendency to blame the last few activities (i.e., "the time it took to arrange for air and ground transportation") for a late delivery, the fact is that every critical path activity contributes to the project's duration. In this case, every activity that had drag of 1 day or more, and that might somehow have been shortened through additional resources or expense, could have saved the lives of those 28 soldiers.

The USS Monitor: A Happy Story of Limiting Drag Cost

When news reached the U.S. Navy in late 1861 that the Confederate Navy was working to convert the former USS Merrimack into an ironclad warship, an emergency order went out for the design of a Union ironclad. John Ericsson's model of "a cheesebox on a raft" was selected, and on October 4, 1861, Continental Iron Works and DeLamater Iron Works, both of New York, were contracted to build the Union ironclad. Ericsson had no project management software, and had never read an article about critical path drag. But he was an engineering genius managing an urgent project. Under his direction, the USS Monitor was launched in Brooklyn and began preparations for combat on January 30, 1862, just 118 days after the Navy's order was submitted.

On March 6, the process of towing Monitor down the Atlantic Coast to Chesapeake Bay began. Late on March 8, the former Merrimack, now rebuilt into the ironclad CSS Virginia, attacked the Union squadron blockading Hampton Roads and sank USS Cumberland and USS Congress. At dusk, the Virginia returned to port, intending to finish the job the next morning. But that night the Monitor arrived, and on March 9 the two ironclads fought their famous battle to a draw, leaving the Union blockade in place. The cost of the one extra day it took for Monitor to arrive was high, but 2 days would have cost far more! Had Ericsson had software to help him eliminate one more day on his critical path, the lives lost on the two Union warships might have been saved. Conversely, had he not so brilliantly shortened the project schedule as much as he did, the blockade would probably have been broken and the Union might have lost the war.

Using Drag to Accelerate the Schedule of a Subdeliverable

A few years ago, a client called to see if I could help with a scheduling issue on a large project: the customer had requested that delivery of a certain component be accelerated by 5 weeks. Part of the problem was that the component was not on the critical path of the 3-year project; it had over 200 days of float. The earliest it could be completed, according to the master schedule, was 5 weeks later than the customer now needed it. And the program manager did not know where to start. In such cases, it is crucial to have a "clean" schedule: with up-to-date progress information, correct dependency links, and no activities performed out-of-sequence (the bane of schedule analysis!). It took a while to "scrub" the data. After 3 or 4 hours, we felt that we had an accurate schedule from the current date forward. Then

1. We targeted the component delivery, making it our last or "sink" activity.

2. We identified the target's "ancestors," that is, all earlier activities on the same logical path: predecessors, predecessors' predecessors, and so on.
3. Next, we eliminated all activities that were *not* ancestors to get a subset of only those activities that were ancestors of the targeted activity.
4. We identified the critical path to the targeted activity, and computed the drags.
5. Finally, we pulled in the component's delivery date just as we would the end of a project, by fast tracking or crashing the durations of those activities with the most drag, recalculating activity drags as the critical path changed.

The adjustments made the component's desired delivery date achievable.

References

Devaux, S. A. 1999. *Total Project Control: A Manager's Guide to Integrated Project Planning, Measuring and Tracking*, John Wiley & Sons, New York, NY.

Devaux, S. A. 2012. The drag efficient: The missing quantification of time on the critical path, *Defense AT&L*, January–February 2012, 18–24.

GAO (General Accounting Office). 1992. http://www.fas.org/spp/starwars/GAO/im92026.htm

7

Decision Tools for Project Management in the Oil and Gas Industry

Education is what you get when you read the fine print; experience is what you get when you don't.

Pete Seeger

The oil market, like the crude it harbors, is very volatile. A comprehensive decision analysis is essential to govern all facets of operations in the oil and gas industry. The industry is highly process oriented, thus requiring process improvement strategies. Understanding processes so that they can be improved by means of a systematic approach requires the knowledge and application of tools and techniques. The effective use of these tools and techniques requires their application within the context of ongoing projects within a practical setting. This chapter presents a collection of common tools and techniques for process improvement in project management, with specific focus on oil and gas projects.

Process Operational Definition

A key part of improving a process is to understand the process in a way that can be communicated to everyone without the risk of ambiguity. This requires an operational definition and assessment framework. The definition must be communicable in a way that presents the same consistent meaning to everyone, ranging from vendors and operators to the customers. Operational definition enables all of the people involved in a transaction to use and understand a term in exactly the same way every time. Many times, in Six Sigma initiatives, we focus too much on the product itself rather than on the understanding of the people involved in running, operating, and managing the production facility. Using an operational definition allows us to include all the entities involved in a comprehensive systematic way. This becomes especially important when collecting data, which is the basis for assessing and improving a process. Each person on a project team may have a different idea about what constitutes a defect. But if an operational definition has been established, the criteria for what is acceptable and the test procedure for separating what is acceptable from what is defective, all team members can determine what is and is not a defect. Operational definitions are relevant only within the particular prevailing operating scenario of interest. For this purpose, the concept of a *clean table* is useful. A clean table refers to establishing a fresh and organized framework for doing what is about to be done. This means starting fresh. For example, if you are using the table as a workbench, then *clean* may only mean that it is free of clutter and has all the necessary tools. On the other hand, if it is a lunch table, you would want some level of cleanliness, which is achieved by using a mild detergent. If it is a medical operating room table, it would have to be antiseptically *clean* to prevent the spread of infection. If it is a writer's work table, it would need to have all the relevant references and writing instruments. The operational definition of *clean* is quite different for each of these examples. So, the context of the definition is important. A sample worksheet for developing an operational definition is presented in Figure 7.1.

The elements of an operational definition are summarized as follows:

Criterion: A standard or metric against which to evaluate the results of a test of the process.

Test: The procedure for measuring a characteristic of the process.

Decision: A determination of whether the test results show that the characteristic meets the criterion.

For example, in a process to assess the wool fibers in a piece of pipe insulation material, the criteria might be as follows for a situation where it is desired to have 50% wool insulation:

Criteria: (a) Wool fibers are evenly distributed; (b) wool fibers comprise half of the blanket's weight.

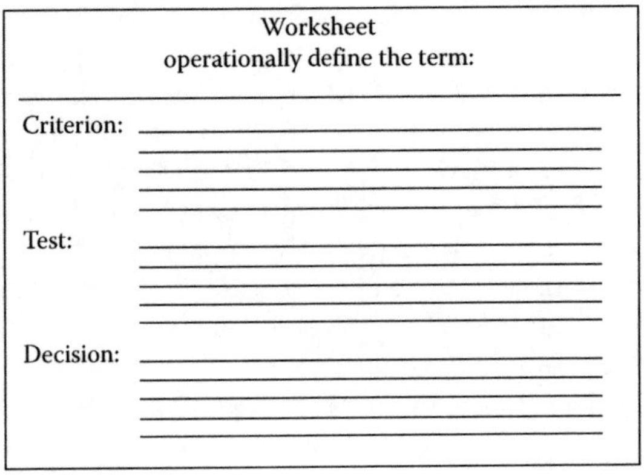

FIGURE 7.1
Template for operational definition.

Test: Analysis of samples to measure distribution and proportion of fibers.

Decision: Goal is met if test shows wool fibers are evenly distributed and comprise half of the insulation weight.

As another example, consider a situation where we desire a rust-free pipe. The operational definition would be done as follows:

Criterion: The absence of any visible oxidation on the surface of the pipe.

Test: Under good lighting conditions, an inspector visually examines the surface of the metal for evidence of oxidation.

Decision: If no oxidation is observed, the criterion is met. The conclusion is that the pipe surface is rust-free.

Although the above are simple examples, the same framework would be applied to more complex operating scenarios, such as oil lubrication in a piece of drilling equipment, applying medical first-aid after an accident, or assessing the flow rate in a pipeline. Figure 7.2 presents a generic environment for applying the technique of operational definition. Control is an essential part of process improvement. A good control system is needed to ensure that appropriate control actions can be taken if a process goes out of

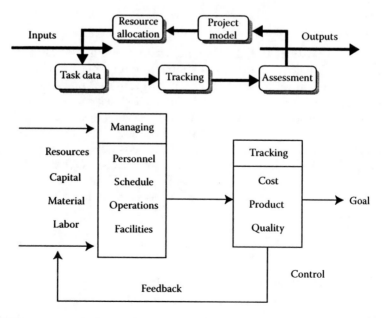

FIGURE 7.2
Input–output scenario for applying operational definition.

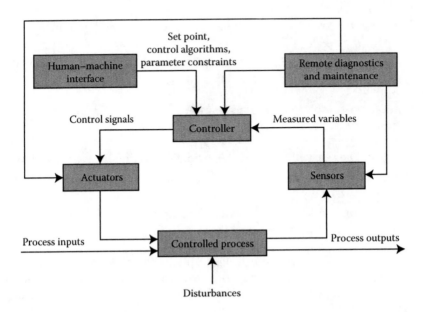

FIGURE 7.3
Components of a control system.

control. Figure 7.3 shows examples of the components of a control system in an oil and gas type of operation.

Process Mapping

One of the initial steps to understand or improve a process is process mapping. By gathering information about the process, we can construct a "dynamic" pictorial representation of the activities that make up a process. Process maps are useful communication tools that help improvement teams understand the process and identify opportunities for improvement. Process mapping provides a common framework, discipline, and language that facilitate a systematic way of working. Complex interactions can be represented in a logical, highly visible, and objective way. It defines where issues, bottlenecks, or kinks exist and provides improvement teams with a common decision-making framework. The steps to constructing a process map are summarized below:

- Brainstorm all activities that routinely occur within the scope of the process
- Group the activities into 4–6 key subprocesses
- Identify the sequence of events and links between the subprocesses
- Define as a high-level process map and subprocess maps using ICOR (inputs, controls, outputs, and resources)

Process maps provide a dynamic view of how an organization can deliver improved business value. *What if* scenarios can be quickly developed by comparing maps of the process *As is* with the process *To be*.

Associated with process mapping is the technique of process flowcharting, which is used for graphically recording exactly what is done in a process. If a flowchart cannot be drawn using these symbols, then the process is not fully understood. The purpose of the flowchart is to learn why the current process operates the way it does and to conduct an objective analysis, which will identify problems, weaknesses, unnecessary steps, duplication, and confirm the objectives of the improvement effort.

Once we have established a good operational definition for the improvement goal with a reliable control system framework, we can proceed to implementing one or more of the common tools and techniques of process improvement. Many of the tools are used in combination with one another. For example, Six Sigma improvement teams use the DMAIC methodology to identify and eliminate the causes of defects. The list includes the following options, some of which are discussed in the sections that follow:

- 5s (seiri, seiton, seiso, seiketsu, shitsuke)
- 6s (sort, stabilize, shine, standardize, sustain, safety)
- Bar charts
- Brainstorming
- Cause and effect diagrams
- CEDAC (cause and effect diagram with the addition of cards)
- Check sheets
- Control charts
- CPI (continuous process improvement)
- Cpk (process capability)
- DEJI (design, evaluate, justify, integrate)
- DFSS (design for Six Sigma)
- DMADV (define, measure, analyze, design, verify)
- DMAIC (define, measure, analyze, improve, control)
- Dot plot or tally chart
- DRIVE (define, review, identify, verify, execute)
- Force field analysis
- Histograms
- Kaizen
- Kanban system
- Lean principles

- Matrix analysis
- Pareto analysis
- PDCA (plan, do, check, act)
- PICK chart (possible, implement, challenge, kill)
- Process flowcharting
- Process mapping
- Scatter diagrams
- SIPOC (suppliers, inputs, process, outputs, customers)
- Six Sigma
- SPC (statistical process control)

Cpk Process Capability Index

Industrial process capability analysis is an important aspect of managing industrial projects. The capability of a process is the spread that contains almost all values of the process distribution. It is very important to note that capability is defined in terms of a distribution. Therefore, capability can only be defined for a process that is stable (has distribution) with common cause variation (inherent variability). It cannot be defined for an out-of-control process (which has no distribution) with variation special to specific causes (total variability). Figure 7.4 shows a process capability distribution.

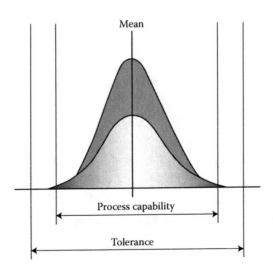

FIGURE 7.4
Process capability distribution.

Capable Process (Cp)

A process is capable (Cp ≥ 1) if its natural tolerance lies within the engineering tolerance or specifications. The measure of process capability of a stable process is $6\hat{\sigma}$, where $\hat{\sigma}$ is the inherent process variability that is estimated from the process. A minimum value of Cp = 1.33 is generally used for an ongoing process. This ensures a very low reject rate of 0.007% and therefore is an effective strategy for prevention of nonconforming items. Cp is defined mathematically as

$$Cp = \frac{USL - LSL}{6\hat{\sigma}}$$

$$= \frac{allowable\,process\,spread}{actual\,process\,spread}$$

where
  USL = upper specification limit
  LSL = lower specification limit

Cp measures the effect of the inherent variability only. The analyst should use R-bar/d_2 to estimate $\hat{\sigma}$ from an R-chart that is in a state of statistical control, where R-bar is the average of the subgroup ranges and d_2 is a normalizing factor that is tabulated for different subgroup sizes (n). We do not have to verify control before performing a capability study. We can perform the study, then verify control after the study with the use of control charts. If the process is in control during the study, then our estimates of capabilities are correct and valid. However, if the process was not in control, we would have gained useful information, as well as proper insights as to the corrective actions to pursue.

Capability Index (Cpk)

Process centering can be assessed when a two-sided specification is available. If the capability index (Cpk) is equal to or greater than 1.33, then the process may be adequately centered. Cpk can also be employed when there is only one-sided specification. For a two-sided specification, it can be mathematically defined as

$$Cpk = Minimum\left\{ \frac{USL - \bar{X}}{3\hat{\sigma}}, \frac{\bar{X} - LSL}{3\hat{\sigma}} \right\}$$

where $\bar{X}$ = overall process average.

However, for a one-sided specification, the actual Cpk obtained is reported. This can be used to determine the percentage of observations out of specification. The overall long-term objective is to make Cp and Cpk as large as possible by continuously improving or reducing process variability, $\hat{\sigma}$, for every iteration so that a greater percentage of the product is near the key quality characteristics target value. The ideal is to center the process with zero variability.

If a process is centered but not capable, one or several courses of action may be necessary. One of the actions may be that of integrating designed experiment to gain additional knowledge on the process and in designing control strategies. If excessive variability is demonstrated, one may conduct a nested design with the objective of estimating the various sources of variability. These sources of variability can then be evaluated to determine what strategies to use in order to reduce or permanently eliminate them. Another action may be that of changing the specifications or continuing production and then sorting the items. Three characteristics of a process can be observed with respect to capability, as summarized below:

1. The process may be centered and capable.
2. The process may be capable but not centered.
3. The process may be centered but not capable.

Process Capability Example

Step 1: Using data for the specific process, determine if the process is capable. Let us assume that the analyst has determined that the process is in a state of statistical control. For this example, the specification limits are set at 0 (lower limit) and 45 (upper limit). The inherent process variability as determined from the control chart is

$$\hat{\sigma} = \bar{R}/d_2 = 5.83/2.059 = 2.83$$

The capability of this process to produce within the specifications can be determined as

$$Cp = \frac{USL - LSL}{6\hat{\sigma}} = \frac{45 - 0}{6(2.83)} = 2.650$$

The capability of the process Cp = 2.65 > 1.0, indicating that the process is capable of producing clutches that will meet the specifications between 0 and 45. The process average is 29.367.

Step 2: Determine if the process can be adequately centered. Cpk = minimum $[C_l \text{ and } C_u]$ can be used to determine if a process can be centered.

$$C_u = \frac{USL - \bar{X}}{3\hat{\sigma}} = \frac{45 - 29.367}{3(2.83)} = 1.84$$

$$C_l = \frac{\bar{X} - LSL}{3\hat{\sigma}} = \frac{29.367 - 0}{3(2.83)} = 3.46$$

Therefore, the capability index, Cpk, for this process is 1.84. Since Cpk = 1.84 is greater than 1.33, then the process can be adequately centered. The potential applications of process capability index are summarized below:

- *Communication*: Cp and Cpk have been used in industry to establish a dimensionless common language useful for assessing the performance of production processes. Engineering, quality, manufacturing, and so on can communicate and understand processes with high capabilities.

- *Continuous improvement*: The indices can be used to monitor continuous improvement by observing the changes in the distribution of process capabilities. For example, if there were 20% of processes with capabilities between 1 and 1.67 in a month, and some of these improved to between 1.33 and 2.0 the next month, then this is an indication that improvement has occurred.

- *Audits*: There are so many various kinds of audits in use today to assess the performance of quality systems. A comparison of in-process capabilities with capabilities determined from audits can help establish problem areas.

- *Prioritization of improvement*: A complete printout of all processes with unacceptable Cp or Cpk values can be extremely powerful in establishing the priority for process improvements.

- *Prevention of nonconforming product*: For process qualification, it is reasonable to establish a benchmark capability of Cpk = 1.33, which will make nonconforming products unlikely in most cases.

In spite of its several possible applications, process capability index has some potential sources of abuse as summarized below:

- Cpk can increase without process improvement even though repeated testing reduces test variability. The wider the specifications, the larger the Cp or Cpk, but the action does not improve the process.

- Analysts tend to focus on number rather than on process.

- *Process control*: Analysts tend to determine process capability before statistical control has been established. Most people are not aware that capability determination is based on process common cause variation and what can be expected in the future. The presence of special causes of variation makes prediction impossible and capability index unclear.

- *Nonnormality*: Some processes result in nonnormal distribution for some characteristics. Since capability indices are very sensitive to departures from normality, data transformation may be used to achieve approximate normality.
- *Computation*: Most computer-based tools do not use $\bar{R}/d_2$ to calculate σ.

Lean Principles

When analytical and statistical tools are coupled with sound managerial approaches, an organization can benefit from a robust implementation of improvement strategies. One approach that has emerged as a sound managerial principle is "lean," which has been successfully applied to many industrial operations. Lean means the identification and elimination of sources of *waste* in operations. Recall that Six Sigma involves the identification and elimination of source of *defects*. When Lean and Six Sigma are coupled, an organization can derive the double benefit of reducing waste and defects in operations, which leads to what is known as Lean Six Sigma. Consequently, the organization can achieve higher product quality, better employee morale, better satisfaction of customer requirements, and more effective utilization of limited resources. The basic principle of "lean" is to take a close look at the elemental compositions of a process so that non-value-adding elements can be located and eliminated. In order to identify value-adding elements of a lean project, the component tasks must be ranked and comparatively assessed. The method below applies relative ratings to tasks. It is based on the distribution of a total point system. The total points available to the composite process or project are allocated across individual tasks. The steps are explained below:

Steps:

1. Let T be the total points available to tasks.
2. $T = 100(n)$, where n = number of raters on the rating team.
3. Rate the value of each task on the basis of specified output (or quality) criteria on a scale of 0 to 100.
4. Let x_{ij} be the rating for task i by rater j.
5. Let m = number of tasks to be rated.
6. Organize the ratings by rater j as shown below:

Rating for Task 1: x_{ij}
Rating for Task 2: x_{2j}
.   .

.
.
.
        .
        .

$$\text{Rating for Task } m: \quad \frac{x_{mj}}{100}$$
$$\text{Total rating points}$$

7. Tabulate the ratings by the raters as shown in Table 7.1 and calculate the overall weighted score for each Task i from the expression below:

$$w_i = \frac{1}{n} \sum_{j=1}^{n} x_{ij}$$

The w_i are used to rank order the tasks to determine the relative value-added contributions of each. Subsequently, using a preferred cut-off margin, the low or noncontributing activities can be slated for elimination. In terms of activity prioritization, a comprehensive lean analysis can identify the important versus unimportant and urgent versus not urgent tasks. It is within the unimportant and not urgent quadrant that one will find "waste" task elements that should be eliminated. Using the familiar Pareto distribution format, Table 7.2 presents an example of task elements within a 20% waste elimination zone.

It is conjectured that activities that fall in the "not important" and "not urgent" zone run the risk of generating points of waste in any productive

TABLE 7.1

Lean Task Rating Matrix

	Rating by Rater $j = 1$	Rating by Rater $j = 2$	...	...	Rating by Rater n	Total Points For Task i	w_i
Rating for Task $i = 1$							
Rating for Task $i = 2$							
...							
...							
Rating for Task m							
Total Points from Rater j	100	100	...	...	100	100n	

TABLE 7.2

Pareto Analysis of Unimportant Process Task Elements

	Urgent	Not Urgent
Important	20%	80%
Not Important	80%	20%

undertaking. That zone should be the first target of review for tasks that can be eliminated. Granted that there may be some "sacred cow" activities that an organization must retain for political, cultural, or regulatory reasons, attempts should still be made to categorize all task elements of a project. The long-established industrial engineering principle of time-and-motion studies is making a comeback due to the increased interest in eliminating waste in lean initiatives. Lean and Six Sigma use analytical tools as the basis for pursuing their goals. But the achievement of those goals is predicated on having a structured approach to the activities of production. If proper project management is practiced at the outset on an industrial endeavor, it will pave the way for achieving Six Sigma results and realizing lean outcomes. The key in any project endeavor is to have a structured design of the project so that diagnostic and corrective steps can be easily pursued. If the proverbial "garbage" is allowed to creep into a project, it would take much more time, effort, and cost to achieve a Lean Six Sigma cleanup.

Kaizen

By applying the Japanese concept of *Kaizen*, which is a compound word meaning "change" (*kai*) "for the better" (*zen*) an organization can redesign its processes to be lean and devoid of excesses. That is, change for improvement. That implies taking apart a process and making it better. In a mechanical design sense, this can be likened to finite element analysis, which identifies how the component parts of a mechanical system fit together. It is by identifying these basic elements that improvement opportunities can be easily and quickly recognized. It should be recalled that the process of work breakdown structure in project management facilitates the identification of task-level components of an endeavor. Consequently, using a project management approach facilitates the achievement of the objectives of "lean." In the context of quality management, a process decomposition hierarchy may help identify elemental characteristic that may harbor waste, inefficiency, and quality impedance. The functional relationships (*f*) are summarized as shown below:

Task = f(activity)

Subprocess = f(task)

Process = f(subprocess)

Quality system = f(process)

Thus, quality improvement can be achieved by hierarchically improving a process and all the elements contained therein. Kaizen focuses on continuous improvement throughout all aspects of life. When applied to the workplace, kaizen continually improves all functions of a business, from manufacturing to management and from the chief executive to the assembly line workers. Kaizen is a daily activity, the purpose of which goes beyond simple productivity improvement. It is a process that, when applied correctly, humanizes the workplace, eliminates overly hard work (muri), and teaches people how to perform experiments on their work using the scientific method and how to learn to spot and eliminate waste in business processes. To be most effective, kaizen must operate with three principles in place:

1. Consider the process and results (not results only) so that actions to achieve effects are surfaced.
2. Systematic thinking of the whole process and not just that immediately in view (i.e., big picture, not solely the narrow view) in order to avoid creating problems elsewhere in the process.
3. A learning, nonjudgmental, nonblaming (because blaming is wasteful) approach and intent to allow the reexamination of the assumptions that resulted in the current process.

In the context of historical recollection, after World War II, to help restore Japan, American occupation forces brought in American experts to help with the rebuilding of Japanese industry while the Civil Communications Section (CCS) developed a Management Training Program that taught statistical control methods as part of the overall material. This course was developed and taught by Homer Sarasohn and Charles Protzman in 1949–1950. Sarasohn recommended Dr. W. Edwards Deming for further training in statistical methods. The Economic and Scientific Section (ESS) group was also tasked with improving Japanese management skills and Edgar McVoy was instrumental in bringing Lowell Mellen to Japan to properly install the Training Within Industry (TWI) programs in 1951. Prior to the arrival of Mellen in 1951, the ESS group had a training film to introduce the three TWI "J" programs (job instruction, job methods, and job relations). The film was titled "Improvement in 4 Steps" (Kaizen eno Yon Dankai). This was the original introduction of "Kaizen" to Japan. For the pioneering, introduction, and implementation of Kaizen in Japan, the Emperor of Japan awarded the 2nd Order Medal of the Sacred Treasure to Dr. Deming in 1960. Consequently, the Union of Japanese Science and Engineering (JUSE) instituted the annual Deming Prizes for achievement in quality and dependability of products. On October 18, 1989, JUSE awarded the Deming Prize to Florida Power & Light Company for its exceptional accomplishments in process and quality control management. It was the first company outside Japan to win the Deming Prize. This example demonstrates that continuous improvement can, indeed, be accomplished in the energy-related industry. Projects in the oil and gas industry can benefit from using tools and techniques of process improvement.

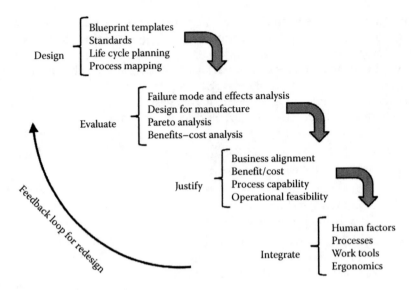

FIGURE 7.5
DEJI model for product development process.

DEJI

Figure 7.5 illustrates the DEJI product decision model. The model (Badiru, 2010) is unique among process improvement tools and techniques because it explicitly calls for a justification of the product within the process improvement cycle. This is important for the purpose of determining when a program should be terminated even after going into production. If the program is justified, it must then be integrated and "accepted" within the ongoing business of the enterprise. Giachetti (2010) emphasizes the importance of integrated design and redesign of an enterprise as it goes through its life cycle.

Military Plane Case Example

The DEJI model facilitates such a recursive design–evaluate–justify–integrate process for enterprise feedback looping. Integration is crucial in highly technical projects not only for the current operation but also for future operations in a dynamic environment. A case example for this is the 2012 revelation by the U.S. Government Accountability Office (GAO) that the U.S. Air Force would spend $9.7 billion over 20 years to upgrade the capabilities of its F-22A Raptor as a result of the service's failure to anticipate the plane's long-term need for modernization. This is integration gone awry. Applying a technique such as DEJI would have called attention to integration needs on the future continuum of new technological developments. The biggest challenge for any project management endeavor is coordinating and integrating the multiple facets that affect the final outputs of a project, where a specific output may be a physical product, a service, or a desired result. Addressing the challenges of project execution from a systems perspective increases the likelihood of success. The DEJI model can facilitate project success through structural implementation. Although originally developed for product development projects, the model is generally applicable to all types of projects as every project goes through the stages of process design, evaluation of parameters, justification of the project, and integration of the project into the core business of the organization. The model can be applied across the spectrum of the following elements of an organization:

1. People
2. Process
3. Technology

Design Stage of DEJI

Product or process design should be structured to follow point-to-point transformation. A good technique to accomplish this is the use of state-space transformation, with which we can track the evolution of a project from concept stage to final product stage. For the purpose of project management, we adopt the general definitions and characteristics of state-space modeling. A state is a set of conditions that describe a process at a specified point in time. A formal definition of *state* in the context of the proposed research is presented below:

> The *state* of a project refers to a performance characteristic of the project which relates input to output such that knowledge of the input time function for $t \geq t_0$ and state at time $t = t_0$ determines the expected output for $t \geq t_0$.

A project *state-space* is the set of all possible states of the project life cycle. State-space representation can solve project design problems by moving from an initial state to another state, and eventually to a goal state. The movement from state to state is achieved by means of actions. A goal is a description of an intended state that has not yet been achieved. The process of solving a project problem involves finding a sequence of actions that represents a solution path from the initial state to the goal state. A state-space model consists of state variables that describe the prevailing condition of the project. The state variables are related to inputs by mathematical relationships. Examples of potential project state variables include schedule, output quality, cost, due date, resource, manpower utilization, and productivity level. For a process described by a system of differential equations, the state-space representation is of the form:

$$\dot{z} = f(z(t), x(t))$$

$$y(t) = g(z(t), x(t))$$

where f and g are vector-valued functions. For linear systems, the representation is

$$\dot{z} = Az(t) + Bx(t)$$
$$y(t) = Cz(t) + Dx(t)$$

where $z(t)$, $x(t)$, and $y(t)$ are vectors and A, B, C, and D are matrices. The variable y is the output vector while the variable x denotes the inputs. The state vector $z(t)$ is an intermediate vector relating $x(t)$ to $y(t)$. The state-space representation of a discrete-time linear project design system is represented as

$$z(t + 1) = Az(t) + Bx(t)$$
$$y(t) = Cz(t) + Dx(t)$$

In generic terms, a project is transformed from one state to another by a driving function that produces a transitional equation given by

$$S_s = f(x \mid S_p) + \varepsilon$$

where
S_s = subsequent state
x = state variable
S_p = the preceding state
ε = error component

The function f is composed of a given action (or a set of actions) applied to the project. Each intermediate state may represent a significant milestone in the project. Thus, a descriptive state-space model facilitates an analysis of what actions to apply in order to achieve the next desired product state.

Design Transformation due to Technology Changes

Project objectives are achieved by state-to-state transformation of project phases. Figure 7.6 shows a product development example involving the transformation from one state to another through the application of action. This simple representation can be expanded to cover several components within the product information framework. Hierarchical linking of product elements provides an expanded transformation structure. The product state can be expanded in accordance with implicit requirements. These requirements might include grouping of design elements, precedence linking (both technical and procedural), required communication links, and reporting requirements. The actions to be taken at each state depend on the prevailing

(INPUTS)		State Transformations	(OUTPUTS)
S_0	Initial condition		
T_1	Planning	➜ $S_1 = T_1(S_0)$	Product specs
T_2	Defining	➜ $S_2 = T_2(S_1)$	Problem statement
T_3	Formulating	➜ $S_3 = T_3(S_2)$	Overall function
T_4	Synthesizing	➜ $S_4 = T_4(S_3)$	Sub-function structure
T_5	Abstracting	➜ $S_5 = T_5(S_4)$	Basic operation
T_6	Varying effects	➜ $S_6 = T_6(S_5)$	Effect variants
T_7	Varying effectors	➜ $S_7 = T_7(S_6)$	Effector variants
T_8	Representing principles	➜ $S_8 = T_8(S_7)$	Solution principles
T_9	Combining	➜ $S_9 = T_9(S_8)$	Assembly variants
T_{10}	Combining	➜ $S_{10} = T_{10}(S_9)$	System variants
T_{11}	Varying forms	➜ $S_{11} = T_{11}(S_{10})$	Varying forms
T_{12}	Laying out	➜ $S_{12} = T_{12}(S_{11})$	Qualitative layout
T_{13}	Dimensioning	➜ $S_{13} = T_{13}(S_{12})$	Scale layout
T_{14}	Analyzing	➜ $S_{14} = T_{14}(S_{13})$	Preliminary layout
T_{15}	Elaborating	➜ $S_{15} = T_{15}(S_{14})$	Final layout
T_{16}	Detailing	➜ $S_{16} = T_{16}(S_{15})$	Detail drawing
T_{17}	Production preparation	➜ $S_{17} = T_{17}(S_{16})$	Production documents
T_{18}	Producing	➜ $S_{18} = T_{18}(S_{17})$	Product
T_{19}	Marketing	➜ $S_{19} = T_{19}(S_{18})$	**Delivery to Market**

FIGURE 7.6
Design transformation due to technology changes.

product conditions. The nature of subsequent alternate states depends on what actions are implemented. Sometimes there are multiple paths that can lead to the desired end result. At other times, there exists only one unique path to the desired objective. In conventional practice, the characteristics of the future states can only be recognized after the fact, thus, making it impossible to develop adaptive plans. In the DEJI implementation, adaptive plans can be achieved because the events occurring within and outside the product state boundaries can be taken into account.

If we describe a product by P state variables s_i, then the composite state of the product at any given time can be represented by a vector S containing P elements. That is

$$S = \{s_1, s_2, \dots, s_P\}$$

The components of the state vector could represent either quantitative or qualitative variables (e.g., cost, energy, color, time). We can visualize every state vector as a point in the M-dimensional state-space. The representation is unique since every state vector corresponds to one and only one point in the state-space. Suppose we have a set of actions (transformation agents) that we can apply to the product information so as to change it from one state to another within the project state-space. The transformation will change a state vector into another state vector. A transformation may be a change in raw material or a change in design approach. Suppose we let T_k be the kth type of transformation. If T_k is applied to the product when it is in state **S**, the new state vector will be $T_k(\mathbf{S})$, which is another point in the state-space. The number of transformations (or actions) available for a product may be finite or countably infinite. We can construct trajectories that describe the potential states of a product evolution as we apply successive transformations. Each transformation may be repeated as many times as needed. Given an initial state $\mathbf{S}_0$, the sequence of state vectors is represented by the following:

$$\mathbf{S}_1 = T_1(\mathbf{S}_0)$$

$$\mathbf{S}_2 = T_2(\mathbf{S}_1)$$

$$\mathbf{S}_3 = T_3(\mathbf{S}_2)$$

$$\dots\dots\dots\dots\dots$$

$$\mathbf{S}_n = T_n(\mathbf{S}_{n-1})$$

The final state, $\mathbf{S}_n$, depends on the initial state **S** and the effects of the actions applied.

Evaluation Stage of DEJI

A project can be evaluated on the basis of cost, quality, and performance. In this section, learning curve modeling is used as the evaluation basis of a project with respect to the concept of growth and decay. The half-life extension (Badiru, 2010) of the basic learning curve presented earlier in Chapter 5 is applicable for the evaluation stage of DEJI. In today's technology-based operations, retention of learning may be threatened by fast-paced shifts in operating requirements. Thus, it is of interest to evaluate the half-life properties of learning curves. Information about the half-life can tell us something about the sustainability of learning-induced performance. This is particularly useful for designing products whose life cycles stretch into the future in a high-tech environment.

Figure 7.7 shows a graphical representation of performance as a function of time under the influence of forgetting (i.e., performance decay). Performance decreases as time progresses. Our interest is to determine when performance has decayed to half of its original level. With half-life computations, a comparative analysis of different learning curves models can be made.

Half-Life Analysis

The basic log–linear model is the most popular learning curve model. It expresses a dependent variable (e.g., production cost) in terms of some independent variables (e.g., cumulative production). The model states that the

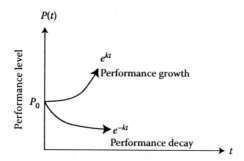

FIGURE 7.7
Concept of learning curve growth and decay.

improvement in productivity is constant (i.e., it has a constant slope) as output increases. That is

$$C(x) = C_1 x^{-b}$$

where
   $C(x)$ = cumulative average cost of producing x units
   C_1 = cost of the first unit
   x = cumulative production unit
   b = learning curve exponent

The percent productivity gain, p, due the effect of learning is computed as

$$p = 2^{-b}$$

The application of half-life analysis to learning curves can help address questions such as:

- How fast and how far can system performance be improved?
- What are the limitations to system performance improvement?
- How resilient is a system to shocks and interruptions to its operation?
- Are the performance goals that are set for the system achievable?

Figure 7.8 shows a pictorial representation of the basic log–linear model, with the half-life point indicated as $x_{1/2}$. The half-life of the log–linear model is computed as follows: Let

C_0 = Initial performance level
$C_{1/2}$ = Performance level at half-life

$$C_0 = C_1 x_0^{-b} \quad \text{and} \quad C_{1/2} = C_1 x_{1/2}^{-b}$$

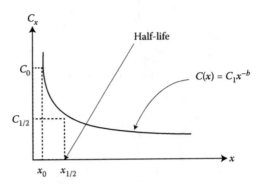

FIGURE 7.8
Profile of a learning curve with half-life point.

But $C_{1/2} = (1/2)C_0$. Therefore, $C_1 x_{1/2}^{-b} = (1/2)C_1 x_0^{-b}$, which leads to $x_{1/2}^{-b} = (1/2)x_0^{-b}$, which, by taking the $(-1/b)$th exponent of both sides, simplifies to yield the following expression as the general expression for the standard log–linear learning curve model:

$$x_{1/2} = \left(\frac{1}{2}\right)^{-(1/b)} x_0, \quad x_0 \geq 1$$

where $x_{1/2}$ is the half-life and x_0 is the initial point of operation. We refer to $x_{1/2}$ as the *first-order half-life*.

The *second-order half-life* is computed as the time corresponding to half of the preceding half. That is

$$x_{1/2(2)} = \left(\frac{1}{2}\right)^{-(2/b)} x_0$$

Similarly, the third-order half-life is

$$x_{1/2(3)} = \left(\frac{1}{2}\right)^{-(3/b)} x_0$$

In general, the kth-order half-life for the log–linear model is represented as

$$x_{1/2(k)} = \left(\frac{1}{2}\right)^{-(k/b)} x_0$$

Justification Stage of DEJI

We need to justify a program on the basis of quantitative value assessment. The systems value model (Troxler and Blank, 1989) is a good quantitative technique that can be used here for project justification on the basis of value. The model provides a heuristic decision aid for comparing project alternatives. It is presented here again for the present context. Value is represented as a deterministic vector function that indicates the value of tangible and intangible attributes that characterize the project. It is represented as

$$V = f(A_1, A_2, \ldots, A_p)$$

where $V = value$, $A = (A_1, \ldots, A_n)$ = vector of quantitative measures or attributes, and $p =$ number of attributes that characterize the project. Examples of project attributes are quality, throughput, capability, productivity, and cost performance. Attributes are considered to be a combined function of factors, x_1, expressed as

$$A_k(x_1, x_2, \ldots, x_{m_k}) = \sum_{i=1}^{m_k} f_i(x_i)$$

where $\{x_i\}$ = set of m factors associated with attribute $A_k(k = 1, 2, \ldots, p)$ and $f_i =$ contribution function of factor x_i to attribute A_k. Examples of factors are market share, reliability, flexibility, user acceptance, capacity utilization, safety, and design functionality. Factors are themselves considered to be composed of indicators, v_i, expressed as

$$x_i(v_1, v_2, \ldots, v_n) = \sum_{j=1}^{n} z_i(v_i)$$

where $\{v_j\}$ = set of n indicators associated with factor $x_i (i = 1, 2, \ldots, m)$ and $z_j =$ scaling function for each indicator variable v_j. Examples of indicators are debt ratio, project responsiveness, lead time, learning curve, and scrap volume. By combining the above definitions, a composite measure of the value of a project is given by the expression below:

$$PV = f(A_1, A_2, \ldots, A_p)$$

$$= f\left\{ \left[\sum_{i=1}^{m_1} f_i\left(\sum_{j=1}^{n} z_j(v_j) \right) \right]_1, \left[\sum_{i=1}^{m_2} f_i\left(\sum_{j=1}^{n} z_j(v_j) \right) \right]_2, \ldots, \left[\sum_{i=1}^{m_k} f_i\left(\sum_{j=1}^{n} z_j(v_j) \right) \right]_p \right\}$$

where PV is the composite project value and m and n may assume different values for each attribute. A weighting measure to indicate the decision maker's preferences may be included in the model by using an attribute weighting factor, w_i, as shown below:

$$PV = f(w_1 A_1, w_2 A_2, \ldots, w_p A_p)$$

where

$$\sum_{k=1}^{p} w_k = 1, \quad (0 \leq w_k \leq 1)$$

In addition to the quantifiable factors, attributes, and indicators that impinge upon overall project value, the human-based subtle factors should also be included in assessing overall project value. Some of such factors are:

- Project communication
- Project cooperation
- Project coordination

Integration Stage of DEJI

Without being integrated, a system will be in isolation and it may be worthless. We must integrate all the elements of a system on the basis of alignment of functional goals. The overlap of systems for integration purposes can conceptually be viewed as projection integrals by considering areas bounded by the common elements of subsystems.

$$A = \iint_{A_y A_x} z(x, y) dy dx$$

$$B = \iint_{B_y B_x} z(x, y) dy dx$$

In Figure 7.9, the projection of a flat plane onto the first quadrant is represented as area A while Figure 7.10 shows the projection on an inclined plane as area B. The net projection encompassing the overlap of A and B is represented as area C in Figure 7.11 and computed as

$$C = \iint_{C_y C_x} z(x, y) dy dx$$

Notice how each successful net projection area decreases with increase in the angle of inclination of the project plane. The fact is that in actual project

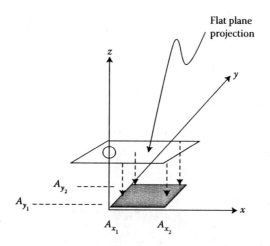

FIGURE 7.9
Flat plane projection for systems integration.

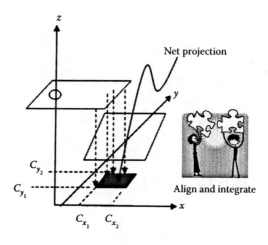

FIGURE 7.10
Inclined plane projection for subsystem alignment and integration.

execution, it will be impractical or impossible to model subsystem scenarios as double integrals. But the concept, nonetheless, demonstrates the need to consider where and how project elements overlap for a proper assessment of integration. For mechanical and electrical systems, one can very well develop mathematical representation of systems overlap and integration boundaries. For the purpose of further explanation, double integrals arise in several technical applications. Some examples are

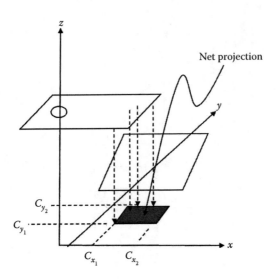

FIGURE 7.11
Reduced net projection area due to steep incline.

- Calculation of volumes
- Calculation of the surface area of a two-dimensional surface (e.g., a plane surface)
- Calculation of a force acting on a two-dimensional surface
- Calculation of the average of a function
- Calculation of the mass or moment of inertia of a body
- Consider the surface area given by the integral

$$A(x) = \int_c^d f(x,y)dy$$

The variable of integration is y, and x is considered a constant. The cross-sectional area depends on x. Thus, the area is a function of x. That is, $A(x)$. The volume of the slice between x and $x + dx$ is $A(x)dx$. The total volume is the sum of the volumes of all the slices between $x = a$ and $x = b$. That is

$$V = \int_a^b A(x)dx$$

If substitute for $A(x)$, we obtain

$$V = \int_a^b \left[\int_c^d f(x,y)dy \right] dx = \int_a^b \int_c^d f(x,y)dydx$$

This is an example of an *iterated* integral. One integrates with respect to y first, then x. The integrals with respect to y and x are called the inner and outer integrals, respectively. Alternatively, one can make slices that are parallel to the x axis. In this case, the volume is given by

$$V = \int_c^d \left[\int_a^b f(x,y)dx \right] dy = \int_c^d \int_a^b f(x,y)dxdy$$

The inner integral corresponds to the cross-sectional area of a slice between y and $y + dy$. The quantities $f(x,y)dydx$ and $f(x,y)dxdy$ represent the value of the double integral in the infinitesimally small rectangle between x and $x + dx$ and y and $y + dy$. The length and width of the rectangle are dx and dy, respectively. Hence, $dydx$ (or $dxdy$) is the area of the rectangle. Thus, the change in area is $dA = dydx$ or $dA = dxdy$.

Computational Example

Consider the double integral:

$$V = \iint_R (x^2 + xy^3)dA,$$

where R is the rectangle $0 \le x \le 1, 1 \le y \le 2$. Suppose we integrate with respect to y first. Then

$$V = \int_0^1 \int_1^2 (x^2 + xy^3)dydx$$

The inner integral is

$$V = \int_1^2 (x^2 + xy^3)dy = \left[x^2y + x\frac{y^4}{4} \right]_{y=1}^{y=2}$$

Note that we treat x as a constant as we integrate with respect to y. The integral is equal to

$$x^2(2) + x(2^4/4) - x^2 - x/4 = x^2 + (15/4)x$$

We are now left with the following integral:

$$\int_0^1 \left(x^2 + \frac{15}{4}x \right) dx = \left(\frac{x^3}{3} + \frac{15}{8}x^2 \right)_{x=0}^{x=1} = \frac{1}{3} + \frac{15}{8} = 2.2083$$

Alternatively, we can integrate with respect to x first and then y. We have

$$V = \int_1^2 \int_0^1 (x^2 + xy^3)dxdy$$

which should yield the same computational result. As a recap of the application of the DEJI model, systems integration is the synergistic linking together of the various components, elements, and subsystems of a system, where the system may be a complex project, a large endeavor, or a large enterprise. Activities that are resident within the system must be managed from both the technical and managerial standpoints. Any weak link in the system, no matter how small, can be the reason that the overall system fails. In this regard, every component of a project is a critical element that must be nurtured and controlled. Embracing the systems integration stage of the DEJI model will increase the likelihood of project success. Figure 7.12 summarizes the flow process, tools, and techniques of the four stages of the model. Postintegration assessment provides feedback inputs that go back into the design stage. The idea is not to prescribe specific tools, but to offer a consistent structure for applying the various tools and techniques that have been presented. Most organizations already have in-house tools and techniques for their processes. Putting those into use in a structured and consistent framework is what the DEJI model offers.

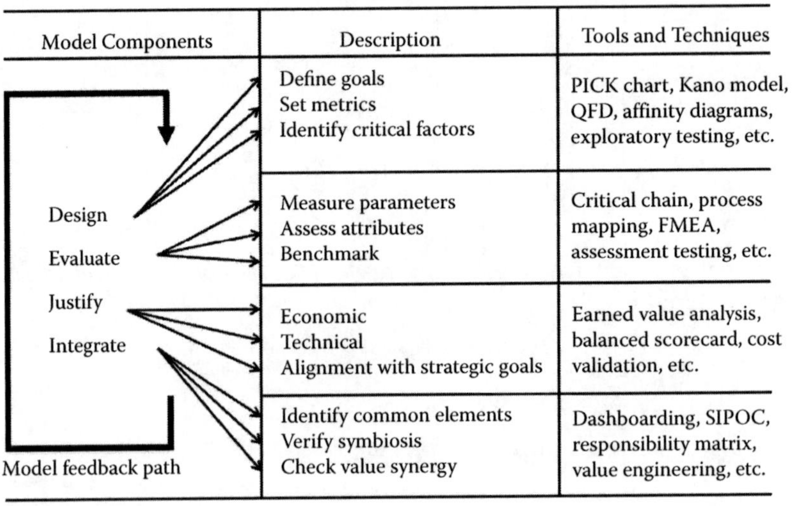

Model Components	Description	Tools and Techniques
	Define goals Set metrics Identify critical factors	PICK chart, Kano model, QFD, affinity diagrams, exploratory testing, etc.
Design	Measure parameters Assess attributes Benchmark	Critical chain, process mapping, FMEA, assessment testing, etc.
Evaluate		
Justify	Economic Technical Alignment with strategic goals	Earned value analysis, balanced scorecard, cost validation, etc.
Integrate		
	Identify common elements Verify symbiosis Check value synergy	Dashboarding, SIPOC, responsibility matrix, value engineering, etc.
Model feedback path		

FIGURE 7.12
Flow diagram of the four stages of the DEJI model.

DMAIC

Figure 7.13 illustrates an application of the DMAIC (design, measure, analyze, improve, and control) technique for acquisitions process improvement. DMAIC is a basic component of the *Six Sigma* methodology for the reduction of variability and it complements the *Lean Approach*, which focuses on eliminating waste in work processes. Variability reduction is achieved through the identification and elimination of sources of defects. Applying DMAIC to acquisitions programs can ensure that a project covers all the elements defined in the scope statement and only the elements defined in the scope. The define stage of DMAIC puts acquisition in the context of a specific military business case. The measure stage of DMAIC lays the ground work for measurement of the metrics of acquisitions performance. In this stage, accurate measurements must be made and relevant data must be collected and analyzed. We must be able to measure a metric before we can control or improve it. The analyze stage of DMAIC is very important to determine the relationships and factors of causality in the acquisitions process. If the focus is to generate products, services, or results, then we must understand what causes what and how the relationships can be enhanced. The improve stage of DMAIC outlines how to plan, pursue, and achieve improvement in acquisitions with an appropriate recognition of organizational structures and impediments. The control stage ensures that any variances that stand out undergo corrective actions before they can adversely influence the end result of an acquisition program. The operational components of DMAIC are explained below and Figure 7.14 shows a typical implementation flowchart for the methodology:

D—Define a problem or improvement opportunity

M—Measure process performance

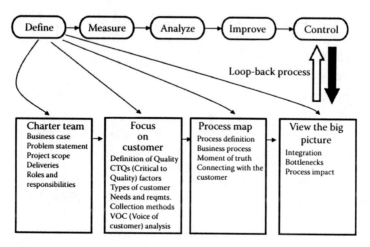

FIGURE 7.13
Application of the DMAIC technique for acquisitions process improvement.

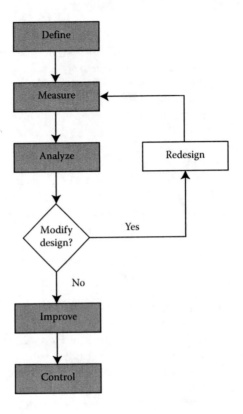

FIGURE 7.14
DMAIC flowchart.

A—Analyze the process to determine the root causes of poor performance (determine whether the process can be improved or should be redesigned)

I—Improve the process by addressing root causes

C—Control the improved process to sustain the improvements

The Six Sigma approach allows for no more than 3.4 defects per million parts in manufactured goods or 3.4 mistakes per million activities in a service operation. To explain the effect of the Six Sigma approach, consider a process that is 99% perfect. That process will produce 10,000 defects per million parts. With Six Sigma, the process will need to be 99.99966% perfect in order to produce only 3.4 defects per million. Thus, Six Sigma is an approach that moves a process toward perfection. Six Sigma, in effect, reduces variability among products produced by the same process.

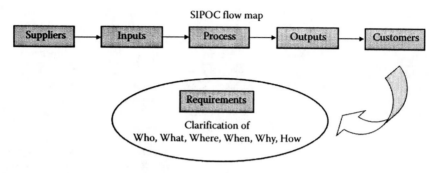

FIGURE 7.15
Application of the SIPOC technique for acquisitions process improvement.

SIPOC

Figure 7.15 illustrates an adaptation of the SIPOC (suppliers, inputs, process, outputs, customers) technique in an acquisitions environment. The diagram is used to identify all performance elements relevant for improvement before the acquisitions program starts. The process improvement team may also add requirements at the end of the SIPOC diagram to identify the specific customer requirements that are to be satisfied. This helps to obtain clarifications of what, who, what, why, and how of improvement efforts. SIPOC can help define a complex acquisition program to ensure that the product will be in alignment with the organizational goal. SIPOC is often applied at the measure stage of the DMAIC methodology. SIPOC complements and provides additional details for the usual process mapping and input–output scoping processes of an acquisitions program. As a case example, SIPOC was applied to an Environmental Safety and Occupational Health (ESOH) project for the purpose of acquisition of lab chemicals at the Air Force Institute of Technology (AFIT) at Wright Patterson Air Force Base. The supplier to customer flow map is shown in Table 7.3. Notice that there are many stakeholders in such a chemical purchase. The information in the table is used later in the example on quantification of the PICK chart.

DRIVE

DRIVE (define, review, identify, verify, and execute) is an approach to problem solving and analysis that can be used as part of the process improvement effort.

Define the scope of the problem and the criteria by which success will be measured and agree on the deliverables and the success factors.

TABLE 7.3

SIPOC Chart for ESOH Improvement Project

Suppliers	Inputs	Process	Outputs	Customers
Consultants	Training	Value stream	Safe working	Local, state, and
Faculty	Purchase process	maps	environment	federal
Chemical vendors	Inventory		Compliance with	agencies
Equipment vendors	Personal		local, state, and	Defense
Base system	protective		federal	financial
(physical	equipment		requirements	accounting
plant, chemical	Laboratory		Properly trained	(invoices, pmts)
school	survey		students	Users (students,
management	Price quotes		Students perform	faculty, external
system, supply/	Government		excellent	visitors)
disposal)	purchase card		Research &	Research
Management	School		Development	sponsors
Students/research	regulations		Student	Maintenance
asst	Federal and		education	staff
Comp support	local law		Useable product	Compliance
Funding agencies	Time to complete		for sponsor	managers
Local business	forms		(equipment,	Facility manager
Contractors	Research		publication,	Internal and
Collaboration with	proposal		information)	external
other colleges	approvals		Safety culture	leadership
Local inventor	Equipment		Degrees	
Funding source	Expertise		Contracts	
Base laser safety	Sponsor		Reports to	
Inspectors	requirements		external	
	Defense		groups,	
	department		contractors, etc.	
	guidelines		Excess item	
			disposal	

Review the current situation, understand the background, identify and collect information, include performance metrics, identify problem areas, improvements, and "low-hanging fruits."

Identify improvements or solutions to the problem and the required changes to enable and sustain the improvements.

Verify and check that the improvements will bring about benefits that meet the defined success criteria; prioritize and pilot the improvements.

Execute the implementation of the solutions and improvements; plan a review; gather feedback and review.

ICOR

ICOR (inputs, controls, outputs, and resources) is an internationally accepted process analysis methodology for process mapping. It allows processes to be broken down into simple, manageable, and more easily understandable units. The maps define the inputs, controls, outputs, and resources for both the high-level process and the subprocesses. This provides hierarchical relationship linking between suppliers and customers. Figure 7.16 shows the layout of ICOR. Figure 7.17 presents an illustration of an implementation of ICOR. Notice the external controls in the example. The oil and gas industry is highly subject to external regulations and legislation. The resources available to the organization include skills, experience, and knowledge of the employees.

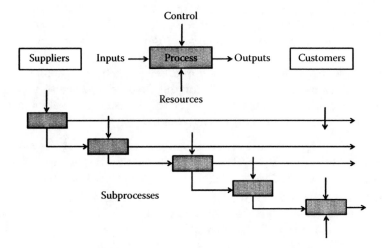

FIGURE 7.16
Framework for ICOR implementation.

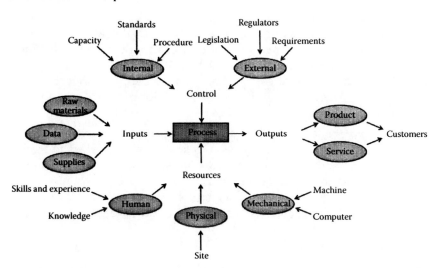

FIGURE 7.17
Implementation example of ICOR.

6s/5s

The methodology known as "6s" is an extension of the "5s" approach, which is a method of workplace organization and visual controls popularized by Hiroyuki Hirano. 6s adds safety to the 5s approach. The five "Ss" refer to five Japanese words—seiri, seiton, seiso, seiketsu, and shitsuke. Seiri means to separate needed and unneeded materials and to remove the latter. Seiton means to neatly arrange (stabilize or straighten) and identify needed materials for ease of use. Seiso means to conduct a cleanup campaign. Seiketsu means to do seiri, seiton, and seiso at frequent intervals and to standardize your 5S procedures. Shitsuke means to form the habit of always following the first four Ss. The origin of 5S comes from the works of two American pioneers, Frederick W. Taylor's Scientific Management (1911) and Henry Ford (1922), who were studied extensively and emulated by the Japanese. Ford's CANDO program (Cleaning up, Arranging, Neatness, Discipline, Ongoing improvement), which builds on Taylor's work, appears as the obvious origin for 5S. The six elements of 6s are shown below:

Sort (Seiri)—Distinguish between what is needed and not needed and remove the latter. The tools and materials in the workplace are sorted out. The unwanted tools and materials are placed in the tag area.

Stabilize (Seiton)—Enforce a place for everything and everything in its place. The workplace is organized by labeling. The machines and tools are labeled with their names and all the sufficient data required. A sketch with exact scale of the work floor is drawn with grids. This helps in achieving a better flow of work and easy access of all tools and machines.

Shine (Seiso)—Clean up the workplace and look for ways to keep it clean. Periodic cleaning and maintenance of the workplace and machines are done. The wastes are placed in a separate area. The recyclable and other wastes are separately placed in separate containers. This makes it easy to know where every components are placed. The clean look of the work place helps in a better organization and increases flow.

Standardize (Seiketsu)—Maintain and monitor adherence to the first three Ss. This process helps to standardize work. The work of each person is clearly defined. The suitable person is chosen for a particular work. People in the workplace should know who is responsible for what. The scheduling is standardized. Time is maintained for every work that is to be done. A set of rules is created to maintain the first 3S's. This helps in improving efficiency of the workplace.

Sustain (Shitsuke)—Follow the rules to keep the workplace 6S-right— "maintain the gain." Once the previous 4S's are implemented some rules are developed for sustaining the other S's.

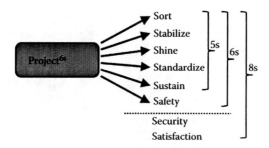

FIGURE 7.18
Graphics representation of 5s/6s process improvement.

Safety—Eliminate hazards. The sixth "S" is added so that focus could be directed at Safety within all improvement efforts. This is particularly essential in the high-risk accident-prone environment of oil and gas operations. This sixth extension is often debated as a separate entity because safety should be implicit in everything we do. Besides, the Japanese word for Safety is "Anzen," which does not follow the "S" rhythm. Going further out on a limp, some practitioners even include additional "S's". So, we could have 8s with the addition of Security and Satisfaction.

- Security (e.g., job security, personal security, mitigation of risk, capital security, intellectual security, property security, information security, asset security, equity security, product brand security, etc.)
- Satisfaction (e.g., employee satisfaction, morale, job satisfaction, sense of belonging, etc.)

Figure 7.18 summarizes the elements contained within the 5s/6s process improvement technique. The figure, developed explicitly for this book, introduces the concept of Project 6s as an equation representation for mnemonic reference purposes.

PICK Chart

PICK chart is a very effective Lean Six Sigma tool used to categorize process improvement ideas. The chart uses a 2 × 2 grid (representing four categories) drawn on a white board or a large flip-chart. Ideas that were written on sticky notes by team members are placed on the grid based on the payoff and difficulty level. The acronym comes from the labels for each of the quadrants of the grid:

Possible (easy, low payoff)   →   Third quadrant
Implement (easy, high payoff)   →   Second quadrant

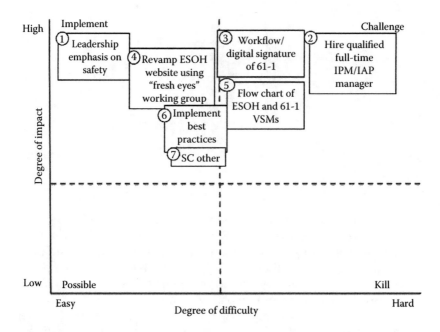

FIGURE 7.19
PICK chart example for ESOH improvement project.

Challenge (hard, high payoff) → First quadrant
Kill (hard, low payoff)   → Fourth quadrant

Figure 7.19 illustrates an example of a PICK chart application to the ESOH project described earlier under the SIPOC technique. When faced with multiple improvement ideas, a PICK chart may be used to determine the most useful one to pick. The vertical axis, representing ease of implementation, would typically include some assessment of the cost to implement the category. More expensive actions can be said to be more difficult to implement.

Quantification of the PICK Chart

The placement of items into one of the four categories in a PICK chart is done through expert ratings, which are often subjective and nonquantitative. In order to put some quantitative basis to the PICK chart analysis, this chapter presents a new methodology of dual numeric scaling on the impact and difficulty axes. Suppose each project is ranked on a scale of 1 to 10 and plotted accordingly on the PICK chart. Then, each project can be evaluated on a binomial pairing of the respective rating on each scale. For our ESOH example, let x represents level of impact and let y represents rating along the axes of difficulty. Note that a high rating along x is desirable while a high rating along y is not desirable. Thus, a composite rating involving x and y

must account for the adverse effect of high values of y. A simple approach is to define $y' = (11 - y)$, which is then used in the composite evaluation. If there are more factors involved in the overall project selection scenario, the other factors can take on their own lettered labeling (e.g., a, b, c, z, etc.). Then, each project will have an n-tuple assessment vector. In its simplest form, this approach will generate a rating such as the following:

$$PICK_{R,i}(x, y') = x + y'$$

where
$PICK_{R,i}(x, y) = $ PICK rating of project i ($i = 1, 2, 3, ..., n$)
$n = $ number of project under consideration
$x = $ rating along the impact axis ($1 \leq x \leq 10$)
$y = $ rating along the difficulty axis $1 \leq y \leq 10$)
$y' = (11 - y)$

If $x + y'$ is the evaluative basis, then each project's composite rating will range from 2 to 20, 2 being the minimum and 20 being the maximum possible. If $(x)(y)$ is the evaluative basis, then each project's composite rating will range from 1 to 100. In general, any desired functional form may be adopted for the composite evaluation. Another possible functional form is

$$PICK_{R,i}(x, y'') = f(x, y'')$$
$$= (x + y'')^2$$

where y'' is defined as needed to account for the converse impact of the axes of difficulty. The above methodology provides a quantitative measure for translating the entries in a conventional PICK chart into an analytical technique to rank the improvement alternatives, thereby reducing the level of subjectivity in the final decision. The methodology can be extended to cover cases where a project has the potential to create negative impacts, which impede organizational advancement. Referring back to the PICK chart for our ESOH example, we develop the numeric illustration shown in Table 7.4.

As expected, the highest $x + y'$ composite rating (i.e., 18) is in the second quadrant, which represents the "implement" region. The lowest composite rating is 12 in the first quadrant, which is the "challenge" region. With this type of quantitative analysis, it becomes easier to select, justify, and implement improvement projects. This facilitates a more rigorous analytical technique compared to the traditional subjective arm-waving approaches.

Kanban

Kanban is a signaling system to trigger action in production operations. Kanban is used as a part pull system. It is demand scheduling, which means

TABLE 7.4

Numeric Evaluation of PICK Chart Rating for ESOH Project

Improvement Project	x Rating	y Rating	$y' = 11 - y$	$x + y'$	xy'
Leadership emphasis	9	2	9	18	81
Full-time issue manager	9	10	1	10	9
Workflow digital signature	9	6	5	14	45
Workgroup process	8	3	8	16	64
Workflow chart VSM	7	6	5	12	35
Implement best practices	7	4	7	14	49
Support center other	6	4	7	13	42

it only produces products to replace the products consumed by its customers, and it only produces products based on signals sent by its customers. Kanban replaces the daily scheduling activities necessary to operate the production process, and the need for production planners and supervisors to continuously monitor schedule status to determine the next item to run and when to change over. Kanban scheduling reduces inventory, improves flow, prevents overproduction, places control at operational level, and creates visual scheduling and management of process.

Quality Circle

The concept of quality circle is based on human resources management, which is considered as one of the key factors in the improvement of product quality and productivity. It implies the development of skills, capabilities, confidence, and creativity of the people through cumulative process of education, training, work experience, and participation. The quality circle concept has three major attributes:

1. It is a form of participative management.
2. It is a human resources development technique.
3. It is a problem-solving technique.

A quality circle is a small group of volunteers (usually 3–12 employees) doing similar work. They meet regularly under the leadership of their immediate supervisor, or someone chosen among the circle to identify problems, set priorities, discover causes, and propose solutions. A quality circle environment is a good example of the actualization of the Chinese quote below:

"Tell me and I forget;
Show me and I remember;
Involve me and I understand."

Direct participation to facilitate understanding is a key benefit of quality circles, which may concern product quality, work productivity, safety, job structure, work process flow, project control, aesthetics of the work areas, and so on.

Poka Yoke

Poka yoke is a method of mistake-proofing a process to preempt the occurrence of defects. It is a way to manufacture or assemble products with minimum or zero defects by practicing zero quality control (ZQC), which is based on the principle that defects are prevented by controlling the performance of a process so that there cannot be defects in the product. A poka yoke system uses sensors or other devices installed in processing equipment or machines to detect errors that are missed by the operators. Poka yoke systems are used to carry out two key elements of ZQC: 100% inspection and immediate feedback. Poka yoke systems are used in source inspection to catch errors before the production process creates a defective product. A poka yoke system regulates the production and prevents defects using one of the following approaches. A control system stops the machine when an irregularity occurs. A warning system signals the operator to stop the machine when error is identified. There are three main methods for using poka yoke systems:

1. Contact methods—This method works by detecting whether a product makes physical or energy contact with a sensing device.
2. Fixed value method—This method is used when a fixed number of parts to be attached to the product or when a fixed number of repeated operations need to be done at a particular work station. The device-fixed system counts the number of repetitions accomplished and releases the product when the fixed count is reached.
3. Motion-step method—This method is used to sense whether a motion or step in the process has been carried out within a certain time. It can also be used to sort and order the sequence in which the process should continue so as to avoid errors.

Because of its error-preemptive nature, the poka yoke system is particularly suitable for the oil and gas production environment.

Culture and Attitude

In spite of all the tools and techniques available, it often happens that the task of process improvement will boil down to prevailing culture and attitudes within the workforce. These are two attributes that are not easy to change. This is basically what hampers many improvement efforts.

Quantifying Operational Efficiency

Operational efficiency, effectiveness, and productivity usually go hand in hand. An integrated definition of each is essential to identify where military organizational improvement can be pursued. The existing techniques for improving efficiency, effectiveness, and productivity are quite amenable for military adaptation. Efficiency refers to the extent to which a resource (time, money, effort, etc.) is properly utilized to achieve an expected outcome. The goal, thus, is to minimize resource expenditure, reduce waste, eliminate unnecessary effort, and maximize output. The ideal (i.e., the perfect case) is to have 100% efficiency. This is rarely possible in practice. Usually expressed as a percentage, efficiency (e) is computed as output divided by input:

$$e = \frac{\text{output}}{\text{input}} = \frac{\text{result}}{\text{effort}}$$

Effectiveness is an ambiguous evaluative term that is difficult to quantify. It is primarily concerned with achieving objectives. To model effectiveness quantitatively, we can consider the fact that an "objective" is essentially an "output" related to the numerator of the efficiency equation above. Thus, we can assess the extent to which the various objectives of an organization are met with respect to the available resources. Although efficiency and effectiveness often go hand in hand, they are, indeed, different and distinct. For example, one can forego efficiency for the sake of getting a particular objective accomplished. Consider the statement "if we can get it done, money is no object." The military, by virtue of being mission driven, often operates this way. If, for instance, our goal is to go from point A to point B to hit a target, and we do hit the target, no matter what it takes, then we are effective. We may not be efficient based on the amount of resources expended to hit the target. For the purpose of this chapter, a cost-based measure of effectiveness is defined as

$$ef = \frac{s_o}{c_o}$$

where
ef = measure of effectiveness
s_o = level of satisfaction of the objective (rated on a scale of 0 to 1)
c_o = cost of achieving the objective (expressed in pertinent cost basis: money, time, measurable resource, etc.)

If an objective is fully achieved, its satisfaction rating will be 1. If not achieved at all, it will be zero. Thus, having the cost in the denominator gives a measure of achieving the objective per unit cost. If the effectiveness measures of achieving several objectives are to be compared, then the denominator (i.e., cost) will need to be normalized to a uniform scale. Overall system effectiveness can be computed as a summation as follows:

$$ef_c = \sum_{i=1}^{n} \frac{s_0}{c_0}$$

where
  ef_c = composite effectiveness measure
  n = number of objectives in the effectiveness window

Because of the potential for the effectiveness measure to be very small based on the magnitude of the cost denominator, it is essential to scale this measure to a scale of 0 to 100. Thus, the highest comparative effectiveness per unit cost will be 100 while the lowest will be 0. The above quantitative measure of effectiveness makes most sense when comparing alternatives for achieving a specific objective. If the effectiveness of achieving an objective in absolute (noncomparative) terms is desired, it would be necessary to determine the range of costs, minimum to maximum, applicable for achieving the objective. Then, we can assess how well we satisfy the objective with the expenditure of the maximum cost versus the expenditure of the minimum cost. By analogy, killing two birds with one stone is efficient. By comparison, the question of effectiveness is whether we kill a bird with one stone or kill the same bird with two stones, if the primary goal is to kill the bird nonetheless. In technical terms, systems that are designed with parallel redundancy can be effective, but not necessarily efficient. In such cases, the goal is to be effective (get the job done) rather than to be efficient. Productivity is a measure of throughput per unit time. The traditional application of productivity computation is in the production environment with countable or measurable units of output in repetitive operations. Manufacturing is a perfect scenario for productivity computations. Typical productivity formulas include the following:

$$P = \frac{Q}{q} \quad \text{and} \quad P = \frac{e}{u}$$

where P is productivity, Q is output quantity, q is input quantity, e is efficiency, and u is utilization. However, the military environment is a nonmanufacturing setting, for which productivity analysis is still of interest. The military organization is composed, primarily, of knowledge workers, whose productivity must be measured in alternate terms, perhaps through work rate analysis.

Project Decision Analysis

Decision analysis facilitates a proper consideration of the essential elements of decisions in a project systems environment. These essential elements include the problem statement, information, performance measure, decision model, and an implementation of the decision. The recommended steps are enumerated below:

Step 1. Problem statement: A problem involves choosing between competing, and probably conflicting, alternatives. The components of problem solving in project management include:

- Describing the problem (goals, performance measures)
- Defining a model to represent the problem
- Solving the model
- Testing the solution
- Implementing and maintaining the solution

Problem definition is very crucial. In many cases, *symptoms* of a problem are more readily recognized than its *cause* and *location*. Even after the problem is accurately identified and defined, a benefit/cost analysis may be needed to determine if the cost of solving the problem is justified.

Step 2. Data and information requirements: Information is the driving force for the project decision process. Information clarifies the relative states of past, present, and future events. The collection, storage, retrieval, organization, and processing of raw data are important components for generating information. Without data, there can be no information. Without good information, there cannot be a valid decision. The essential requirements for generating information are:

- Ensuring that an effective data collection procedure is followed
- Determining the type and the appropriate amount of data to collect
- Evaluating the data collected with respect to information potential
- Evaluating the cost of collecting the required data

For example, suppose a manager is presented with a recorded fact that says, "Sales for the last quarter are 10,000 units." This constitutes ordinary data. There are many ways of using the above data to make a decision depending on the manager's value system. An analyst, however, can ensure the proper use of the data by transforming it into information, such as, "Sales of 10,000 units for last quarter are within x percent of the targeted value." This type of information is more useful to the manager for decision making.

Step 3. Performance measure: A performance measure for the competing alternatives should be specified. The decision maker assigns a perceived worth or value to the available alternatives. Setting measures of performance is crucial to the process of defining and selecting alternatives. Some performance measures commonly used in project management are project cost, completion time, resource usage, and stability in the workforce.

Step 4. Decision model: A decision model provides the basis for the analysis and synthesis of information and is the mechanism by which competing alternatives are compared. To be effective, a decision model must be based on a systematic and logical framework for guiding project decisions. A decision model can be a verbal, graphical, or mathematical representation of the ideas in the decision-making process. A project decision model should have the following characteristics:

- Simplified representation of the actual situation
- Explanation and prediction of the actual situation
- Validity and appropriateness
- Applicability to similar problems

The formulation of a decision model involves three essential components:

Abstraction: Determining the relevant factors

Construction: Combining the factors into a logical model

Validation: Assuring that the model adequately represents the problem

The basic types of decision models for project management are described next.

Descriptive models. These models are directed at describing a decision scenario and identifying the associated problem. For example, a project analyst might use a CPM network model to identify bottleneck tasks in a project.

Prescriptive models. These models furnish procedural guidelines for implementing actions. The Triple C approach (Badiru, 2008), for example, is a model that prescribes the procedures for achieving communication, cooperation, and coordination in a project environment.

Predictive models. These models are used to predict future events in a problem environment. They are typically based on historical data about the problem situation. For example, a regression model based on past data may be used to predict future productivity gains associated with expected levels of resource allocation. Simulation models can be used when uncertainties exist in the task durations or resource requirements.

Satisficing models. These are models that provide trade-off strategies for achieving a satisfactory solution to a problem within given constraints. Goal programming and other multicriteria techniques provide good satisficing solutions. For example, these models are helpful in cases where time limitations, resource shortages, and performance requirements constrain the implementation of a project.

Optimization models. These models are designed to find the best available solution to a problem subject to a certain set of constraints. For example, a linear programming model can be used to determine the optimal product mix in a production environment.

In many situations, two or more of the above models may be involved in the solution of a problem. For example, a descriptive model might provide insights into the nature of the problem; an optimization model might provide the optimal set of actions to take in solving the problem; a satisficing model might temper the optimal solution with reality; a prescriptive model might suggest the procedures for implementing the selected solution; and a predictive model

Step 5. Making the decision: Using the available data, information, and the decision model, the decision maker will determine the real-world actions that are needed to solve the stated problem. A sensitivity analysis may be useful for determining what changes in parameter values might cause a change in the decision.

Step 6. Implementing the decision: A decision represents the selection of an alternative that satisfies the objective stated in the problem statement. A good decision is useless until it is implemented. An important aspect of a decision is to specify how it is to be implemented. Selling the decision and the project to management requires a well-organized persuasive presentation. The way a decision is presented can directly influence whether or not it is adopted. The presentation of a decision should include at least the following: an executive summary, technical aspects of the decision, managerial aspects of the decision, resources required to implement the decision, cost of the decision, the time frame for implementing the decision, and the risks associated with the decision.

Group Decision Making

Systems decisions are often complex, diffuse, distributed, and poorly understood. No one person has all the information to make all decisions accurately. As a result, crucial decisions are made by a group of people. Some organizations use outside consultants with appropriate expertise to make recommendations for important decisions. Other organizations set up their own internal consulting groups without having to go outside the organization. Decisions can be made through linear responsibility, in that case one person makes the final decision based on inputs from other people. Decisions can also be made through shared responsibility, in that case a group of people share the responsibility for making joint decisions. The major advantages of group decision making are listed below:

1. Facilitation of a systems view of the problem environment.
2. Ability to share experience, knowledge, and resources. Many heads are better than one. A group will possess greater collective ability to solve a given decision problem.
3. Increased credibility. Decisions made by a group of people often carry more weight in an organization.
4. Improved morale. Personnel morale can be positively influenced because many people have the opportunity to participate in the decision-making process.
5. Better rationalization. The opportunity to observe other people's views can lead to an improvement in an individual's reasoning process.
6. Ability to accumulate more knowledge and facts from diverse sources.
7. Access to broader perspectives spanning different problem scenarios.
8. Ability to generate and consider alternatives from different perspectives.
9. Possibility for a broader-base involvement, leading to a higher likelihood of support.
10. Possibility for group leverage for networking, communication, and political clout.

In spite of the much-desired advantages, group decision making does possess the risk of flaws. Some possible disadvantages of group decision making are listed below:

1. Difficulty in arriving at a decision.
2. Slow operating time frame.
3. Possibility for individuals' conflicting views and objectives.
4. Reluctance of some individuals in implementing the decision.
5. Potential for power struggle and conflicts among the group.
6. Loss of productive employee time.
7. Too much compromise may lead to less than optimal group output.
8. Risk of one individual dominating the group.

9. Overreliance on group process may impede agility of management to make decision fast.

10. Risk of dragging feet due to repeated and iterative group meetings.

Brainstorming

Brainstorming is a way of generating many new ideas. In brainstorming, the decision group comes together to discuss alternate ways of solving a problem. The members of the brainstorming group may be from different departments, may have different backgrounds and training, and may not even know one another. The diversity of the participants helps create a stimulating environment for generating different ideas from different viewpoints. The technique encourages free outward expression of new ideas no matter how far-fetched the ideas might appear. No criticism of any new idea is permitted during the brainstorming session. A major concern in brainstorming is that extroverts may take control of the discussions. For this reason, an experienced and respected individual should manage the brainstorming discussions. The group leader establishes the procedure for proposing ideas, keeps the discussions in line with the group's mission, discourages disruptive statements, and encourages the participation of all members. After the group runs out of ideas, open discussions are held to weed out the unsuitable ones. It is expected that even the rejected ideas may stimulate the generation of other ideas that may eventually lead to other favored ideas. Guidelines for improving brainstorming sessions are presented as follows:

- Focus on a specific decision problem.
- Keep ideas relevant to the intended decision.
- Be receptive to all new ideas.
- Evaluate the ideas on a relative basis after exhausting new ideas.
- Maintain an atmosphere conducive to cooperative discussions.
- Maintain a record of the ideas generated.

Delphi Method

The traditional approach to group decision making is to obtain the opinion of experienced participants through open discussions. An attempt is made to reach a consensus among the participants. However, open group discussions are often biased because of the influence of subtle intimidation from dominant individuals. Even when the threat of a dominant individual is not present, opinions may still be swayed by group pressure. This is called the "bandwagon effect" of group decision making.

The Delphi method attempts to overcome these difficulties by requiring individuals to present their opinions anonymously through an intermediary.

The method differs from the other interactive group methods because it eliminates face-to-face confrontations. It was originally developed for forecasting applications, but it has been modified in various ways for application to different types of decision making. The method can be quite useful for project management decisions. It is particularly effective when decisions must be based on a broad set of factors. The Delphi method is normally implemented as follows:

1. *Problem definition.* A decision problem that is considered significant is identified and clearly described.

2. *Group selection.* An appropriate group of experts or experienced individuals is formed to address the particular decision problem. Both internal and external experts may be involved in the Delphi process. A leading individual is appointed to serve as the administrator of the decision process. The group may operate through the mail or gather together in a room. In either case, all opinions are expressed anonymously on paper. If the group meets in the same room, care should be taken to provide enough room so that each member does not have the feeling that someone may accidentally or deliberately observe their responses.

3. *Initial opinion poll.* The technique is initiated by describing the problem to be addressed in unambiguous terms. The group members are requested to submit a list of major areas of concern in their specialty areas as they relate to the decision problem.

4. *Questionnaire design and distribution.* Questionnaires are prepared to address the areas of concern related to the decision problem. The written responses to the questionnaires are collected and organized by the administrator. The administrator aggregates the responses in a statistical format. For example, the average, mode, and median of the responses may be computed. This analysis is distributed to the decision group. Each member can then see how his or her responses compare with the anonymous views of the other members.

5. *Iterative balloting.* Additional questionnaires based on the previous responses are passed to the members. The members submit their responses again. They may choose to alter or not to alter their previous responses.

6. *Silent discussions and consensus.* The iterative balloting may involve anonymous written discussions of why some responses are correct or incorrect. The process is continued until a consensus is reached. A consensus may be declared after five or six iterations of the balloting or when a specified percentage (e.g., 80%) of the group agrees on the questionnaires. If a consensus cannot be declared on a particular point, it may be displayed to the whole group with a note that it does not represent a consensus.

In addition to its use in technological forecasting, the Delphi method has been widely used in other general decision making. Its major characteristics of anonymity of responses, statistical summary of responses, and controlled procedure make it a reliable mechanism for obtaining numeric data from subjective opinion. The major limitations of the Delphi method are:

1. Its effectiveness may be limited in cultures where strict hierarchy, seniority, and age influence decision-making processes.
2. Some experts may not readily accept the contribution of nonexperts to the group decision-making process.
3. Since opinions are expressed anonymously, some members may take the liberty of making ludicrous statements. However, if the group composition is carefully reviewed, this problem may be avoided.

Nominal Group Technique

The nominal group technique is a silent version of brainstorming. It is a method of reaching consensus. Rather than asking people to state their ideas aloud, the team leader asks each member to jot down a minimum number of ideas, for example, five or six. A single list of ideas is then written on a chalkboard for the whole group to see. The group then discusses the ideas and weeds out some iteratively until a final decision is made. The nominal group technique is easier to control. Unlike brainstorming where members may get into shouting matches, the nominal group technique permits members to silently present their views. In addition, it allows introversive members to contribute to the decision without the pressure of having to speak out too often. In all of the group decision-making techniques, an important aspect that can enhance and expedite the decision-making process is to require that members review all pertinent data before coming to the group meeting. This will ensure that the decision process is not impeded by trivial preliminary discussions. Some disadvantages of group decision making are:

1. Peer pressure in a group situation may influence a member's opinion or discussions.
2. In a large group, some members may not get to participate effectively in the discussions.
3. A member's relative reputation in the group may influence how well his or her opinion is rated.
4. A member with a dominant personality may overwhelm other members in the discussions.
5. The limited time available to the group may create a time pressure that forces some members to present their opinions without fully evaluating the ramifications of the available data.

6. It is often difficult to get all members of a decision group together at the same time.

Despite the noted disadvantages, group decision making definitely has many advantages that may nullify the shortcomings. The advantages as presented earlier will have varying levels of effect from one organization to another. The Triple C principle presented in Chapter 2 may also be used to improve the success of decision teams. Team work can be enhanced in group decision making by adhering to the following guidelines:

1. Get a willing group of people together.
2. Set an achievable goal for the group.
3. Determine the limitations of the group.
4. Develop a set of guiding rules for the group.
5. Create an atmosphere conducive to group synergism.
6. Identify the questions to be addressed in advance.
7. Plan to address only one topic per meeting.

For major decisions and long-term group activities, arrange for team training that allows the group to learn the decision rules and responsibilities together. The steps for the nominal group technique are

1. Silently generate ideas, in writing.
2. Record ideas without discussion.
3. Conduct group discussion for clarification of meaning, not argument.
4. Vote to establish the priority or rank of each item.
5. Discuss vote.
6. Cast final vote.

Interviews, Surveys, and Questionnaires

Interviews, surveys, and questionnaires are important information gathering techniques. They also foster cooperative working relationships. They encourage direct participation and inputs into project decision-making processes. They provide an opportunity for employees at the lower levels of an organization to contribute ideas and inputs for decision making. The greater the number of people involved in the interviews, surveys, and questionnaires, the more valid the final decision. The following guidelines are useful for conducting interviews, surveys, and questionnaires to collect data and information for project decisions:

1. Collect and organize background information and supporting documents on the items to be covered by the interview, survey, or questionnaire.

2. Outline the items to be covered and list the major questions to be asked.

3. Use a suitable medium of interaction and communication: telephone, fax, electronic mail, face-to-face, observation, meeting venue, poster, or memo.

4. Tell the respondent the purpose of the interview, survey, or questionnaire, and indicate how long it will take.

5. Use open-ended questions that stimulate ideas from the respondents.

6. Minimize the use of yes or no type of questions.

7. Encourage expressive statements that indicate the respondent's views.

8. Use the who, what, where, when, why, and how approach to elicit specific information.

9. Thank the respondents for their participation.

10. Let the respondents know the outcome of the exercise.

Multivote

Multivoting is a series of votes used to arrive at a group decision. It can be used to assign priorities to a list of items. It can be used at team meetings after a brainstorming session has generated a long list of items. Multivoting helps reduce such long lists to a few items, usually three to five. The steps for multivoting are

1. Take a first vote. Each person votes as many times as desired, but only once per item.

2. Circle the items receiving a relatively higher number of votes (i.e., majority vote) than the other items.

3. Take a second vote. Each person votes for a number of items equal to one-half the total number of items circled in step 2. Only one vote per item is permitted.

4. Repeat steps 2 and 3 until the list is reduced to three to five items depending on the needs of the group. It is not recommended to multivote down to only one item.

5. Perform further analysis of the items selected in step 4, if needed.

The tools, techniques, and concepts presented in this chapter provide practical guidance for applying decision tools for oil and gas project management. To improve a project is to improve the project's underlying processes. A process encompasses the steps and decisions involved in the way that work is accomplished. Every oil and gas project environment can benefit from the illustrative examples presented in the chapter. To summarize, Figure 7.20 presents a flowchart of process improvement steps with critical decision points embedded.

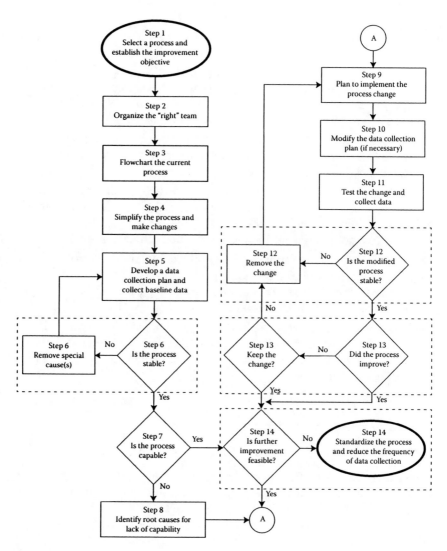

FIGURE 7.20
Flowchart of process improvement steps.

Supplier Selection Decision

Supplier selection decision is a frequent problem in oil and gas project management. A technique, such as the PICK chart, can be adapted for selecting suppliers based on qualitative or subjective analysis. For a more rigorous selection approach, quantitative methods may be necessary. The supplier selection problem is very much like an outsourcing problem, and they both can benefit from rigorous analytical selection tools and techniques. Some of the commonly used techniques for vendor selection include the following:

- Total cost approach: In this approach, the quoted price from each vendor is taken as the starting point and each constraint under consideration is replaced iteratively by a cost factor. The contract is awarded to the vendor with the lowest unit total cost.
- Multiattribute utility theory (MAUT): In this approach, multiple, and possibly conflicting, attributes are fed into a comprehensive mathematical model. This approach is useful for global contracting applications.
- Multiobjective programming: In this approach, flexibility and vendor inclusiveness are achieved by allowing a varying number of vendors into the solution such that suggested volume of allocation to each vendor is recommended by the mathematical model.
- Total cost of ownership: In this philosophy-based approach, the selection process looks beyond price of purchase to include other purchase-related costs. This is useful for demonstrating vendor buy-in and overall involvement in project success.
- Analytic hierarchy process: In this approach, pair-wise comparison of vendors is conducted in a stage-by-stage decision process. This is useful for cases where qualitative considerations are important for the decision process.

Wadhwa–Ravindran Supplier Selection Technique

Several other mathematical models are available in the literature. One comprehensive quantitative technique that uses multicriteria modeling is presented by Wadhwa and Ravindran (2007). They present a multicriteria formulation of the vendor selection problem with multiple buyers and multiple vendors under price discounts. This is applicable to cases where different divisions of an organization buy through one central purchasing department. The number of buyers in this scenario is set equal to the number of divisions buying through the central purchasing office. The model is also applicable for a case where the number of buyers is equal to one. The formulation considers the least restrictive case where any of the buyers can acquire one or more products from any vendors. The potential set of vendors chosen by an organization is constrained by the following:

- Quality level of the products from different vendors
- Lead time of the supplied products
- Production capacity of the vendors

The Wadhwa–Ravindran model helps any organization to select a subset of the most favorable vendors for various outsourced components and to determine the respective quantities to order from each of the chosen most favorable vendors; with the objective of meeting project needs. The model uses the following notations:

I = set of products to be purchased

J = set of buyers who procure multiple units in order to fulfill some demand

K = potential set of vendors

M = set of incremental price breaks

p_{ikm} = cost of acquiring one unit of product i from vendor k at price level m

b_{ikm} = quantity at which incremental price breaks occurs for product i by vendor k

F_k = fixed ordering cost associated with vendor k

d_{ij} = demand of product i for buyer j

l_{ijk} = lead time of vendor k to produce and supply product i to buyer j

q_{ik} = quality that vendor k maintains for product i (measured in percent of defects)

L_{ij} = lead time that buyer j requires for product i

Q_j = minimum quality level that buyer j requires for all vendors to maintain (percent rejection)

CAP_k = production capacity of vendor k

N = maximum number of vendors that can be selected

X_{ijkm} = number of units of product i supplied by vendor k to buyer j at price level m

Z_k = decision variable denoting whether or not a particular vendor is chosen (1 or 0)

Y_{ijkm} = decision variable indicating whether or not price level m is used (1 or 0)

The objective of the model is to simultaneously minimize price, lead time, and rejects. The mathematical representations of these multiple objectives are presented below for price, lead time, and quality:

Total purchasing cost = total variable cost + total fixed cost

$$= \sum_i \sum_j \sum_k \sum_m p_{ikm} X_{ijkm} + \sum_k F_k Z_k$$

Total lead time = summation over all products, buyers, and vendors

$$= \sum_i \sum_j \sum_k \sum_m l_{ijk} X_{ijkm}$$

Quality = sum of rejects over all products, buyers, and vendors

$$= \sum_i \sum_j \sum_k \sum_m q_{ijk} X_{ijkm}$$

The constraints in the model are expressed in terms of capacity constraint, demand constraint, maximum number of vendors, linearization, and nonnegativity. These are expressed as follows:

$$\text{Capacity constraint:} \sum_i \sum_j \sum_m X_{ijkm} \le CAP_k Z_k \quad \forall k$$

$$\text{Demand constraint:} \sum_k \sum_m X_{ijkm} = d_{ij} \quad \forall i,j$$

$$\text{Maximum number of vendors:} \sum_k Z_k \le N$$

Because of price discounts, the objective function will be nonlinear. Linearizing constraints are needed to convert the nonlinear objective function to a linear function. These constraints are expressed as

$$X_{ijkm} \le (b_{ikm} - b_{ikm-1}) * Y_{ijkm} \quad \forall i,j,k; \quad 1 \le m \le m_k$$

$$X_{ijkm} \ge (b_{ikm} - b_{ikm-1}) * Y_{ijkm+1} \quad \forall i,j,k; \quad 1 \le m \le m_k - 1$$

Note that price breaks occur at the following sequence of quantities:

$$0 = b_{i,k,0} < b_{i,k,1} < \cdots < b_{i,k,m}$$

The unit price of ordering X_{ijkm} units from vendor k at price level m is given by p_{ikm}, if $b_{i,k,m-1} < X_{ijkm} \le b_{i,k,m}$ ($1 \le m \le m[k]$).

The linearizing constraints force quantities in the discount range for a vendor to be incremental. Because the "quantity" is incremental, if the order quantity lies in discount interval m, that is, $Y_{ijkm} = 1$, then the quantities in the interval 1 to $m - 1$, should be at the maximum of those ranges The first of the two constraints also assures that a quantity in any range is no greater than the width of the range. The nonnegativity and binary constraint is expressed as

$$X_{ijkm} \geq 0; \quad Z_k, Y_{ijkm} \in (0,1).$$

The above formulations present the general structure of the Wadhwa–Ravindran model. Interested readers should consult Wadhwa and Ravindran (2007) for the full exposition of the model as well as a numeric example of the model. Several different methods are available for solving multiobjective optimization problems. Wadhwa and Ravindran (2007) cover the following solution methods:

1. Weighted objective method
2. Goal programming method
3. Compromise programming method

Weighted Objective Method

Weighing the objectives to obtain an efficient or Pareto-optimal solution is a common multiobjective solution technique. Under the weighted objective approach, the vendor selection problem is transformed to the following single-objective optimization problem:

$$\text{Min } w_1 \left[\sum_i \sum_j \sum_k \sum_m p_{ikm} X_{ijkm} + \sum_k F_k \right] + w_2 \left[\sum_i \sum_j \sum_k \sum_m l_{ijk} X_{ijkm} \right]$$

$$+ w_3 \left[\sum_i \sum_j \sum_k \sum_m q_{ik} X_{ijkm} \right]$$

where w_1, w_2, and w_3 are the weights on each of the objectives. The optimal solution to the weighted problem is a noninferior solution to the multiobjective problem as long as all the weights are positive. The weights can be systematically varied to generate several efficient solutions. This is generally not a good method for finding an exact representation of the efficient set. It is often used to approximate the efficient solution set.

Goal Programming

Goal programming approach views a decision problem as a set of goals to be accomplished subject to a set of *soft constraints* representing the targets to be achieved. Typical optimization techniques assume that the decision constraints are *hard constraints* that cannot be violated. Goal programming relaxes that strict requirement by focusing on compromises that can be accommodated in favor of achieving an overall improvement in the set of goals. The compromises are modeled as deviations from the goals. Goal programming attempts to minimize the set of deviations from the specified goals. The goals are considered simultaneously, but they are weighted in accordance with their relative importance to the decision maker. Goal programming is a three-step approach:

Step 1: The decision maker provides the goals and targets to be achieved for each objective. Because the goals are not hard constraints, some of the goals may not be achievable. Let us consider an objective f_i with a target value of b_i. The goal constraint will be written as

$$f_i(x) + d_i^- - d_i^+ = b_i$$

where

d_i^- = underachievement of goal

d_i^+ = overachievement of goal

Step 2: The decision maker provides his/her preference on achieving the goals. This can be done as ordinal (preemptive rank order), cardinal (absolute weights), or hybrid measure.

Step 3: Find a solution that will come as close as possible to the stated goal in the specified preference order. As an illustration, preemptive weights are used in the model presented here. Priority order is assigned to the goals. Goals with higher priorities are satisfied before lower-priority goals are considered. For the example below, price is the highest-priority goal, followed by lead time, and then quality. The formulation is represented as shown below:

$$\text{Min } Z = P_1 d_1^+ + P_2 d_2^+ + P_3 d_3^+$$

Subject to

$$\sum_i \sum_j \sum_k \sum_m p_{ikm} \cdot X_{ijkm} + \sum_k F_k \cdot Z_k + d_1^- - d_1^+; \quad \text{for price goal}; \quad \forall i, j$$

$$\sum_k \sum_m q_{ik} \cdot X_{ijkm} + d_2^- - d_2^+ \quad \text{for quality goal}; \quad \forall i, j$$

$$\sum_k \sum_m l_{ijk} \cdot X_{ijkm} + d_3^- - d_3^+ \quad \text{for lead time goal;} \quad \forall i, j$$

where p_1, p_2, and p_3 are the preemptive priorities assigned to each criterion.

Compromise Programming

Compromise programming (CP) is an approach that sets the identification of an ideal solution as a point where each attribute under consideration achieves its optimum value and seeks a solution that is as close as possible to the ideal point. Comparative weights are used as measures of relative importance of the attributes in the CP model. Although weights representing relative importance are used as the preference structure in CP, the mathematical basis for applying CP is superior to conventional weighted-sum methods for locating efficient solutions, or the so-called Pareto points. CP is very useful for collective decision making, such as procurement selection. It is a methodology for approaching the *ideal solution* as closely as possible within the decision sphere. An ideal solution corresponds to the best value that can be achieved for each objective, ignoring other objectives, subject to the overall constraints. Since the objectives are conflicting, the ideal solution cannot be achieved, but it can be approached as closely as possible. "Closeness," in this regard, is represented by a distance metric, L_p, defined as follows:

$$L_p = \left[\sum_{i=1}^{k} \lambda_i^p (f_i - f_i^*)^p \right]^{1/p}, \quad \text{for } p = 1, 2, \ldots, \infty$$

where the variables, f_1, f_2, ..., f_k, are the different objectives. The factor, $f_i^* = \min(f_i)$, ignoring other criteria, is called the ideal value for the ith objective. The weights given to the various criteria are the λ_i values. In general, using w_i's as the relative weights, we have the following relationship:

$$\lambda_i = \frac{w_i}{f_i^*}$$

A *compromise solution* is identified as any point that minimizes the L_p function for the following conditions:

$$\lambda_i > 0$$
$$\sum \lambda_i = 1$$
$$1 \leq p \leq \infty$$

The compromise solution is always nondominated in the optimization sense. As p increases, larger deviations are assigned higher weights. For $p = \infty$, the largest of the deviations completely dominates the distance determination. For the vendor selection application, the compromise-programming approach will proceed as follows:

Step 1: Obtain the ideal solution by optimizing the problem separately for each objective. The ideal values for each of the three objectives price, lead time, and quality are denoted, respectively, by p_i^*, l_i^*, and q_i^*.

Step 2: Obtain a compromise solution by using an appropriate distance measure.

Thus, we have the following mathematical expression for the vendor selection problem:

$$
\text{Min} \left[
\begin{array}{l}
w_1 \left\{ \dfrac{\left(\left(\displaystyle\sum_i \sum_j \sum_k \sum_m p_{ijkm} * X_{ijkm} + \sum_k F_k * Z_k \right) - p_i^* \right)^p}{p_i^*} \right\} \\[2em]
+ w_2 \left\{ \dfrac{\left(\left(\displaystyle\sum_i \sum_j \sum_k \sum_m l_{ijk} * X_{ijkm} \right) - l_i^* \right)^p}{l_i^*} \right\} \\[2em]
+ w_3 \left\{ \dfrac{\left(\left(\displaystyle\sum_i \sum_j \sum_k \sum_m q_{ik} * X_{ijkm} \right) - q_i^* \right)^p}{q_i^*} \right\}
\end{array}
\right]^{1/p}
$$

Typical values used for p are 1, 2, and ∞. By changing the value of the parameter p, different efficient solutions can be obtained from the above expression. Several optimization software tools are available for solving the models and formulations discussed above.

References

Badiru, A. B. 2008, *Triple C Model of Project Management: Communication, Cooperation, and Coordination*, Taylor & Francis/CRC Press, Boca Raton, FL.

Badiru, A. B. 2010, Half-life of learning curves for information technology project management, *International Journal of IT Project Management*, 1(3), 28–45.

Giachetti, R. E. 2010, *Design of Enterprise Systems: Theory, Architecture, and Methods*, CRC Press/Taylor & Francis, Boca Raton, FL.

Troxler, J. W. and Blank, L. 1989, A comprehensive methodology for manufacturing system evaluation and comparison, *Journal of Manufacturing Systems*, 8(3), 176–183.

Wadhwa, V. and Ravindran, A. R. 2007, Vendor selection in outsourcing, *Computers & Operations Research*, 34, 3725–3737.

8

Project Schedule Forecast and Control: Reliable Schedule Forecasting in Federal Design–Build Facility Procurement*

Introduction

Construction engineering managers participate in a multifaceted process riddled with technical and social pressures. According to the Accreditation Board for Engineering and Technology (ABET), engineering management programs must prepare graduates to understand the relationships between planning, organizing, leading, and controlling (ABET Engineering Accreditation Commission 2010). Overall, these four tasks facilitate the structuring and execution of work. In this manner, scheduling is an important process that network tasks in order to communicate what should happen in the future.

> Schedules are those outputs of work structuring that link directly with production control. (Ballard et al. 2002)

Federal design–build (DB) facility procurement involves several factors that influence the scheduling process. First, government agencies must maintain fair and competitive bidding of DB contracts in accordance with the Federal Acquisition Reform Act of 1996 (Public Law 104-106) (American Society of Civil Engineers 2010). Consequently, public construction operates in a uniquely regulated acquisition environment in pursuit of transparency and equity. This pressures a construction schedule to provide reliable records of contract progress and payments.

Next, in terms of planning and project delivery methods, schedulers deal with more unknowns at the beginning of a DB project than a traditional design–bid–build process since DB contracts typically present no more than

---

* Reprinted with permission from Gannon T., Feng P., and William S. 2012, *Lean Construction Journal*, 1–14. A publication of the Lean Construction Institute.

35% design in a proposal. DB contracts feature concurrent development within the design and construction phases, which can generate technical and behavioral tendencies described as the "90% syndrome" and the "Liar's club" that contribute to schedule degradation (Ford and Sterman 2003a, 2003b). This is the condition of stagnating around a 90% completion plateau on a project and the associated behavior of hiding rework or fault in the hold. Although DB proponents may laud the synergy created in coupling processes and responsibilities under one contractor, social factors can play a major role in project performance. Accordingly, a 2010 construction management literary review conducted by Xue et al. finds that success in collaborative working within the construction industry predominately hinges on two factors: the business environment and human behavior (Xue et al. 2010).

By design, DB projects tend to shift more risk and liability to the general contractor (GC) and potentially forego a degree of owner participation (Agostini 1996). In this way, DB can offer a quicker contracting solution with possibly less end-user coordination. Although this method may produce a reduction of owner control, careful schedule and cost review practices are measures noted to "bridge the gap" needed in federal management oversight between owners and contractors (Rookard-Everett 2009). Overall, the federal government pursues DB contracts to most expediently allocate funds and maximize budget execution. As such, schedule communication remains a critical management process.

Schedule communication fundamentally rests on the problems of planning. Production theory addresses these problems through a comparison of pushing and pulling methods (Spearman and Zuzanis 1992). In this research, pull systems reveal advantages in control and reliability for manufacturing. However, obstacles exist in implementing pull techniques to construction, especially with design components (Ballard 1999). DB construction management may present challenges to combine the techniques to facilitate realistic schedule forecasting. In order to understand what is happening on real projects in regard to reliable schedules, we investigated three different cases.

This research considers the following military construction projects on Wright Patterson Air Force Base, Dayton, Ohio:

- Project 1: DB addition/alteration of Signature Technology Laboratory (new construction attached to existing)
- Project 2: DB alteration of Acquisition Management and Materials Laboratory Facility (renovation of two separated buildings)
- Project 3: DB addition/alteration of Sensors Directorate Laboratory (new construction and renovation of multiple facilities)

United States Army Corps of Engineers (USACE) provides construction management services for nearly all Air Force construction over $750,000 (USD). For these observed projects, USACE uses contract specifications to outline requirements of a detailed activity-based schedule. USACE mandates a standardization of activity codes for schedule submittals and links the pay application to reported progress per submittal. The initial schedule, required no later than 40–42 days after the notice to proceed, serves as the baseline for monitoring this progress (United States Army Corps of Engineers 2007). Thus, managing scheduled activities is USACE's basis of maintaining contractor accountability.

The question considered in the paper is: Do initial schedules specified in federal DB facility procurements provide reliable forecasting for project control?

Following a review of the objective and limitations, this paper provides case project descriptions to identify stakeholders, scopes, and complexities. We then present the research question and methods. Next, the analysis and results expand on our findings of schedule variance and shortfalls. The recommendation section then provides a concept to address the variance. Finally, the conclusions section discusses the overall schedule shortfalls and impacts.

Objective

The objective is to understand how the scheduling process performs for each of the three projects and identify how project change underscores schedule uncertainty. The metrics of total cost, total duration, and activity count allow us to analyze data from the initial schedules versus the subsequent schedule updates. These metrics illustrate the forecasting shortfalls in the activity-based scheduling currently used in public sector construction management.

Limitations

Federal facility procurement is a massive industry for which we only examine three project case studies. This research is limited to Wright Patterson Air Force Base, Dayton, Ohio. The project contracts are all DB, which is most common for projects of this magnitude. We monitored project progress and scheduling issues with construction managers on an average of 2 times a month over the course of 14 months (September 2009–November 2010). The Primavera P6 XER schedule data files were available for a total 61 updates between the three projects combined. Human input errors inevitably exist in these files as well.

Case Study Project Descriptions

The prime contractor for all the case study projects is Butt Construction Company (BCC) with award dates between June 2008 and April 2009 after competitive bidding processes. The individual project descriptions below are in chronological order according to award date. Coincidently, this is also the order of contract award price, smallest to largest, ranging from $8.5 million to $36.2 million. The facilities are all located within a 1 km radius and managed by the local USACE construction services office, which is located within this radius as well.

The USACE and BCC management personnel vary on the three projects with some overlap. Following are the staffing differences and similarities for key positions between the projects:

- Different USACE project managers (although some overlap due to transfers)
- Different USACE construction management project engineers
- Different USACE quality assurance representatives
- Different BCC project managers
- Different BCC quality control and schedule managers
- Same USACE resident engineer
- Same USACE senior project controller
- Same BCC project engineer

The Department of Defense (DoD) Base Realignment and Closure (BRAC) 2005 process spurred the funding of all three projects. These BRAC facility procurements each support a high-priority movement of a diverse group of Air Force personnel and operations upon completion. Therefore, the motive of construction across the projects is similar, although the end-users are different. The descriptions below provide further overview of each project's contract requirements and challenges.

Project 1

This DB contract for the addition/alteration to the Signature Technology Laboratory, awarded on June 12, 2008, was $8,540,000 with an original performance period of 540 calendar days. The contract consisted of new construction of a three-story office building attached to an existing facility along with new parking. In adherence to security requirements, the new building had no windows and entailed multiple Sensitive Compartmented Information Facility (SCIF) rooms. The new constructed area totaled approximately 3700 square meters (40,000 square feet (SF)). Near the completion of the project, the only major contract modification included the $300,000 change order for finishing floor three.

Project 2

This DB contract for the alteration of Acquisition Management and Materials Laboratory Facility, awarded on September 22, 2008, was $18,539,000 with an original performance period of 570 calendar days. This renovation entailed new structural, electrical, and HVAC systems for two separate buildings both built in the late 1920s. The acquisition management facility renovation incorporate about 6050 square meters (65,000 SF) of office space while the materials lab was nearly 2790 square meters (30,000 SF). One of the main challenges was to update the buildings to the DoD Anti-Terrorism/Force Protection standards. This included new window, wall, and structural support designs to mitigate blast hazards. The materials laboratory also required specialized equipment and air quality standards for experimental use. The largest change to the contract was the addition of finishes to the bottom floor and office furniture installation for all three floors of office space in the acquisition management facility.

Project 3

This DB addition/alteration of Sensors Directorate Laboratory, awarded on April 27, 2009, was $36,212,000 with an original performance period of 690 calendar days. The contract required new construction of an office building, storage warehouses, and sensors testing range along with eight different renovation areas for laboratory and office space totaling 13,750 square meters (148,000 SF). The project entailed integrating detailed laboratory needs into the final designs of the renovations and relocating personnel into temporary office space during construction. This expanded interface with the end-users created a challenge to deliver requirements and execute ongoing refinements.

Research Question and Methods

This exploratory research asks the question: do initial schedules provide reliable forecasting for project control? In order to understand this inquiry in the context of federal DB facility procurement, we first investigated the general scheduling process used by a local USACE office. We used specifications and the schedule data from periodic updates to extract the details of cost, duration, and activities to evaluate consistency between projects.

We used comparative analysis of this schedule data to understand the variability encountered in the updating process. Our approach to capturing project information entailed many conversations with management personnel from both USACE and BCC along with visits to the project sites. The core of the research relies on the examination of Primavera P6 schedule files, schedule narratives, contract schedule specifications, and presentations on USACE scheduling requirements. In the pursuit of organizing this information into applicable findings, we performed the following steps:

1. Outlined and characterized scheduling process (using value stream mapping)
2. Gathered schedule data
3. Analyzed change and trends in cost, duration, and activities
4. Identified timing of changes relative to percent schedule and design complete
5. Employed qualitative root cause analysis on the changes

As described by Creswell (2003), the framework of our inquiry rests on a mixed method of quantitative and qualitative strategies. Using concurrent procedures of research, the observed schedule metrics merges with the gathered observations from project managers to form an understanding of the overall results.

Analysis and Results

We discovered that the three projects had consistent requirements. The contract specifications clearly set the same expectations. However, the process as a whole involved multiple handoffs using redundant information systems. The value stream map revealed a possible problem with the information exchange for schedule updating. In this exchange, the GC first produces schedules in Primavera P6 software, but then must upload schedules to USACE's Quality Control System/Resident Management System (QCS/RMS) in order to complete a pay application. Even though the USACE project manager primarily uses the QCS/RMS information to verify project status, the GC must still submit a hard and soft copy of the Primavera P6 schedule file along with a schedule narrative for the review by the project controller. Waste, therefore, exists in the maintenance of multiple lines of schedule communication. Opportunity for inconsistent data using multiple incompatible systems is a documented challenge (Rasdorf et al. 2009). The information and communication technology study conducted by Lam et al. (2010) reveals a similar redundancy of electronic and hard copies used by multidisciplinary teams throughout the construction industry (Lam et al. 2010).

In all three projects, the GC managers also meet challenges in integrating potential modifications or options in the schedule updates. As per the schedule specifications, contract modifications cannot be included into the official schedule until approved. This drives the official schedule to carry unsound planning of cost and activity logic in several instances. In terms of work execution, GC project managers commented, "we really need to keep two schedules: one for USACE to show no changes, and one for us to implement the items necessary to complete the changes and stay on schedule." This conflict creates a chance to introduce waste and error in the data. The intent for the GC was to use the schedule as a management tool versus a reporting device. However, the demand to communicate compensation for unapproved work led to a variety of schedule approaches by both the USACE and BCC managers. For example, Project 1 began tracking a major change order on a separate schedule and later included it on the final schedule updates. On the other hand, Project 2 rearranged sequencing in the schedule and then delayed a correction of the cost loading until given approval of the change. Project 3 initially reported unapproved modifications on the schedule as floating activities without finalizing sequencing logic until USACE officially awarded the options. Despite the unique circumstances, the friction in developing a schedule update to acknowledge a cost incurred for an unapproved change is a recurring issue in each project.

Change in these projects originates from the following sources:

- Owner/USACE-driven contract modifications
- Definition and sequencing clarification/correction
- Delays from weather and material delivery
- Hidden rework from uncertainty ("Liar's club")
- Process learning

Further schedule analysis quantifies the amount of change incurred in terms of cost, duration, and activity metrics. Table 8.1 summarizes this analysis of the project schedule data. For all three projects, the summary shows a cumulative growth using each metric of cost, duration, and activity count when comparing the schedule updates to the initial schedule. The calculation for the percentage change is total change reported in the schedule updates divided by the original value. For instance, the 7% total cost growth for Project 1 is calculated by subtracting the original contract cost ($8,540,000) from the last scheduled cost ($9,104,448) and dividing by the original contract amount ($8,540,000) and multiplying by 100. The values for original cost and performance period come directly from the awarded contract, whereas the initial number of activities originates from the initial schedule created by BCC in accordance with USACE schedule requirements.

Note that cost growth and duration growth are not synonymous with cost overrun or behind schedule, respectively. Government change in the contract drives most of the variability. Meanwhile, the activity count parameter captures the evolving nature of the scheduled events. The increase in activity count reveals the detailing of the schedule and indicates a new formation

TABLE 8.1

Summary of Schedule Analysis

	Project 1	Project 2	Project 3
Original project amount	$8,540,000	$18,539,000	$36,212,000
Total % cost growth (((last scheduled total cost − original cost)/original cost) * 100%)	7%	4%	41%
Original performance period (cal. days)	540	570	690
Total % duration growth (((last scheduled total duration − original duration)/ original duration) * 100%)	29%	10%	7%
Initial number of activities	331	544	1084
Total % activity growth (((last scheduled activity count − original activity count)/ original activity count) * 100%)	31%	31%	44%

in planning and project change. Although the Lean community views this process of detailing as a favorable way of including production level tasks, the public sector maintains that the initial schedule is a contractual anchor point. Thus, activity count variation is problematic.

Project 1 and 2 schedules report an overall 7% and 4% increase in scheduled cost, respectively. Although contract modifications justify these increases, they are still within a feasible contingency budget estimate of 7.5% as predicted by a recent Air Force construction cost model (Thal et al. 2010). The Project 1 and 2 cost growths also fall within one standard deviation of another cost model of public construction developed in Jordan (Hammad et al. 2010). The 41% cost increase from Project 3 is a result of cumulative options as well as major modifications. The schedule originally removed the numerous options then added them back following each official approval of change. At the time this case study was compiled, Project 3 was still in progress and undergoing further change in the overall scope and budget.

The positive duration growth may indicate a noncompliance to the contracted performance period. However, owner modifications to the contracts have created extensions to the must-finish dates for all projects. Project 1 reported an increase of nearly 160 days to complete the finishes of an additional floor. Contract changes, including office furniture and floor finishes, attribute for most of the growth in Project 2's duration. Finally, the 7% increase in time on the Project 3 schedule is a result of executed options and durations linked to new activities.

Finally, activity growth seen in Table 8.1 signifies an increase in the number of tasks tracked in the progression of schedule updates. The table reports a task expansion of 31% for Project 1 and 2 and 44% for Project 3. According to the USACE DB contract specifications, the remaining construction activities are to be included with cost loading by the completion of the design phase. Within this time, managers can anticipate a degree of change. However, timing analysis of activity change indicates that the majority of activity growth occurs after 100% design.

Figure 8.1 shows the amount of cumulative activity growth in relation to the schedule percent complete throughout the performance period. The horizontal axis represents time in months after the notice to proceed. Using the left vertical axis as its reference, the solid line represents the schedule percent complete according to the progress updates through time. The right vertical axis provides the reference for the cumulative activity growth graphically depicted by the dashed line in the figure. Together, the figure shows the relative timing of activity growths for each project. The activity growth in Project 1 shows an increase of over 80 activities in the last months of the project. These activities relate to the finish of floor three; however, the approval of the change order delayed the actual inclusion of this work in the official project schedule. The GC manager instead tracked the work on a separate schedule to avoid misrepresenting the contract. In turn, the initial schedule does not include a means to monitor the progress of this final phase. Consequently, the schedule exhibits a plateau of the schedule percent complete just below 100% for the last 6 months.

Activity count variance from the initial schedule in Project 2 occurs mostly before the design is complete. Even so, the cumulative growth shows another increase approximately 6 months after the 100% design. The GC manager also reports zero schedule completion for the first 5 months. This anomaly is an error and a result of a manual update of the schedule submittals into the USACE QCS program instead of tracking correctly in the P6 files. The activity growth is a result of both a fleshing out of the schedule during the design and incorporating owner changes during the contract performance. Project 2 is unique from the others in displaying a small drop in the growth at month 15. This indicates a removal of activities from the schedule. Even so, growth continues and the initial schedule becomes more unreliable in monitoring progress.

Finally, the cumulative schedule activity growth for Project 3 indicates considerable deviation from the number of activities planned in the initial schedule. Sixteen percent of cumulative activity growth occurs within the design phase. The recorded project shows an additional 29% activity growth after the design completes. Note that the project is still in progress and is prone to further changes to the activity count based on the trend. The Project 3 graph indicates growth in every periodic schedule update provided. Again, options awarded within the performance period of Project 3 help shape the changes of activity counts.

Despite the differences in project requirements and management personnel, the scheduling process is similar. Missing schedule submittals reveal gaps in the percent schedule complete and activity count trends seen in Figure 8.1. Yet the graphs still clearly present evidence that the number of activities increases throughout the project and that the design development accounts for only a fraction of this change. The growth in total activities requires additional effort from the GC to maintain and USACE to review. The upward trend of activities in all three projects indicates that schedules

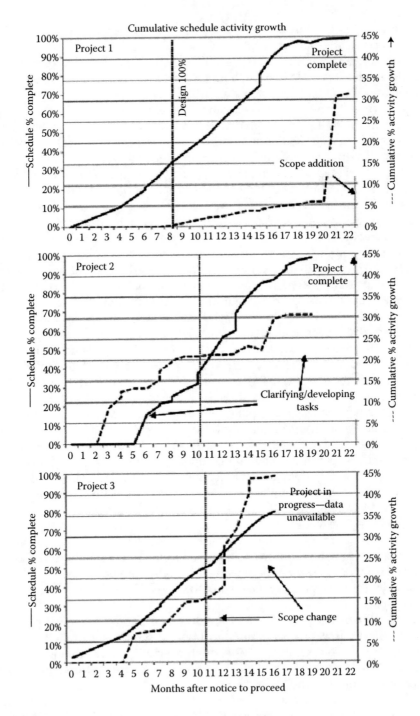

FIGURE 8.1
Plot of comparison of percent complete and cumulative activity schedule.

transform throughout the projects despite the establishment of a single base-line specified in the beginning. Although this initial schedule is required within the first 2 months, these DB contracts do not reach 100% design until the 8–11 month point. Moreover, a third or more of the activity growth occurs months beyond the 100% design. Consequently, these project schedules appear dynamic in the attempt to capture unexpected modifications through-out the performance period.

Recommendation

Given the unintended dynamics observed in activity-based scheduling, our recommendation is to change specifications in federal facility procurement contracts in pursuit of more reliable forecasting. The goal is to allow a sched-ule to adjust according to expected uncertainty while maintaining control. As seen in the cone of uncertainty in Figure 8.2, the variability early in a DB project at 35% design is much greater than later at 100% design.

The target finish date on the horizontal axis acts as a surrogate for any target schedule metric such as cost, duration, or activity count. Thus, when a design is only at 35%, a project manager can expect the variability around a target metric to be large. Based on the case studies, the accuracy of the ini-tial schedule at roughly 35% design ranges 4–44% depending on the metric. However, as more design is completed, the cone narrows shaping a reduc-tion in the level of project uncertainty. In the cases explored, the schedules at the 100% design mark show a reduction of 50% in the variability. Production changes still occur due to unexpected planning and this needs to be known and worked by all stakeholders.

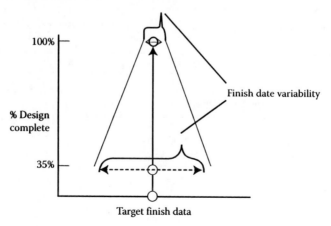

FIGURE 8.2
Cone of uncertainty.

The timing in establishing a phase baseline determines the amount of uncertainty the schedule will carry and potential for rework in actual activity execution. Accordingly, Feng et al. (2008) demonstrate how rework timing affects a project as a whole; by delaying final plan and work in order to resolve unknowns, the overall time required for negative rework decreases.

As a foundation to the change, production theory and lean thinking offers an innovative perspective to the construction industry by promoting a simultaneous adherence to the principles of transformation, flow, and value (TFV) (Koskela 1992; Ballard 2000; Ballard et al. 2002). In particular, the Last Planner™ System (LPS) focuses on these TFV goals to provide production control in the scheduling process (Ballard and Howell 1998, 2003; Ballard 2000). LPS also embodies a "management by means" foundation of thought by addressing internal goals and metrics through "percent planned complete" of weekly work (Kim and Ballard 2010). Kim and Ballard discuss how the LPS concept thus better suits an operational level of work such as the daily construction management endeavors where "each task is highly interdependent." LPS incorporates four levels of planning as seen in Figure 8.3.

Using these four levels or planning, management can structure work using the most recent information and provide reliable workflow with pull techniques and active conflict resolution. Planning therefore integrates changes into the schedule updates. As the time of execution nears, details explode and the basis for measured progress is a current set of promises or goals. The Last Planner system provides an alternative scheduling method applicable to the public sector but must be carefully implemented in DB projects.

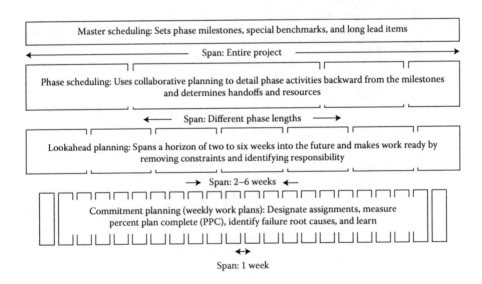

FIGURE 8.3
Alternate levels of schedules.

More specifically, our recommendation involves using relational contracting to establish progressive phase schedules aimed at target value designs. Relational contracting provides a way to share risk and commitment through strategies of target value design, collaboration, holistic thinking, and learning (Lichtig 2005). In this way, the schedule specifications could still call for a master schedule to provide a target value design and needed end-date and important milestones to the end-user. The master schedule is the skeleton of the baseline. However, the payouts to GCs coincide with progress in the more flexible set of phase schedule baselines that correspond with the development of the design and execution planning. The GC would withhold phase schedules until the design approvals at 65%, 95%, and 100% and develop progressive phase baselines aimed to include all changes in the early stages. In doing so, we defer more decisions until the last responsible moment and consequently strengthen the reliability of the schedule for the remainder of all planning, organizing, leading, and controlling tasks in management.

Several obstacles still exist in a progressive baseline approach. Without the correct incentives and contractual conditions, liability and transparency are a concern for the government. The perceived risk is higher if the government accepts an incomplete or flexible-type schedule in the beginning of the project. The transformation of the phased baseline schedules could provide GCs with an opportunity to make unaccounted changes. Yet, the implementation of the bottom tiers of the lookahead and commitment planning offers the connective tissue most important to the monitoring of the plans and production. Scheduling, in this way, becomes a pull system controlled by the production team to support the hard constraints of the project. Contractual completion dates can therefore be set while internal execution is more fluid. The four levels of schedules used together reinforce the trust and learning needed to execute positive control and ultimately provide more reliable forecasting.

Conclusions

Uncertainty challenges construction managers throughout the scheduling process. The initial activity-based schedules from the case study projects reveal shortfalls in forecasting:

- Final project cost
- Final project duration
- Total activity count

The growth in these metrics indicates that uncertainty in the beginning of the projects is unavoidable. Although creating a baseline early may establish an indicator of project plan and scope, the encountered modifications can quickly deem the efforts obsolete. Since change happens, it should be incorporated progressively. The activity growth in particular warrants concern of GC managers since USACE expects them to justify deviation from the initial schedule and convey positive control. The schedule specification mandates a submission of reasoning and solution with any behind-schedule activities (United States Army Corps of Engineers 2007). Accordingly, the added work of explaining reported activities does not contribute to any of the lean goals of transformation, flow, or value.

> Pursuing lean goals in public facility procurement could make a major impact on the entire construction industry.

Although initial activity-based schedules seemingly provide a comprehensive and networked plan in which to monitor project progression, a different approach to capture change appears necessary. Detailing work breakdowns and critical paths in the beginning of the project does not provide a reliable baseline. If the project does use initial schedules as baselines, change disturbs efforts to monitor realistic outputs. In this way, updates deviating from the baseline schedule require continual justification. At the same time, known contract changes cannot be included in the schedule until official approval, which pushes the uncertainty into the future. Ultimately, the scheduling required by USACE only facilitates as supporting documentation for payment applications but is misaligned for project control.

Balancing the financial investments and risks of a facility project against the progression of completed work is a key management mechanism for those overseeing federal procurement. On the other side, general contractors are obliged to show a plan to accomplish work and receive compensation through an initial schedule. However, if the initial schedule fails as a suitable baseline because of change, change ought to be integrated into project management from both sides of the contract. Together, the government and contractors can work to pull scheduling into place rather than push. Since the unknowns for DB projects are unavoidable, the four-tiered planning approach of the Last Planner system may provide the only viable option. At the least, schedule specifications need to acknowledge a demand for a progressive baseline that responds quickly to change. Overall, controlling uncertainty can provide more reliable schedule forecasting and project control.

References

ABET Engineering Accreditation Commission. 2010, *Criteria for Accrediting Engineering Programs*. ABET, Inc., Baltimore, MD.

Agostini, J. 1996, *Legal Aspects and Risk Management of Design/Build Contracts.* Architects/ Engineers Professional Network, http://www.aepronet.org/pn/vol9-no1.html.

American Society of Civil Engineers. 2010, *Policy Statement 400—Design-Build Procurement.* http://www.asce.org/Content.aspx?id=8503 (December, 2010).

Ballard, G. 1999, Can pull techniques be used in design management, *Conference on Concurrent Engineering in Construction,* August 26–27, Helsinki, Finland, pp. 1–18.

Ballard, G. 2000, *The Last Planner(tm) System of Production Control,* PhD thesis, University of Birmingham, School of Civil Engineering, Birmingham, UK.

Ballard, G. and Howell, G. 1998, Shielding production: Essential step in production control, *Journal of Construction Engineering and Management;* ASCE, 124(1), 11–17.

Ballard, G. and Howell, G. A. 2003, An update on last planner, *11th Annual Conference of the International Group for Lean Construction,* July 22–24, Blacksburg, VA, pp. 1–13.

Ballard, G., Tommelein, I., Koskela, L., and Howell, G. 2002, Lean construction tools and techniques. *Design and Construction: Building in Value,* R. Best and G. De Valence, eds., Butterworth-Heinemann, Oxford, pp. 227–255.

Creswell, J. W. 2003, *Research Design: Qualitative, Quantitative, and Mixed Method Approaches,* Sage Publications Inc., Thousand Oaks, CA.

Feng, P. P., Tommelein, I. D., and Booth, L. 2008, Modeling the effect of rework timing: Case study of a mechanical contractor, P. Tzortzopoulos and M. Kagioglou, eds., *Proceedings of the 16th Annual Conference of the International Group for Lean Construction (IGLC 16),* July 16–18, Manchester, UK.

Ford, D. N. and Sterman, J. D. 2003a, The Liar's club: Concealing rework in concurrent development, *Concurrent Engineering: Research and Applications,* 11(3), 211–219.

Ford, D. N. and Sterman, J. D. 2003b, Overcoming the 90% syndrome: Iteration management in concurrent development projects, *Concurrent Engineering: Research and Applications,* 11(3), 177–186.

Gannon T., Feng P., and William S. 2012, Reliable schedule forecasting in federal design-build facility procurement. *Lean Construction Journal,* 8(1), 1–14.

Hammad, A. A. A., Ali, S. M. A., Sweis, G. J., and Sweis, R. J. 2010, Statistical analysis on the cost and duration of public building projects, *Journal of Management in Engineering,* 26(2), 105–112.

Kim, Y. W. and Ballard, G. 2010, Management thinking in the earned value method system and the last planner system, *Journal of Management in Engineering,* 26(4), 223–228.

Koskela, L. 1992, *Application of the New Production Philosophy to Construction,* Rep. No. Technical Report #72, Center for Integrated Facility Engineering, Department of Civil Engineering, Stanford University, CA.

Lam, P. T. I., Wong, F. W. H., and Tse, K. T. C. 2010, Effectiveness of ICT for construction information exchange among multidisciplinary project teams, *Journal of Computing in Civil Engineering,* 24(4), 365–376.

Lichtig, W. 2005, Sutter health: Developing a contracting model to support lean project delivery. *Lean Construction Journal,* 2(1), 105–112.

Rasdorf, W., Hummer, J. E., Harris, E. A., and Sitzabee, W. E. 2009, IT issues for the management of high-quantity, low-cost assets, *Journal of Computing in Civil Engineering,* 23(2), 91–99.

Rookard-Everett, T. 2009, Bridging the oversight gap. *The Military Engineer,* 101(662), 65–66.

Spearman, M. L. and Zuzanis, M. A. 1992. Push and pull production systems: Issues and comparisons. *Operations Research,* 40(3), 521–532.

Thal Jr, A. E., Cook, J. J., and White III, E. D. 2010, Estimation of cost contingency for Air Force construction projects, *Journal of Construction Engineering and Management*, 136(11), 1181–1187.

United States Army Corps of Engineers. 2007, General Requirement Section 01 32 01.00 10 Project Schedule. *Unified Facilities Guide Specifications*, pp. 1–18.

Xue, X., Shen, Q., and Ren, Z. 2010, Critical review of collaborative working in construction projects: Business environment and human behaviors, *Journal of Management in Engineering*, 26(4), 196–208.

9

*Multiattribute Drilling System Selection**

This chapter is an application case study of a methodology to select optimal onshore environmentally friendly drilling (EFD) systems at Green Lake near McFaddin, Texas. The chapter describes sensitivity analysis procedures, which will help decision makers examine the robustness of the optimal solution to changes in input parameters in system selection decisions, where the system is made up of components (drilling technologies) that can be combined in many different ways. Two different sensitivity analysis techniques are presented. One is a sensitivity analysis for weighting factors of each attribute and the other is a sensitivity analysis for uncertainty of the overall attribute inputs.

Introduction

Input data used in multiattribute decision making (MADM) problems are often perceived to be imprecise by decision makers because they are generally based on expert assessments. Because of this, the results of a MADM model tend to not be fully trusted by decision makers due to the inherent uncertainty. As a result, an important step in many applications of MADM is to perform a sensitivity analysis on the input data to help decision makers understand how in which regions of the input data space they can be most confident in the recommended decisions and where improved input information is most needed. Earlier work by the authors developed a multiattribute decision model for helping decision makers select an optimal oil and gas drilling system for a specific site with the objective of balancing environmental burden, public opinion, and drilling cost (herein, a system is defined as a particular drilling set of technologies). In this chapter, we build from this previous work to show how a sensitivity analysis can be conducted for this type of problem.

Two different approaches for conducting a sensitivity analysis for selection of onshore EFD systems are presented in this chapter. Sensitivity analysis for multiattribute utility problems can be categorized based on the number

---

* Reprinted with permission from Yu, O.-Y., Guikema, S., Briaud, J.-L., and Burnett, D. 2012, Sensitivity analysis for multi-attribute system selection problems in onshore environmentally friendly drilling (EFD), *Systems Engineering*, 15(2), 153–171.

of times an optimization routine needs to be run to analyze sensitivity [1]. If various individuals have distinct weight combinations for multiattribute utility problems, each combination could be given as a discrete weight combination to the optimization routine and any resulting change in the technology selected would indicate sensitivity to an individual's choice of weight combination. In this case, not only do relatively few optimization need to be run, but also relatively little postprocessing of the optimization results is needed to evaluate sensitivity [1]. The sensitivity analysis for discrete weight combinations of multiattribute utility problems has been addressed many times in the literature. Call and Merkhofer [2], for example, developed one approach to sensitivity analysis using predefined weight combinations (i.e., high and low for each attribute).

On the other hand, if decision makers do not feel confident enough in their assessments to specify precise values, uncertainties in input parameters such as the weights of each attribute in multiattribute utility problems can arise. In this case, the proper values can lie anywhere within a possibly wide range of values specified by the decision makers. For this type of sensitivity analysis, multiple optimizations need to be run and the breakpoints in the resulting recommended decisions become important. In this chapter, for example, the breakpoints where the optimal drilling systems change are very important aspect. This type of sensitivity analysis is more difficult and time consuming than discrete sensitivity analysis. Less research has focused this type of sensitivity analysis in the literature than for the discrete sensitivity analysis.

This chapter focuses on extending the types of sensitivity analysis that can be conducted for MADM problems. Yu et al. [3,4] developed a system evaluation protocol to incorporate a number of current and emerging onshore EFD technologies into an optimal drilling system by minimizing cost and environmental impact, and by maximizing public perception. They also conducted a case study at Green Lake at McFaddin in Texas to illustrate the applicability of their approach. This chapter introduces the case study conducted by Yu et al. [3,4] and demonstrates methods for conducting a sensitivity analysis for multiattribute system selection problems in onshore EFD.

Green Lake Case Study for System Selection Problems

Yu et al. [3,4] introduced the use of multiattribute utility theory to develop a quantitative decision tool for helping decision makers select an optimal drilling system for a specific onshore oil and gas drilling site, properly balancing the inherent trade-off among cost, environmental impacts, perceptions, and safety based on a formal quantification of preferences. The proposed decision model has been developed over the last 5 years as part of a comprehensive academic–industry program funded by the U.S. Department of Energy (DOE) and Research Partnership to Secure Energy for America (RPSEA) to integrate the key drilling phases [5]. Many subject matter experts from the oil and gas drilling community, including commercial vendors and consultants familiar with the currently available drilling technologies, have assisted in the development of a more reasonable and practical decision model through a series of individual and group meetings and technical workshops. In this chapter, we focus on the sensitivity analysis aspects of this same problem and case study.

For the Green Lake case study, it is assumed that an independent operator is to drill a well on their lease in South Texas in an environmentally sensitive wetland area. The lease extends to the center of Green Lake on the McFaddin Ranch. The formation target is the upper Frio sand [6] at approximately 8500 ft in vertical depth. This section describes the results of the case study that provided a logical and comprehensive approach that balanced the trade-offs between the economic and environmental goals of both the landowner and the oil company leaseholder. The step-by-step procedures to arrive at the optimal drilling system for the site are fully described by Yu et al. [3,4].

Identify Main Subsystems, Subsets, and Available Technologies within Each Subset

The system selection approach proposed by Yu et al. [3,4] includes four main "subsystems" (site access, drill site, rig, and operation) and 13 subsets that have been previously identified through EFD operations (see Figure 9.1). A drilling technology selection example is also presented in Figure 9.1, where technologies indicated within circles through the subset tables represent one example of a possible drilling system. Yu et al. [3,4] developed an approach to evaluate all possible combinations of technologies to find the optimal drilling system for a given site on the basis of elicited preferences over a range of attributes. Three different systems are prespecified by EFD experts in order to identify possible drilling technologies for this case study as given in Table 9.1. Figure 9.2 briefly illustrates the total possible number of systems used in this case study.

Define Attributes and Attribute Scales

In this case study, nine attributes and the corresponding scales are considered for the selection of the EFD system as shown in Figure 9.3. An attribute is defined as one of the parameters considered in the evaluation of the system (e.g., cost, footprint, emission, perception, and safety). Each attribute has an attribute scale used to score the technology on how well it meets the objective for this attribute (e.g., minimizes cost, footprint, emission, and

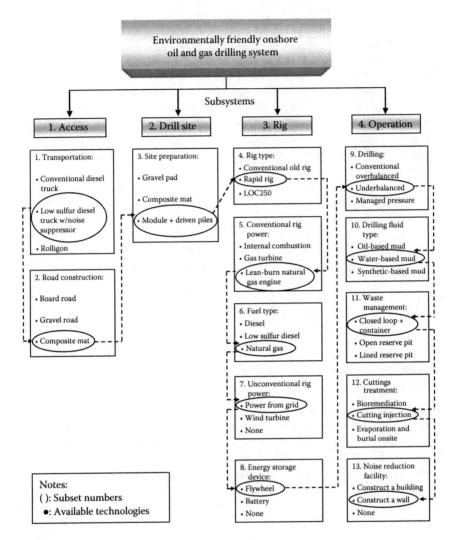

FIGURE 9.1
An example of the EFD system selection.

maximizes positive perception and safety value). In order to evaluate available technologies against each attribute, attribute scales that explicitly reflect the impacts on the system selection process are needed [7]. The nine attributes considered in this case study are briefly described below. These are the same attributes used in Yu et al. [3,4].

1. Total cost (x_1): The total expenditure in dollars during the drilling operation.
2. Footprint (x_2): The total land area used by the entire drilling process.

TABLE 9.1

Prespecified Drilling Systems

Subsets	1. Conventional Drilling	2. Moderately Improved Drilling	3. EFD in 5 Years
1. Transportation	Conventional diesel truck	Low-sulfur diesel truck w/noise suppressor	Low-sulfur diesel truck w/noise suppressor
2. Road construction	Gravel road	Composite mat	Composite mat
3. Site preparation	Gravel pad	Composite mat	Aluminum modules + driven piles
4. Rig type	Traditional older rig	Rapid rig	LOC250 (CWD)
5. Conventional power	Internal combustion engine	Internal combustion engine w/SCR, w/noise suppressor	Lean-burn natural gas engines w/noise suppressor
6. Fuel	Conventional diesel	Low-sulfur diesel	Natural gas
7. Unconventional power	None	Electric power from grid (10%)	Electric power from grid (30%)
8. Energy storage	None	Flywheel	Flywheel
9. Drilling technology	Conventional overbalanced drilling	Underbalanced drilling w/noise suppressor	Managed pressure drilling w/noise suppressor
10. Fluid	Water-based muds	Water-based muds	Water-based muds
11. Waste management	Lined reserve pit + solid control equipment	Closed loop + containers + solid control equipment	Closed loop + containers + solid control equipment
12. Cuttings treatment	Cuttings injection	Cuttings injection	Chemical fixation and solidification (CFS)
13. Noise reduction facility	None	None	None

3. Emissions of air pollutants (x_3): Emissions of three air contaminants (i.e., CO, NO_x, and PM). The relative importance of these contaminants is CO (20%), NO_x (40%), and PM (40%) as shown in Table 9.2, which shows an example of how to calculate the overall air emission score for each technology. First, we estimate the actual emissions level for each of the three contaminants in pounds per operating hour. Second, in order to get an overall air emission score for each technology, we transform each contaminant emission rate into a nondimensional score (normalization) between 0 and 1 using the proportional scoring approach, (x–worst score)/(best score–worst score). In this calculation, the best and worst scores should be obtained among all possible technologies that could be used. Finally,

1. When "<u>Diesel engine</u>" is selected as a conventional power generation

Subsets	Subsystems				Π
	1. Access	2. Drill site	3. Rig	4. Drilling	
1.	2	4**	3	3	72
2.	3*		2***	1	6
3.			2	2	4
4.			1	2	2
5.			1	1	1
				Π	3456

2. When "<u>Natural gas engine</u>" is selected as a conventional power generation

Subsets	Subsystems				Π
	1. Access	2. Drill site	3. Rig	4. Drilling	
1.	2	4**	3	3	72
2.	3*		1	1	3
3.			1	2	2
4.			1	2	2
5.			1	1	1
				Π	864

∴ Total number of possible systems   =   3456
within 1 power allocation scenario   864

Σ   4320

4 different power scenarios (0, 10, 20, 30% of unconventional power) were considered.
∴ Total number of iterations =   4   × 4320   = 17,280

*2 options of composite mat (rent and buy); **2 options of composite mat (rent and buy);
***2 types of diesel engines

FIGURE 9.2
Total number of possible systems used in this case study.

we calculate the overall air emission score of a technology as $\Sigma k_i u_i$, where k_i is a weight factor for each air contaminant and u_i is a non-dimensional score for each contaminant.

4. Emissions of solid and liquid pollutants (x_4): The ordinal scale as constructed in Table 9.3.

5. Emissions of noise (x_5): The 8-h time-weight average sound level (TWA) given in decibels.

6. Perception of government regulators (x_6): The ordinal scale as constructed in Table 9.4.

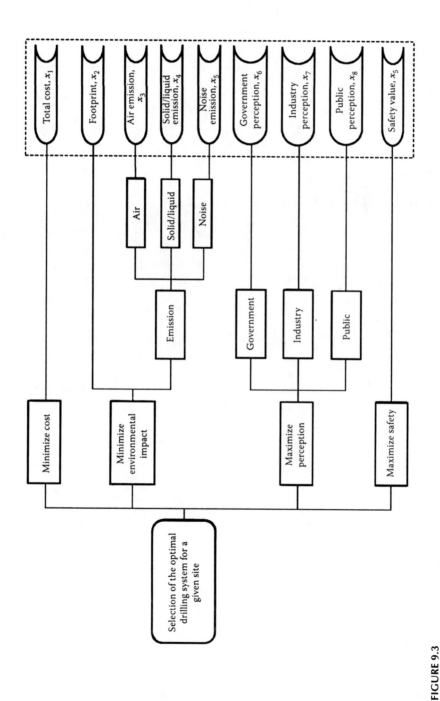

FIGURE 9.3
Fundamental objectives hierarchy and attributes $(x_1 \sim x_9)$ for the EFD project.

TABLE 9.2

An Example of Air Emission Score Calculation

Technologies	Unit	0.2 CO	0.4 NO_x	0.4 PM	Overall Score
Internal combustion engine	(lb/MWh)	6.2	21.8	0.78	0.118
	(lb/h)/unit	6.200	21.800	0.780	
	(lb/h)*portion	6.200	21.800	0.780	
	(lb/operating)	1339.200	4708.800	168.480	
	U-value	0.588	0.000	0.000	
Internal combustion engine w/SCR, w/noise suppressor	(lb/MWh)	6.2	4.7	0.78	0.431
	(lb/h)/unit	6.200	4.700	0.780	
	(lb/h)*portion	6.200	4.700	0.780	
	(lb/operating)	1339.200	1015.200	168.480	
	U-value	0.588	0.784	0.000	
Lean-burn natural gas engines w/noise suppressor	(lb/MWh)	5	2.2	0.03	0.878
	(lb/h)/unit	5.000	2.200	0.030	
	(lb/h)*portion	5.000	2.200	0.030	
	(lb/operating)	1080.000	475.200	6.480	
	U-value	0.668	0.899	0.962	
Power from grid	(lb/MWh)	0	0	0	1.000
	(lb/h)/unit	0.000	0.000	0.000	
	(lb/h)*portion	0.000	0.000	0.000	
	(lb/operating)	0.000	0.000	0.000	
	U-value	1.000	1.000	1.000	

7. Perception of industry as decision maker (x_7).
8. Perception of the general public (x_8).
9. Safety value (x_9).
10. The ordinal scales of x_7 through x_9 are similar to x_6 [3,4].

Assign Scores to All Technologies Using the Attribute Scales

In order to evaluate available technologies in terms of the nine attributes (i.e., x_1 through x_9), EFD subject matter experts' inputs, basic assumptions, and other references are used.

Figure 9.4 shows the basic assumptions used in this case study and key input variables that affect the input values of technologies. The influence

TABLE 9.3

Draft Attribute Scale for Solid and Liquid Emission

Waste Management Technologies	Cuttings Treatment	Solid/Liquid Emission Score
Closed loop	Cutting injection	1.00
—	Bioremediation, composting, *in situ* vitrification, land spreading, plasma arc, microwave technology	0.75
Lined reserve pit	Thermal desorption	0.50
—	Chemical fixation and solidification	0.25
Open reserve pit	Evaporation and burial onsite	0.00

TABLE 9.4

Draft Attribute Scale for Government Perception

Description	Perception Score
Strongly support. All parties will encourage its use and are willing to appropriate funds for the cause.	1.00
Moderate support. There is interest from a majority. Its use will be encouraged, but funds will not be appropriated.	0.75
Neutrality. All parties are indifferent. There is no resistance, but there is also no help.	0.50
Moderate opposition. Some resistance from the majority. Its use may be discouraged, but fines or restrictions will not be imposed.	0.25
Strong opposition. Strong resistance to its use from all parties. Restrictions or fines will be set up to eliminate this option.	0.00

diagram for the drilling site shown in Figure 9.5 should be considered before estimating attribute scores of technologies because attribute scores of a technology can be dependent on key influence variables described in Figure 9.4. For example, different rig type causes the variation of total drilling time and the total drilling time varies the total cost of technologies within many subsets.

Calculate the Overall Attribute Score for Each Attribute

As given in Yu et al. [3,4], after each technology is evaluated in terms of the nine attributes, for each attribute, the overall attribute score of a system is calculated by adding the technology scores of the system or selecting the minimum technology score of the system. The overall scores of cost (x_1), footprint (x_2), and emissions $(x_3$ through $x_5)$ are calculated by summing the scores of technologies selected within each subset as shown in Equation 9.1. The overall scores of perceptions $(x_6$ through $x_8)$ and safety (x_9), however, are calculated by choosing the minimum score among technologies

Basic Assumptions

- Power consumption (peak):   1 MW
- Access road width: 25 ft (2 lanes)
- Access road length: 1 miles
- Width of drilling site: 350 ft (conventional rig + pad)
      300 ft (compact rig + pad)
      200 ft (conventional rig + modules + piles)
      150 ft (compact rig + modules + piles)
- Length of drilling site: 350 ft (conventional rig + pad)
      300 ft (compact rig + pad)
      125 ft (conventional rig + modules + piles)
      100 ft (compact rig + modules + piles)

Key Influence Variables

- Transportation type: Conventional diesel truck
- Rig type: LOC250 (CWD)
- Engine type: Internal combustion engine
- Drilling type: Conventional overbalanced drilling
- Noise reduction type: N/A
- Proportion of unconventional power: 30.0%
- Resale value: 80.0%
- Drilling time: 9.0 days
- Move/rig up: 1.0 days
- Number of wells: 1 wells

FIGURE 9.4
Basic assumptions and key influence variables.

selected within each subset as shown in Equation 9.2 because it is suggested that perception and safety values should be considered on the systems level and not on the individual technology level. The overall score on the ith attribute (X_i) is

$$X_i = \sum_{n=1}^{N} x_{i,n} y_n \quad \text{for attribute } x_1 \text{ through } x_5 \text{ (i.e., } i = 1 \text{ through 5)} \quad (9.1)$$

$$X_i = \text{Min}\left[x_{i,n} y_n \right] \quad \text{for attribute } x_6 \text{ through } x_9 \text{ (i.e., } i = 6 \text{ through 9)} \quad (9.2)$$

where n is the index for possible technologies, N is the number of possible technologies, i is the index for the attributes, $x_{i,n}$ is the score of the nth technology on the ith attribute, and y_n is a binary decision variable that is one if the nth technology is selected, and zero if it is not.
The constraints on the optimization route are

$$\sum_{n=1}^{M} y_n = 1 \quad \text{for each subset except subsets (7), (8), and (13)} \quad (9.3)$$

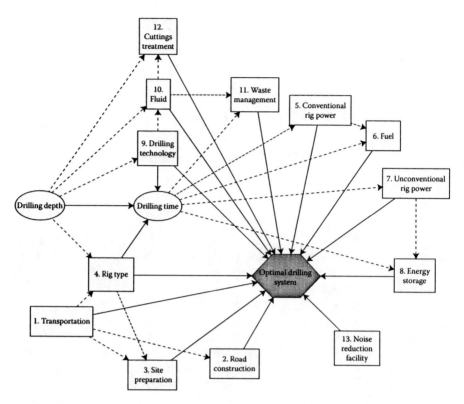

FIGURE 9.5
Influence diagram for the drilling site of the case study.

where n is the index for possible technologies, M is the number of possible technologies within each subset, and y_n is a binary decision variable. This constraint ensures that only one technology will be selected for each subset except subsets (7), (8), and (13) because subsets (7), (8), and (13) are optional (see Figure 9.1). In this case study, for example, the range of unconventional power usage is varied from 0% to 30% of total power usage, so if it is decided not to use an unconventional power generation technology (subset (7)), an energy storage device (subset (8)) is also not considered as a possible subset. More detailed constraints for each subsystem can be found in Yu et al. [4].

Develop a Utility Function for Each Attribute

Figure 9.6 shows an example of single-attribute utility function curves used in this case study. These are the same utility functions used in Yu et al. [3,4]. A utility function is a relationship between the dimensional attribute score (e.g., $, acres, and grades) and a nondimensional number (between 0 and 1) that captures decision maker preferences. The utility function is used to

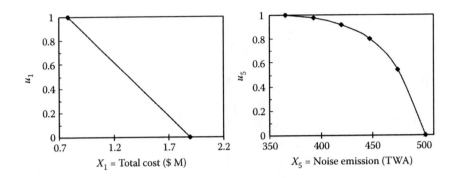

FIGURE 9.6
An example of single-attribute utility function curves.

transform all scores into nondimensional values between 0 and 1. This allows the decision maker to make the overall attribute score for each attribute uniform and comparable.

In situations with multiple evaluation attributes, the term "utility function" is used in the literature to mean a function that encodes trade-offs among the multiple evaluation attributes when either (1) uncertainty is not formally modeled with probabilities, or (2) uncertainty is formally modeled with probabilities and expected utility is used as the decision criterion. We do not formally model uncertainty with probabilities in this chapter. What we call a utility function is sometimes called a value function [7]. In this study, the general shapes of the utility function for each attribute are linear with the exception of the noise attribute. It is very important, before proceeding, to do consistency checks on the reasonableness of the shape of the utility functions [7]. Once each single-attribute utility function $u_i(X_i)$ is derived for its attribute measure, these individual utility values are combined into a final utility value. The multiattribute utility function with the additive form used in this case study is shown in Equation 9.4 [8].

$$U(X_1, X_2, \ldots, X_9) = U\{u_1(X_1), u_2(X_2), \ldots, u_9(X_9)\}$$
$$= k_1 u_1(X_1) + \cdots + k_9 u_9(X_9) = \sum_{i=1}^{9} k_i u_i(X_i) \tag{9.4}$$

where $u_i(X_i)$ is the utility of the ith attribute scaled from 0 to 1, and k_i is the weighting constant for the ith attribute.

Decide on a Weight Factor for Each Attribute

Formally, the weight combination represents the trade-offs between the utility of the different attributes [9]. In general, weight factors are elicited

TABLE 9.5

Assigned Weight Factor for Each Attribute (Base Case)

Attributes	Weights
Total cost (x_1)	0.40
Footprint (x_2)	0.20
Air emission (x_3)	0.20/3
Solid/liquid emission (x_4)	0.20/3
Noise emission (x_5)	0.20/3
Government perception (x_6)	0.05
Industry perception (x_7)	0.05
Public perception (x_8)	0.05
Safety (x_9)	0.05

from a decision maker. For this case study, the base-case weight factors were assessed by two EFD experts who participated in this research as shown in Table 9.5. The weight factors can be assessed by different methods. For example, using a series of trade-offs between different attribute levels can be used for a problem with small numbers of attributes [10]. Another procedure is to assign points to each attribute in proportion to its relative importance within the level of the hierarchy. In this study, direct scoring procedure was used. Two experts were asked to order the attributes from the most importance to the least importance for the Green Lake case study and then to assign score to each attribute within the summation of one. In summary, according to the fundamental objectives hierarchy and attributes (see Figure 9.3), it is assumed that the experts determined that a unit increase in utility in the cost objective and the environmental impact objective is eight times as important as a unit increase in utility in the safety objective, and a unit increase in utility in the perception objective is three times as important as a unit increase in utility in the safety objective. This implies weights of 0.40, 0.40, 0.15, and 0.05 for each objective. The experts also determined that a unit increase in utility in the footprint attribute (x_2) is three times as important as a unit increase in utility in each emission attribute ($x_3 \sim x_5$). Moreover, they determined that the weights of the three emission and perception attributes ($x_3 \sim x_6$ and $x_6 \sim x_8$) are evenly distributed.

The overall base-case weight for the each attribute can be calculated by multiplying its within-objective weight by the weight on the objective it describes above. For example, since a three to one level of importance between footprint attribute (x_2) and each emission attribute ($x_3 \sim x_5$) implies 3/6, 1/6, 1/6, 1/6 weightings, respectively, the overall weighting for the footprint (x_2) and air emission (x_3) attributes are given by

$$\text{Footprint weight} = (0.4) \times (3/6) = 0.2 \tag{9.5}$$

$$\text{Air emission weight} = (0.4) \times (1/6) = 0.2/3 \tag{9.6}$$

The overall weights for other attributes are similarly calculated and presented in Table 9.5.

Find the Best System

As given in Yu et al. [3,4], the exhaustive search optimization model for the system selection problem with nine attributes and the weight factors given in Table 9.5 is used to find the base-case optimal drilling system.

Figure 9.7 shows the final utility value of the best drilling system as well as the overall attribute scores of each attribute. It is noted that attribute scores are not evaluated for the empty cells because those attribute scores are not relevant to the particular subsets, or because these are already included in technologies within other subsets. After the optimization scheme has given the "best" system, a sensitivity analysis should be conducted to help the decision maker understand how robust the best system is to changes in the input parameters such as attribute scores and weight factors. Conducting a sensitivity analysis for the system selection process is an important step because it can give an idea of the range of weights over which certain systems should be selected for a specific site [11].

Sensitivity Analysis for Weight Factors of Each Attribute

The first step in conducting a sensitivity analysis is to examine discrete combinations of input values. For example, these inputs represent the weight factors, single-attribute utility functions, or the attribute scores assessed by different individuals [1]. A sensitivity analysis for different weight combinations is presented in this section. Four different weight scenarios (i.e., different points of view) are defined by EFD experts as shown in Table 9.6, the optimization routine is run for each weight combination, and then the results are compared as shown in Table 9.7. Table 9.6 shows that each weight combination includes five weight components ($W_1 \sim W_5$), and it is noted that three emission attributes ($x_3 \sim x_5$) and three perception attributes ($x_6 \sim x_8$) are grouped in W_4 and W_5, respectively. Each of those weight combinations represents a different point of view for the EFD technology selection problem. In this step, the change in the final utility score of the optimal system is not a good sensitivity measure because the final utility score directly depends on the input parameters being used and there are also many uncertainties in those input values. Instead, it is suggested to look at the changes in the technologies selected for the optimal system because this is the decision that is of most interest to the decision makers.

A sensitivity analysis for uncertainty in the weight combinations is also conducted in this case study. This includes uncertainties in what the

Selected Technologies in Each Subset	40% Total Cost ($)	20% Ecological Footprint (Acres)	6.667% Emissions Air	6.667% Emissions Solid and Liquid	6.667% Emissions Noise (TWA)	5% Perceptions Gov.	5% Perceptions Ind.	5% Perceptions Public	5% Safety Value
1. Transportation: Low sulphur diesel truck w/tier III engine, w/noise suppressor						1.000	0.500	1.000	0.750
2. Road construction: Composite mat (rent)	$147,840	1.515	0.976		64.696	1.000	0.500	1.000	1.000
3. Site preparation: Composite mat (rent)	$100,800	1.033	0.984		62.356	0.750	0.750	0.750	1.000
4. Rig type: LOC250 (CWD)	$173,800		0.985		60.366	1.000	0.500	1.000	1.000
5. Rig power (Conventional): Lean-burn natural gas engines w/noise supperssor	$70,354		0.918		85.603	1.000	0.500	1.000	0.750
6. Fuel type: Natural gas	$25,650					1.000	0.500	1.000	0.750
7. Rig power (Unconventional): Electric power from grid (10%)	$3840	0.000	1.000		0.000	0.500	1.000	1.000	1.000
8. Energy storage: Flywheels	$30,000	0.000				0.500	1.000	1.000	0.750
9. Drilling tech.: Underbalanced drilling w/noise suppressor	$184,500				95.700	0.750	0.750	0.750	0.750
10. Fluid type: Water-based muds	$47,970					1.000	1.000	1.000	1.000
11. Waste mgmt.: Closed loop + containers + solid control equip.*	$27,000	0.000		1.000		1.000	0.500	1.000	0.750
12. Cuttings mgmt.: Cuttings injection	$45,000			1.000		1.000	0.500	1.000	0.750
13. Noise reduction facility: N/A						1.000	0.500	1.000	0.750
Overall attribute scores (Σ or minimum value)	$856,724	2.548	4.863	2.000	368.721	0.500	0.500	0.750	0.750
Single attribute utility values	0.931	0.764	0.986	1.000	0.998	0.500	0.500	0.750	0.750

∴ Multiattribute utility value = 0.849

* Solid control equipment includes shakers, possibly cone centrifuge, desander, desilter, cuttings dryer, and perhaps decanting centrifuge.

FIGURE 9.7
Final utility score of the best system.

TABLE 9.6

Weight Combinations Used in the Sensitivity to Point of View Analysis

Weight No.	Cost (W_1)	Footprint (W_2)	Emissions (W_3)			Perception (W_4)			Safety (W_5)	Note
			Air	S/L	Noise	Govt.	Ind.	Public		
1	0.60	0.25	0.05/3	0.05/3	0.05/3	0.05/3	0.05/3	0.05/3	0.05	Conventional
2	0.40	0.20	0.20/3	0.20/3	0.20/3	0.05	0.05	0.05	0.05	Base case
3	0.27	0.25	0.08	0.08	0.08	0.05	0.05	0.07	0.07	EFD
4	0.12	0.30	0.10	0.10	0.10	0.05	0.05	0.09	0.09	More EFD

proper weight combinations are for the EFD technology selection problem. In order to generate the combinations of weights required to conduct this sensitivity analysis, upper and lower bounds on the parameters need to be assessed. This can be done by asking project staff members and decision makers for absolute bounds for each attribute weight (i.e., the highest and lowest), or by asking for probabilistic bounds (e.g., the 95th and 5th percentiles of a probability distribution for each attribute weight) [1]. In this case study, the lower and upper bounds of each weight component ($W_1 \sim W_5$) are assigned between zero (the possible minimum weight) and one (the possible maximum weight) as shown in Table 9.8. This is primarily because of a lack of expert assessment of this part of the process. Notice that the weights of attributes included in W_3 and W_4 are always evenly distributed in this study. Based on the ranges provided in Table 9.8, this study enumerates all possible weight combinations within these bounds that summed to one in increments of 0.1. Since the weights must sum to one (the standard normalization technique used in decision analysis), as one weight increases, the others must decrease. In this case study, as W_1 increases, the other weights (i.e., W_2, W_3, W_4, and W_5) decrease by the ratio of the weight combination shown in Table 9.6. For example, the weight of the cost attribute can be changed from the initially assigned value of 0.60 in the first set of weights (Weight No. 1) given in Table 9.6. The weights of the other attributes are changed proportionally once the weight of the cost attribute varies. The total number of weight combinations for further consideration is about 410 in this sensitivity analysis.

Deciding on the number of combinations of weights being used in a sensitivity analysis usually involves a trade-off between increased computational time for the analysis and the potential for increased modeling accuracy. This trade-off needs to be made on a case-specific basis [1]. Once the combinations of input parameters are defined, the optimization routine is performed for each combination. This has the potential to consume significant time in the process, especially for problems where a large number of technologies are considered. In this sensitivity analysis where weights are varied, 12 different drilling systems are selected as the optimal systems for at least one of the weight combinations being considered. Table 9.9

TABLE 9.7

Results of the Sensitivity Analysis

Subsets	For Weight #1	For Weight #2	For Weight #3 and #4
1. Transportation	Conventional diesel truck	Low-sulfur diesel truck w/tier III engine, w/noise suppressor	Low-sulfur diesel truck w/tier III engine, w/noise suppressor
2. Road construction	DURA-BASE from composite mat (rent)	DURA-BASE from composite mat (rent)	DURA-BASE from composite mat (rent)
3. Site preparation	DURA-BASE from composite mat (rent)	DURA-BASE from composite mat (rent)	Aluminum modules + driven piles (elevated platform)
4. Rig type	LOC250 (CWD)	LOC250 (CWD)	LOC250 (CWD)
5. Conventional power	Lean-burn natural gas engines w/noise suppressor	Lean-burn natural gas engines w/noise suppressor	Lean-burn natural gas engines w/noise suppressor
6. Fuel	Natural gas	Natural gas	Natural gas
7. Unconventional power	None	Electric power from grid (10%)	Electric power from grid (10%)
8. Energy storage	None	Flywheel	Flywheel
9. Drilling technology	Underbalanced drilling w/noise suppressor	Underbalanced drilling w/noise suppressor	Managed pressure drilling w/noise suppressor
10. Fluid	Water-based muds	Water-based muds	Water-based muds
11. Waste management	Closed loop + containers + solid control equipment	Closed loop + containers + solid control equipment	Closed loop + containers + solid control equipment
12. Cuttings treatment	Cuttings injection	Cuttings injection	Cuttings injection
13. Noise reduction facility	None	None	None

TABLE 9.8

Range of the Allowable Weight Factor for Each Attribute

Weights	Cost (W_1)	Footprint (W_2)	Emissions (W_3)			Perception (W_4)			Safety (W_5)
			Air	S/L	Noise	Govt.	Ind.	Public	
Maximum	1.00	1.00	1/3	1/3	1/3	1/3	1/3	1/3	1.00
Minimum	0.00	0.00	0.00	0.00	0.00	0.00	0.00	0.00	0.00

TABLE 9.9

Proportion of the Optimal Systems for This Case Study

	SET No.											
	1	2	3	4	5	6	7	8	9	10	11	12
Proportion (%)	41.6	35.0	9.1	4.7	2.0	1.5	1.5	1.5	1.5	0.7	0.5	0.5

shows the proportion of the 410 weights for which each of the 12 systems selected for this sensitivity analysis. The fact that SET 1, the most frequently selected optimal solution, is selected for only 42% of the weight combinations emphasizes the need for a sensitivity analysis.

Throughout the sensitivity analysis conducted in this section, the six most frequently selected drilling systems are suggested for further analysis for the Green Lake drilling site as shown in Figure 9.8. Notice that even though SETs 7 through 9 were also selected the same fraction of times as SET 6 according to Table 9.9, only SET 6 was selected for inclusion in the sensitivity analysis because SET 6 has a specific combination of technologies, a conventional drilling system, while the others (SETs 7–9) are similar to SETs 4 and 5. Figure 9.8 shows which technologies are selected for each suggested system. The results of six systems can indicate the potential for further simplification of system selection problems in onshore EFD. In this case, the technologies selected for five subsets (i.e., (2), (4), (10), (12), and (13)) are always same in all suggested systems. Therefore, if sensitivity to weights is the only concern, the optimal decision would revolve around the technologies for only eight subsets of the original 13 subsets. Figure 9.9 shows the comparison of the single-attribute utility values of the six suggested drilling systems described in Figure 9.8.

Figures 9.10 and 9.11 show which system should be selected as W_1 and W_2 are varied, respectively, by the ratio of the base-case weight combination (Weight No. 2) given in Table 9.6. In Figure 9.10, for example, as W_1 increases, the other weights (i.e., W_2 through W_5) decrease by the ratio of the base-case weight combination shown in Table 9.6. Figure 9.10 shows that SET 2 is preferred over SET 1 as W_1 increases, and SET 4, containing 30% of unconventional power usage, is only selected as the optimal system when the cost attribute has a very low weight ($W_1 < 4\%$). This is because currently developed unconventional power

SET 1 (41.6%)	SET 2 (35%)
(1) Transportation: Low sulphur diesel truck w/tier III engine, w/noise suppressor	(1) Transportation: Low sulphur diesel truck w/tier III engine, w/noise suppressor
(2) Road construction: DURA-BASE from Composite Mat (rent)	(2) Road construction: DURA-BASE from Composite Mat (rent)
(3) Site preparation: Aluminum modules + driven piles	(3) Site preparation: DURA-BASE from Composite Mat (rent)
(4) Rig type: LOC250 (CWD)	(4) Rig type: LOC250 (CWD)
(5) Rig power (Conventional): Lean-burn natural gas engines w/noise suppressor	(5) Rig power (Conventional): Lean-burn natural gas engines w/noise suppressor
(6) Fuel type: Natural gas	(6) Fuel type: Natural gas
(7) Rig power (Unconventional): Electric power from grid (10 %)	(7) Rig power (Unconventional): Electric power from grid (10 %)
(8) Energy storage: Flywheels	(8) Energy storage: Flywheels
(9) Drilling tech.: Managed pressure drilling w/noise suppressor	(9) Drilling tech.: Underbalanced drilling w/noise suppressor
(10) Fluid type: Water-based muds	(10) Fluid type: Water-based muds
(11) Waste mgmt.: Closed loop + containers + solid control equip.*	(11) Waste mgmt.: Closed loop + containers + solid control equip.*
(12) Cuttings mgmt.: Cuttings injection	(12) Cuttings mgmt.: Cuttings injection
(13) Noise reduction: N/A	(13) Noise reduction: N/A

SET 3 (9.1%)	SET 4 (4.7%)
(1) Transportation: Coventional diesel truck	(1) Transportation: Low sulphur diesel truck w/tier III engine, w/noise suppressor
(2) Road construction: DURA-BASE from Composite Mat (rent)	(2) Road construction: DURA-BASE from Composite Mat (rent)
(3) Site preparation: DURA-BASE from Composite Mat (rent)	(3) Site preparation: Aluminum modules + driven piles
(4) Rig type: LOC250 (CWD)	(4) Rig type: LOC250 (CWD)
(5) Rig power (Conventional): Lean-burn natural gas engines w/noise suppressor	(5) Rig power (Conventional): Lean-burn natural gas engines w/noise suppressor
(6) Fuel type: Natural gas	(6) Fuel type: Natural gas
(7) Rig power (Unconventional): N/A (0 %)	(7) Rig power (Unconventional): Electric power from grid (30 %)
(8) Energy storage: N/A	(8) Energy storage: Flywheels
(9) Drilling tech.: Underbalanced drilling w/noise suppressor	(9) Drilling tech.: Managed pressure drilling w/noise suppressor
(10) Fluid type: Water-based muds	(10) Fluid type: Water-based muds
(11) Waste mgmt.: Closed loop + containers + solid control equip.*	(11) Waste mgmt.: Closed loop + containers + solid control equip.*
(12) Cuttings mgmt.: Cuttings injection	(12) Cuttings mgmt.: Cuttings injection
(13) Noise reduction: N/A	(13) Noise reduction: N/A

SET 5 (2%)	SET 6 (1.5%)
(1) Transportation: Coventional diesel truck	(1) Transportation: Coventional diesel truck
(2) Road construction: DURA-BASE from Composite Mat (rent)	(2) Road construction: DURA-BASE from Composite Mat (rent)
(3) Site preparation: DURA-BASE from Composite Mat (rent)	(3) Site preparation: DURA-BASE from Composite Mat (rent)
(4) Rig type: LOC250 (CWD)	(4) Rig type: LOC250 (CWD)
(5) Rig power (Conventional): Internal combustion engine	(5) Rig power (Conventional): Internal combustion engine
(6) Fuel type: Low sulphur diesel	(6) Fuel type: Conventional diesel
(7) Rig power (Unconventional): N/A (0 %)	(7) Rig power (Unconventional): N/A (0 %)
(8) Energy storage: N/A	(8) Energy storage: N/A
(9) Drilling tech.: Underbalanced drilling w/noise suppressor	(9) Drilling tech.: Underbalanced drilling w/noise suppressor
(10) Fluid type: Water-based muds	(10) Fluid type: Water-based muds
(11) Waste mgmt.: Closed loop + containers + solid control equip.*	(11) Waste mgmt.: Lined reserve pit + solid control equip.*
(12) Cuttings mgmt.: Cuttings injection	(12) Cuttings mgmt.: Cuttings injection
(13) Noise reduction: N/A	(13) Noise reduction: N/A

FIGURE 9.8
Six systems suggested for Green Lake drilling site.

generation methods and energy storage devices are costly even though they significantly decrease emission rates. Moreover, "conventional diesel truck" is selected for subset (1) rather than "low-sulfur diesel truck with noise suppressor" when W_1 is greater than 60% because "conventional diesel truck" is cheaper than "low-sulfur diesel truck with noise suppressor."

Figure 9.11 shows that an increase in W_2 has little effect on the overall utility score of SET 6.

The main purpose of graphically displaying the results of sensitivity analysis is to help the decision maker clearly understand what the results of a decision support analysis mean [12]. Another way of displaying the results of the sensitivity analysis originally developed by Guikema and Milke [1] is shown in Figures 9.12 and 9.13. These figures focus not on the relative overall utility score of the different systems but on the system selections themselves. This display method is more useful and intuitive when people want to know which system should be selected with a given weight combination. However, the drawback of using this method is that they are only three-dimensional plots, so two remaining weights should be fixed at zero in this case study.

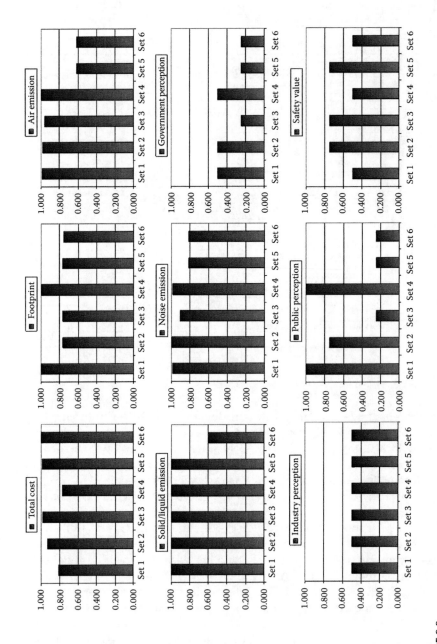

FIGURE 9.9
Comparison of the single-attribute utility.

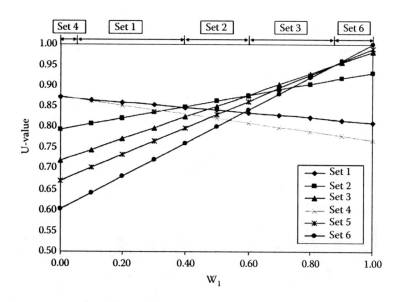

FIGURE 9.10

Optimal utility scores of the suggested systems when W_1 is varied.

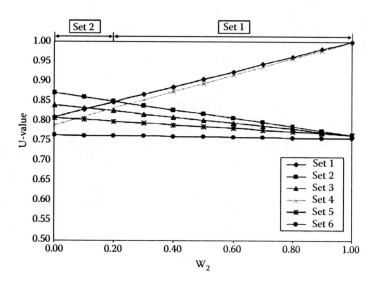

FIGURE 9.11

Optimal utility scores of the suggested systems when W_2 is varied.

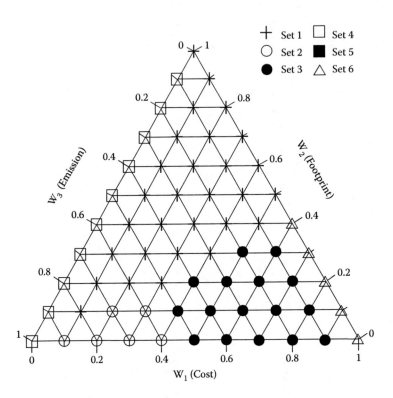

FIGURE 9.12
Optimal system selection as a function of W_1, W_2, and W_3.

Sensitivity Analysis for Uncertainty of Overall Attribute Scores

In order to identify how sensitive the overall utility score is to changes in the input attribute scores, overall attribute scores of two different systems (i.e., SET 1 and SET 6, which are the most and the least suggested optimal system) are varied from the original values with two different discrete weight combinations shown in Table 9.10. The variation of the cost, footprint, and emission attribute scores are ±10% from the original values and the variation of other attribute scores (i.e., perception and safety) are one level upper and lower grade score from the original values. The input scores and the variation of the overall attribute scores being used in this sensitivity analysis are shown in Figures 9.14 and 9.15. It is noted that the overall public perception score of SET 1 is unable to be varied to the upper grade score because the original score of this attribute is one, which is the possible maximum score assigned to this attribute. The possible maximum and minimum score should be considered for attributes using the ordinal scales such as solid/liquid emission, three perceptions, and safety.

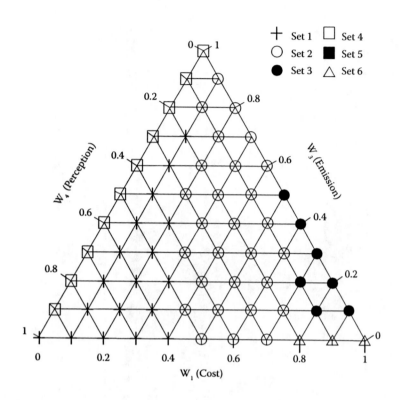

FIGURE 9.13

Optimal system selection as a function of W_1, W_3, and W_4.

TABLE 9.10

Weight Combinations Used in This Section

Weight No.	Cost (W_1)	Footprint (W_2)	Emissions (W_3) Air	S/L	Noise	Perception (W_4) Govt.	Ind.	Public	Safety (W_5)	Note
1	1/9	1/9	1/9	1/9	1/9	1/9	1/9	1/9	1/9	Even
2	0.40	0.20	0.20/3	0.20/3	0.20/3	0.05	0.05	0.05	0.05	Base case

Figures 9.16 through 9.19 show the sensitiveness of input attribute scores for the two systems with two discrete weight combinations given in Table 9.10. In Figure 9.16a, for example, since a steeper slope indicates a more sensitive attribute, the air emission attribute seems to be the most sensitive attribute among the nine attributes. In Figure 9.16b, however, the perception and the safety attributes seem to change the final utility score of SET 1 more than other attributes. This is because the cost, the footprint, and the emission attributes vary by only ±10% from the original values while the perception and the safety attributes vary by about 20 ~ 100% from the original values due to the graded score scale (i.e., 0, 0.25, 0.50, 0.75, 1.00).

Selected Technologies In Each Subset	Total Cost ($)	Ecological Footprint (Acres)	Emissions			Perceptions			Safety Value
			Air	Solid & Liquid	Noise (TWA)	Gov.	Ind.	Public	
(1) Transportation: Low sulphur diesel truck w/tier III engine, w/noise suppressor						1.000	0.500	1.000	0.750
(2) Road construction: DURA-BASE from Composite Mat (rent)	$147,840	1.515	0.976		64.696	1.000	0.500	1.000	1.000
(3) Site preparation: Aluminum modules + driven piles	$226,261	0.005	0.989		76.265	1.000	0.500	1.000	0.500
(4) Rig type: LOC250 (CWD)	$173,800		0.985		60.366	1.000	0.500	1.000	1.000
(5) Rig power (Conventional): Lean-burn natural gas engines w/noise suppressor	$70,354		0.918		85.603	1.000	0.500	1.000	0.750
(6) Fuel type: Natural gas	$25,650					1.000	0.500	1.000	0.750
(7) Rig power (Unconventional): Electric power from grid (10 %)	$3,840	0.000	1.000		0.000	0.500	1.000	1.000	1.000
(8) Energy storage: Flywheels	$30,000	0.000				0.500	1.000	1.000	0.750
(9) Drilling tech.: Managed pressure drilling w/noise suppressor	$193,500				94.100	0.750	0.750	1.000	1.000
(10) Fluid type: Water-based muds	$47,940					1.000	1.000	1.000	1.000
(11) Waste mgmt.: Closed loop + containers + solid control equip.*	$27,000	0.000		1.000		1.000	0.500	1.000	0.750
(12) Cuttings mgmt.: Cuttings injection	$45,000			1.000		1.000	0.500	1.000	0.750
(13) Noise reduction: N/A									
Overall Attribute Scores (Σ or minimum value)	$991,184	1.520	4.868	2.000	381.030	0.500	0.500	1.000	0.500
Single Attribute Utility Values (Original)	0.811	0.966	0.768	1.000	0.988	0.500	0.500	1.000	0.500

Used Upper Bound Scores	$1,090,302.68	1.67	5.35	2.00	419.13	0.75	0.75	1.00	0.75
Used Lower Bound Scores	$892,065.83	1.37	4.38	1.80	342.93	0.25	0.25	0.75	0.25

FIGURE 9.14
Input values and variation of the overall attribute scores of SET 1.

Selected Technologies In Each Subset	Total Cost ($)	Ecological Footprint (Acres)	Emissions			Perceptions			Safety Value
			Air	Solid & Liquid	Noise (TWA)	Gov.	Ind.	Public	
(1) Transportation: Conventional diesel truck						0.250	1.000	0.250	0.750
(2) Road construction: DURA-BASE from Composite Mat (rent)	$132,000	1.515	0.964		82.870	1.000	0.500	1.000	1.000
(3) Site preparation: DURA-BASE from Composite Mat (rent)	$90,000	1.033	0.976		79.945	0.750	0.750	0.750	1.000
(4) Rig type: LOC250 (CWD)	$167,000		0.977		77.458	1.000	0.500	1.000	1.000
(5) Rig power (Conventional): Internal combustion engine	$50,000		0.338		107.998	0.500	1.000	0.500	0.750
(6) Fuel type: Conventional diesel	$45,600					0.500	1.000	0.500	0.500
(7) Rig power (Unconventional): N/A (0 %)	$0	0.000	1.000		0.000	0.250	1.000	0.250	1.000
(8) Energy storage: N/A	$0	0.000				0.250	1.000	0.250	1.000
(9) Drilling tech.: Underbalanced drilling w/noise suppressor	$184,500				95.700	0.750	0.750	0.750	0.750
(10) Fluid type: Water-based muds	$47,940					1.000	1.000	1.000	1.000
(11) Waste mgmt.: Lined reserve pit + solid control equip *	$18,000	0.037		0.500		0.750	0.750	0.750	0.500
(12) Cuttings mgmt.: Cuttings injection	$45,000			1.000		1.000	0.500	1.000	0.750
(13) Noise reduction: N/A									
Overall Attribute Scores (Σ or minimum value)	$780,040	2.585	4.254	1.500	443.971	0.250	0.500	0.250	0.500
Single Attribute Utility Values	0.935	0.756	0.613	0.600	0.820	0.250	0.250	0.250	0.500

	Total Cost ($)	Ecological Footprint	Air	Solid & Liquid	Noise	Gov.	Ind.	Public	Safety
Used Upper Bound Scores	$858,044.00	2.84	4.68	1.65	488.37	0.50	0.75	0.50	0.75
Used Lower Bound Scores	$702,036.00	2.33	3.83	1.35	399.57	0.00	0.25	0.00	0.25

FIGURE 9.15
Input values and variation of the overall attribute scores of SET.

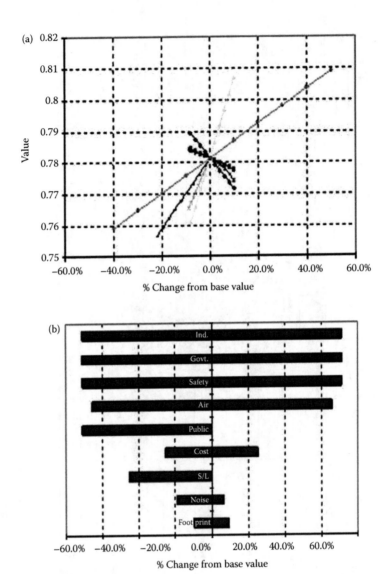

FIGURE 9.16
Results for SET 1 with "even" weight combination in Table 9.10. (a) Spider graph of U-value,
(b) tornado diagram for U-value.

The weight factor assigned to each attribute is a very important element
when identifying the sensitiveness of input attribute scores. In Figure 9.17a, for
example, the cost attribute seems to be the most sensitive attribute for SET
1, which is not the same result shown in Figure 9.16a. This is because the
weight assigned to each attribute is different between these two figures. The

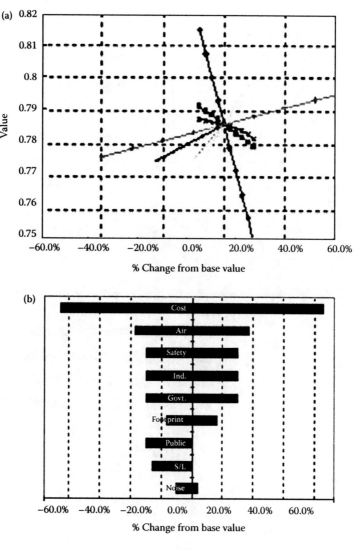

FIGURE 9.17
Results for SET 1 with "base" weight combination in Table 9.10. (a) Spider graph of U-value, (b) tornado diagram for U-value.

weight assigned to the cost attribute is 1/9 in Figure 9.16 while the weight is 0.40 in Figure 9.17. The noise emission attribute seems to be the most sensitive attribute for SET 6 with "even" weight combination as shown in Figure 9.18a. It is also indicated that since the noise attribute utility curve is not linear, the result of the variation (±10%) does not seem to be symmetrical from the original value.

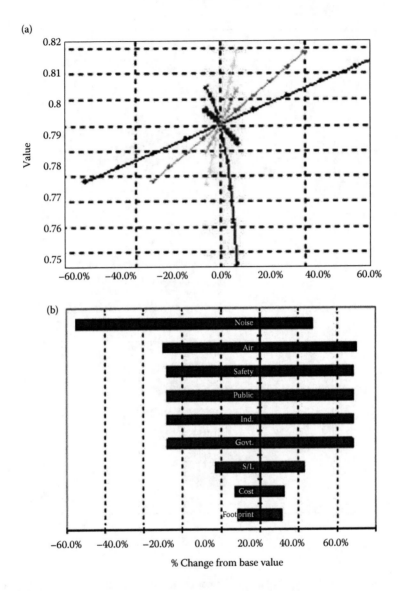

FIGURE 9.18
Results for SET 6 with "even" weight combination in Table 9.10. (a) Spider graph of U-value, (b) tornado diagram for U-value.

In summary, if weight factors are evenly distributed to each attribute, the air emission and the noise emission scores are the most sensitive inputs for SET 1 and SET 6 as shown in Figures 9.16a and 9.18a, respectively. On the other hand, if weight factors are not evenly distributed to each attribute, the most sensitive input attribute can be identified after running sensitivity analysis described in this section.

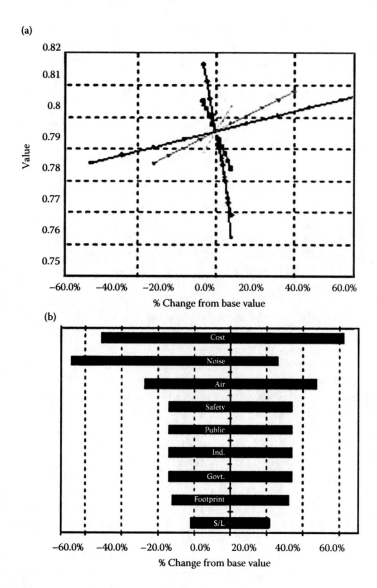

FIGURE 9.19
Results for SET 6 with "base" weight combination in Table 9.10. (a) Spider graph of U-value, (b) tornado diagram for U-value.

Discussion

The knapsack optimization model described below was initially solved for each weight combination in this case study using the Solver tool in Microsoft Excel. However, a critical issue arose while conducting the sensitivity analysis.

Using the branch-and-bound optimization algorithm in Microsoft Excel, we were not always able to find the global optimal solution due to the complexity of the system selection process. In some cases, the Solver tool was trapped at a local optimal solution, a well-known problem with knapsack optimization problems. Therefore, in order to always get the global optimal solution (i.e., retain system with maximum final utility score), an exhaustive search optimization was used in this study. However, we do not recommend using an exhaustive search approach for problems with a larger or more complex search space due to the computation time. One of the future research tasks is to implement existing, more advanced optimization methods that can efficiently search the entire (not truncated) solution space using only standard personal computers. A number of good optimization approaches are available from traditional branch-and-bound methods to more modern heuristic methods such as genetic algorithms, particle swarm optimization, and ant colony optimization. The particular optimization approach used in a given application of the general method presented here will need to be matched to the complexity and size of the application.

Although the focus of the sensitivity analysis presented in this chapter has been on sensitivity to weights and overall input scores, the approach could also be applied to sensitivity to risk attitude (i.e., risk-averse, risk-neutral, and risk-seeking) or to other input parameters. The sensitivity to those unapplied input parameters is an important area for further research to suggest more robust optimal systems, but they involve a trade-off between increased computational time for the analysis and the potential for increased modeling accuracy. This trade-off should be made on a case-specific basis [1].

The authors intend to apply the proposed approach to many different areas containing selection problems in systems engineering. For example, current applications of the same approach include engineering systems for ecosystem management, varying input values on the operation of water utilities, limited budget allocations for potential projects, and so on. However, it is fair to acknowledge that some limitations of this approach include the fact that the computational burden of the procedure may become prohibitive for problems with a large number of decision variables. One possible way to resolve this problem in this research is if the analyst can identify subsets that will always select the same technology for any weight combinations, the elimination of those subsets from further consideration can significantly reduce computational burdens in future steps.

Knapsack Optimization Model

The optimization model for the EFD technology selection problem with nine attributes is given as follows:

$$\text{Max}_j \sum_{i=1}^{9} k_i u_i(X_{ij})$$

where j is the index for systems, i is the index for the attributes, k_i is the weight assigned to the ith attribute (k must sum to 1), X_{ij} is the overall score of the jth system on the ith attribute, and $u_i(X_{ij})$ is the single-attribute utility value for system j on attribute i, scaled from 0 to 1. In order to calculate the overall score of a system on the ith attribute (X_i), Equations 9.1 through 9.3 should be considered.

Conclusion

This chapter has demonstrated a sensitivity analysis method that, when coupled with the multiattribute system selection method proposed by Yu et al. [3,4], can yield strong decision support for selecting systems in onshore oil and gas drilling projects. We have shown that it is possible to suggest a small number of suitable drilling systems that are particularly attractive for the case study drilling site at Green Lake, Texas. Six different drilling systems are suggested for this case study as shown in Figure 9.8. The most frequently suggested optimal drilling system, SET 1, across a wide range of weights is only optimal for about 42% of the weight combinations tested, which implies that different systems would be suggested for 58% of plausible weight combinations. This indicates that the sensitivity analysis conducted in this study is a worthy topic for further investigation.

According to the sensitivity analysis results described in this chapter, sensitiveness of the input attribute scores varies. For example, air emission score has more influence on the final system selection than footprint score (see Figures 9.16 through 9.19). Sometimes, decision makers already have enough sensitivity analysis results, but they do not know how to effectively use them for their decision-making process. Therefore, effective displays of sensitivity analysis for the system selection problems would be crucial as an aid in decision-making process, and also as an aid in explaining the optimal EFD system selections to interested parties in this study. In addition, the display methods chosen in any given situation should be illustrated by the abilities and needs of the decision makers [1]. For example, more complicated displays such as Figures 9.10 and 9.11 can be used for technically trained people while simpler displays such as Figures 9.12 and 9.13 should be used for less technically trained people.

The approach presented in this chapter is designed to help decision makers gain a deeper understanding of the system selection problems in onshore EFD by examining the robustness of the optimal solution to input parameters such as weights and attribute scores. The results of the sensitivity analysis suggest more robust optimal systems for this case study. Even though the system selection process in onshore EFD can be computationally burdensome, it can be very helpful for decision makers to refine their decisions on a more scientific basis.

References

1. Guikema, S. D. and Milke, M. W. 2003, Sensitivity analysis for multi-attribute project selection problems, *Civil Engineering and Environmental Systems*, 20(3), 143–162.
2. Call, H. and Merkhofer, M. 1988, A multi-attribute utility analysis model for ranking superfund sites, *Superfund 88: Proceedings of the 9th National Conference*, Washington, DC, pp. 44–54.
3. Yu, O.-Y., Guikema, S., Bickel, E., Briaud, J.-L., and Burnett, D. 2009, Systems approach and quantitative decision tools for technology selection in environmentally friendly drilling, *The SPE Americas E&P Environmental & Safety Conference*, San Antonio, TX, USA.
4. Yu, O.-Y., Guikema, S., Briaud, J.-L., and Burnett, D. 2011, Quantitative decision tools for system selection in environmentally friendly drilling, *Civil Engineering and Environmental Systems*, 28(3), 185–208.
5. Haut, R., Williams, T., Burnett, D., Theodori, G., and Veil, J. 2010, *The Environmentally Friendly Drilling Systems Program Report*, Houston Advanced Research Center.
6. Hovorka, S. D., Doughty, C., Knox, P. R., Green, C. T., Pruess, K., and Benson, S. M. 2001, Evaluation of brine-bearing sands of the Frio formation, upper Texas Gulf Coast for geological sequestration of CO_2, *First National Conference on Carbon Sequestration*, Washington, DC, May 14–17.
7. Keeney, R. L. and Raiffa, H. 1976, *Decisions with Multiple Objectives: Preferences and Value Tradeoffs*, Cambridge University Press, New York, NY.
8. Clemen, R. T. and Reilly, T. 2001, *Making Hard Decisions with DecisionTools*, Duxbury, Pacific Grove, CA.
9. Hardaker, J. B. 2004, *Coping with Risk in Agriculture*, CABI Publishing, Cambridge, MA.
10. O'Banion, K. 1980, Use of value functions in environmental decisions, *Environmental Management*, 4(1), 3–6.
11. Guikema, S. and Milke, M. 1999, Quantitative decision tools for conservation programme planning: Practice, theory and potential, *Environmental Conservation*, 26(03), 179–189.
12. Insua, D. R. 1990, *Sensitivity Analysis in Multi-Objective Decision Making*, Springer-Verlag, New York, NY.
13. Yu, O.-Y., Guikema, S., Briaud, J.-L., and Burnett, D. 2012, Sensitivity analysis for multi-attribute system selection problems in onshore environmentally friendly drilling (EFD), *Systems Engineering*, 15(2), 153–171.

10

Managing Construction Projects in Oil and Gas*

Introduction

Construction is an integral part of the oil and gas industry. From basic office structure construction to oil rig platform construction, there are unique aspects of the oil and gas business that must be addressed. Oil and gas construction projects are different from that of conventional manufacturing or service enterprises. The oil and gas industry itself does embody its own unique manufacturing and service operations, which also require special management of the associated construction projects. In oil and gas construction project, the focus is on the fulfillment of the owner's requirements with respect to the defined scope of work within the available budget and the specified schedule. The attendant risks that exist in the oil and gas business make the management of construction projects particularly dicey. Management must address the challenging categories of quality, schedule, and cost. Specialized tools are needed, such as plan, do, check, act (PDCA); define, measure, analyze, improve, control (DMAIC); suppliers, inputs, process, outputs, customers (SIPOC); design, evaluate, justify, integrate (DEJI); quality function deployment (QFD); affinity diagrams; flowcharts; Pareto charts; and histograms; many of which are addressed earlier in Chapters 5 through 7.

Project Definition

PMI (2004) defines the word *project* in terms of its distinctive characteristics: "A project is a temporary endeavor undertaken to create a unique product or service." "Temporary" means that every project has a definite beginning and a definite end. "Unique" means that the product or service is different in some distinguishing way from all similar products or services.

It further states that projects are often critical components of the performing organization business strategy. Examples of projects include

---

* Adapted from R. Abdul Razzak, *Quality Management in Construction Projects*, Taylor & Francis/CRC Press, Boca Raton, FL. With permission.

- Developing a new product or service
- Effecting a change in structure, staffing, or style of an organization
- Designing a new transportation vehicle/aircraft
- Developing or acquiring a new or modified information system
- Running a campaign for political office
- Implementing a new business procedure or process
- Constructing a building or facility

The duration of a project is finite; projects are not ongoing efforts, and the project ceases when its declared objectives have been attained. Among other shared characteristics, projects are

1. Performed by people
2. Constrained by limited resources
3. Planned, executed, and controlled

Pyzdek (1999) defined "project" as

1. A plan or proposal; a scheme
2. An undertaking requiring concrete effort

The "plan" is defined as

1. A scheme, program, or method worked beforehand for the accomplishment of an objective; a plan of attack
2. A proposed or tentative projective or course of action
3. A systematic arrangement of important parts

According to Kerzner (2001), a project can be considered to be a set of activities and tasks that

- Have a specified objective to be completed within certain specifications
- Have defined start and end dates
- Have funding limits (if applicable)
- Consume human and nonhuman resources (i.e., money, people, and equipment)
- Are multifunctional (i.e., cut across several lines)

Based on various definitions, the project can be defined as follows: "A project is a plan or program performed by the people with assigned resources to achieve an objective within a finite duration."

Construction Projects

Construction has a history of several thousand years. The first shelters were built from stone or mud and the materials collected from the forests to provide protection against cold, wind, rain, and snow. These buildings were primarily for residential purposes, although some may have had some commercial function.

During the New Stone Age, people introduced dried bricks, wall construction, metal working, and irrigation. Gradually, people developed the skills to construct villages and cities, and considerable skills in building were acquired. This can be seen from the great civilizations in different parts of the world—some 4000–5000 years ago. During the early period of Greek settlement, which was about 2000 BCE, the buildings were made of mud using timber frames. Later, temples and theaters were built from marble. Some 1500–2000 years ago, Rome became the leading center of world culture, which extended to construction.

Marcus Vitruvius Pollo, the first century military and civil engineer, penned in Rome the world's first major treatise on architecture and construction. It dealt with building materials, the styles and design of building types, the construction process, building physics, astronomy, and building machines.

During the Middle Ages (476–1492), improvements occurred in agriculture and artisanal productivity and exploration, and as a consequence, the broadening of commerce took place and in the late Middle Ages, building construction became a major industry. Craftsmen were given training and education in order to develop skills and to raise their status. At this time, guilds came up to identify true craftsmen and set standards for quality.

The fifteenth century brought a "renaissance" or renewal in architecture, building, and science. Significant changes occurred during the seventeenth century and thereafter due to the increasing transformation of construction and urban habitat.

The scientific revolution of the seventeenth and eighteenth centuries gave birth to the great Industrial Revolution of the eighteenth century. After some delay, construction followed these developments in the nineteenth century.

The first half of the twentieth century witnessed the construction industry becoming an important sector throughout the world, employing many workers. During this period, skyscrapers, long-span dams, shells, and bridges were developed to satisfy new requirements and marked the continuing progress of construction techniques. The provision of services such as heating, air conditioning, electrical lighting, water mains, and elevators in buildings became common. The twentieth century has seen the transformation of the construction and building industry into a major economic sector. During the second half of the twentieth century, the construction industry began to industrialize, introducing mechanization, prefabrication, and system building. The design of building services systems changed considerably in the last

20 years of the twentieth century. It became the responsibility of designers to follow health, safety, and environmental regulations while designing any building.

Building and commercial—traditional A&E type—construction projects account for an estimated 25% of the annual construction volume. Building construction is a labor-intensive endeavor. Every construction project has some elements that are unique. No two construction or R&D projects are alike. Though it is clear that many building projects are more routine than research and development projects, some degree of customization is a characteristic of the projects.

Construction projects involve a cross section of many different participants. These both influence and depend on each other in addition to the "other players" involved in the construction process. Figure 10.1 illustrates the concept of the traditional construction project organization.

Traditional construction projects involve three main groups. These are

1. Owners—A person or an organization that initiates and sanctions a project. He/she outlines the needs of the facility and is responsible for arranging the financial resources for creation of the facility.
2. Designers (A&E)—This group consists of one or more architects or engineers and consultants. They are the owner's appointed entities accountable for converting the owner's conception and need

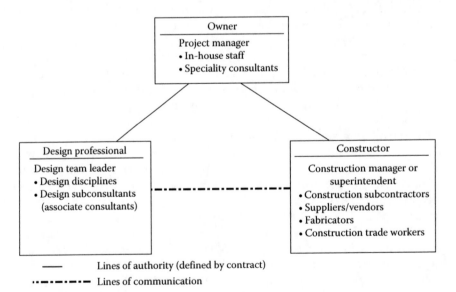

FIGURE 10.1
Traditional construction project organization. (From American Society of Civil Engineers, *Quality in the Constructed Project,* 2000. Reprinted with permission from ASCE.)

into a specific facility with detailed directions through drawings and specifications adhering to the economic objectives. They are responsible for the design of the project and in certain cases its supervision.

3. Contractors—A construction firm engaged by the owner to complete the specific facility by providing the necessary staff, work force, materials, equipment, tools, and other accessories to the satisfaction of the owner/end user in compliance with the contract documents. The contractor is responsible for implementing the project activities and for achieving the owner's objectives.

Construction projects are executed based on a predetermined set of goals and objectives. With traditional construction projects, the owner heads the team, designating a project manager. The project manager is a person/member of the owner's staff or independently hired person/firm with over-all or principal responsibility for the management of the project as a whole.

Oberlender (2000) states that the working environment and culture of a construction project is unique compared to most working conditions. A typical construction project consists of a group of people, normally from several organizations, that are hired and assigned to a project to build the facility. Due to the relatively short life of a construction project, these people may view the construction project as accomplishing short-term tasks. However, the project manager of the construction team must instill in the team the concept that building a long-term relationship is more important in career advancement than trying to accomplish short-term tasks.

In certain cases, owners engage a professional firm, called a construction manager, trained in the management of construction processes, to assist in developing bid documents, and overseeing and coordinating the project for the owner. The basic construction management concept is that the owner assigns a contract to a firm that is knowledgeable and capable of coordinating all the aspects of the project to meet the intended use of the project by the owner. In the construction management type of construction projects, the consultants (architect/engineer) prepare complete design drawings and contract documents, then the project is put for competitive bid and the contract is awarded to the competitive bidder (contractor). Next, the owner hires a third party (construction manager) to oversee and coordinate the construction.

ASCE (2000) categorized two types of construction managers: agency construction managers (ACM) and construction managers-at-risk (CM-at-risk). An ACM functions wholly within the policies, procedures, and practices of the owner's organization. A CM-at-risk typically contracts with the owner in two stages. During the first stage, CM-at-risks act as consultants or even design professionals, and when the design is completed they become involved in the completion of the construction work.

There are numerous types of construction projects:

Process-type projects
    Liquid chemical plants
    Liquid/solid plants
    Solid process plants
    Petrochemical plants
    Petroleum refineries
Nonprocess-type projects
    Power plants
    Manufacturing plants
    Support facilities
    Miscellaneous (R&D) projects
    Civil construction projects
    Commercial/A&E projects

Civil construction projects and commercial/A&E projects can further be categorized into four somewhat arbitrary but generally accepted major types of construction. These are

1. Residential construction
2. Building construction (institutional and commercial)
3. Industrial construction
4. Heavy engineering construction

Residential construction: Residential construction includes single-family homes, multiunit town houses, garden, apartments, high-rise apartments, and villas.

Building construction: Building construction includes structures ranging from small retail stores to urban redevelopment complexes, from grade schools to new universities, hospitals, commercial office towers, theaters, government buildings, recreation centers, warehouses, and neighborhood centers.

Industrial construction: Industrial construction includes petroleum refineries, petroleum plants, power plants, heavy manufacturing plants, and other facilities essential to our utilities and basic industries.

Heavy engineering construction: Heavy engineering construction includes dams and tunnels, bridges, railways, airports, highways and urban rapid transit system, ports and harbors, water treatment and distribution, sewage and storm water collection, treatment and disposal system, power lines, and communication network.

Table 10.1 shows a brief classification of projects/characteristics.

TABLE 10.1

Classification of Projects/Characteristics

			Type of Project/Industry			
	In-House R&D	Small Construction	Large Construction	Aerospace/ Defense	MIS	Engineering
Need for interpersonal skills	Low	Low	High	High	High	Low
Importance of organizational structure	Low	Low	Low	Low	High	Low
Time management difficulties	Low	Low	High	High	High	Low
Number of meetings	Excessive	Low	Excessive	Excessive	High	Medium
Project manager's supervision	Middle management	Top management	Top management	Top management	Middle management	Middle management
Project sponsor present	Yes	No	Yes	Yes	No	No
Conflict intensity	Low	Low	High	High	High	Low
Cost control level	Low	Low	High	High	High	Low
Level of planning/ scheduling	Milestones only	Milestones only	Detailed plan	Detailed plan	Milestones only	Milestones only

Source: H. Kerzner. *Project Management*. 2001. Copyright Wiley-VCH Verlag GmbH & Co. KGaA. Reprinted with permission.

Construction and Manufacturing

Construction has unique problems compared to manufacturing. A few of these are listed as follows:

- The construction is a custom rather than a routine, repetitive business and differs from manufacturing.
- Quality in manufacturing passes through series of processes. The output is monitored by inspection and testing at various stages of production.
- Construction is different from both that of mass production and batch (lot) production manufacturing.
- In construction projects, the scenario is not the same as that of manufacturing. If anything goes wrong, the nonconforming work is very difficult to rectify and remedial action is sometimes not possible. Quality costs play an important role in construction projects.
- In construction, an activity may be repeated at various stages, but it is done only one time for a specific work. Therefore, it has to be right from the onset.
- In manufacturing, the buyer does not enter the scene until the product comes into being, whereas in construction the buyer is involved from beginning to end. Even during the construction phase, it is likely that certain modifications may take place.
- The owner is deeply involved in the construction process, while the purchaser of manufactured goods is not. Buyers of the usual manufactured products seldom have access to the plant where they are made, nor do they deal directly with factory managers.
- Most projects or their individual work phases are of relatively short duration. One consequence is that management teams and possibly the work force must be assembled quickly and cannot be shaken out or restructured before the project or work phase is completed.
- To a great extent, each project has to be designed and built to serve a specific need and therefore it is necessary to make certain modifications in the system process to fit the particular conditions of each construction project and its specific problems.
- The location of construction projects varies widely. In a manufacturing plant a given operation is assigned to and carried out in one place. In contrast, specialized construction crews progress from location to location.
- Operations are commonly conducted out of doors and are subject to all the interruptions and variation in conditions and the other difficulties that rain, snow, heat, and cold can introduce.

- The final product is usually of unique design and differs from workstation to workstation so that no fixed arrangement of equipment or aids such as jigs and fixtures are possible as is in the case of manufacturing.
- The construction is a preliminary step leading to a completed facility; the layout and arrangements may make access for construction difficult and permanent provisions for safety impossible.
- The construction often needs highly skilled craftsmen rather than unskilled workers; individual crews, whether union or nonunion, usually do specialized operations.
- Construction involves installation and integration of various materials, equipment, systems, or other components to complete the facility.
- Construction focuses mainly on overall performance of the project or facility in which a product(s) or a system(s) is a part and assembled/installed to achieve the objectives.
- Construction projects work against defined scope, schedules, and budget to achieve the specified result.
- Performance of construction projects can be evaluated only after it is completed and put into use/operation.

Quality Cost

Introduction

Quality has an impact on the costs of products and services. The cost of poor quality is the annual monetary loss of products and processes that are not achieving their quality objective.

According to Gryna (2001) the concept of quality costs emerged during the 1950s, and different people assigned different meaning to the term. Some people equated quality costs with the costs of attaining quality; some people equated the term with the extra costs incurred because of poor quality. He further states that

> The cost of poor quality is the annual monetary loss of products and processes that are not achieving their quality objectives. The main components of the cost of low quality are
>
> 1. Cost of nonconformities
> 2. Cost of inefficient processes
> 3. Cost of loss opportunities of sales revenue

Juran and Godfrey (1999) also state that "the term *quality costs* has different meanings to different people. Some equate quality costs with the cost of poor quality (mainly the costs of finding and correcting defective work); others equate the term with the costs to attain quality; still others use the term to mean the costs of running the quality department."

Categories of Costs

Costs of poor quality are the costs associated with providing poor-quality products or services. These are costs that would not be incurred if things were done right from the start time and at every stage thereafter in order to achieve the quality objective. There are four categories of costs:

1. Internal failure costs. (The costs associated with defects found before the customer receives the product or service. It also consists of cost of failure to meet customer satisfaction and needs and cost of inefficient processes.)

2. External failure costs. (The costs associated with defects found after the customer receives the product or service. Also includes lost opportunity for sales revenue.)

3. Appraisal costs. (The costs incurred to determine the degree of conformance to quality requirements.)

4. Prevention costs. (The costs incurred to keep failure and appraisal costs to minimum.)

Thomas Pyzdek (1999) has detailed these costs as follows:

1. Prevention costs: Costs incurred to prevent the occurrence of nonconformance in the future. Examples of prevention costs include

   Quality planning

   Process control planning

   Design review

   Quality training

   Gage design

2. Appraisal costs: Costs incurred in measuring and controlling concurrent production to assure conformance to requirements. Examples of appraisal costs are

   Receiving inspection

   Laboratory acceptance testing

   In-process inspection

   Outside endorsements (e.g., UL approval)

   Calibration

   Inspection and test equipment

   Field testing

3. Internal failure costs: Costs generated before a product is shipped as a result of nonconformance to equipment. Examples of internal failure costs include

Scrap

Rework

Process troubleshooting

Vendor-caused scrap or work

Material review board activity

Reinspection or retest

Downgrading

4. External failure costs: Costs generated after a product is shipped as a result of nonconformance to requirement. Examples of external failure costs include

Processing of customer complaints

Service

Unplanned field repair

Recalls

Processing of returned materials

Warranty

These cost categories allow the use of quality cost data for a variety of purposes. Quality costs can be used for measurement of progress, for analyzing the problem, or for budgeting. By analyzing the relative size of the cost categories, the company can determine if its resources are properly allocated.

CII product no. EM-4A (1994) states that "the cost of quality is the penalty paid for an imperfect world. It is the costs of all the extra work we do beyond merely doing a task correctly the first time to meet the requirements and expectations. The simple formula for defining the cost of quality is Cost of Quality = Cost of Prevention and Appraisal + Cost of Deviation Correction."

It has further elaborated the components of costs of quality as follows:

Prevention and appraisal: All measures taken to assure that requirements are met, such as quality control systems, inspection, work checking, design review, constructability or maintainability review, shop inspection, and auditing.

Deviation correction: Work done more than once because it did not meet requirements the first time.

Reasons for Poor Quality

According to the survey carried out by the Construction Industry Institute (CII), the primary reasons for poor quality are mainly due to poor management and are illustrated in Figure 10.2.

As per Ireland (1991), the cost of quality is the total price of all efforts to achieve product or service quality. This includes all work to build a product or service that conforms to the requirements as well as work resulting from nonconformance to the requirements. The general areas of costs for a quality system are illustrated in Table 10.2.

Figure 10.3 shows the expected results of the total quality management system on quality cost. It shows that increasing prevention costs, that is, doing things that will prevent problems, reduces the cost of appraisal and failure and gain a net cost benefit to the organization. CII has made the following recommendations to reduce the rework:

1. Reduce the number of design changes
2. Implement a quality management program
3. Adopt the standard set of quality related terminology
4. Develop and implement system to establish a database
5. Implement a QPMS

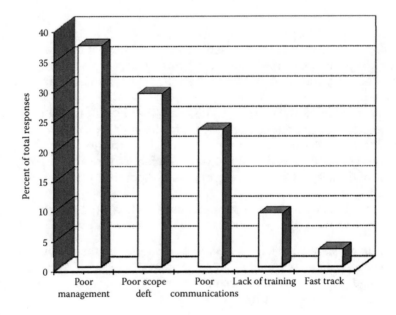

FIGURE 10.2
Primary reasons for poor quality (139 responses). (From CII, Source 79. Reprinted with permission of CII, University of Texas.)

TABLE 10.2

Conformance versus Nonconformance Costs

Cost of Conformance	Cost of Nonconformance
Planning	Scrap
Training and indoctrination	Rework
Field testing	Expedition
Product design validation	Additional material or inventory
Process validation	Warranty repairs or service
Test and evaluation	Complaint handling
Quality audits	Liability judgments
Maintenance and calibration	Product recalls
Other	Productive corrective actions

Source: L. R. Ireland, 1991. *Quality Management for Projects & Programs.* Reprinted with permission from PMI.

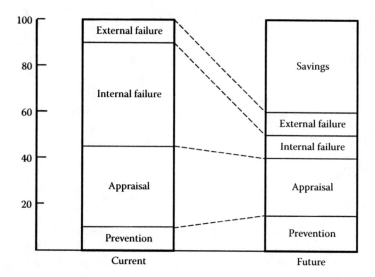

FIGURE 10.3
Total quality cost. (H. Kerzner. *Project Management.* 2001. Copyright Wiley-VCH Verlag GmbH & Co. KGaA. Reprinted with permission.)

Chung (1999) has quoted (Robert, 1991) that "quality does not cost—it pays." Figure 10.4 summarizes the quality-related costs expressed as a percentage of total construction costs. He further states that through the implementation of a proactive quality system that costs about 1% of the project value (the prevention cost), the expenditure as a result of repair, and so forth (failure cost) drops from 10% to 2% representing a saving of 7%. These categories of

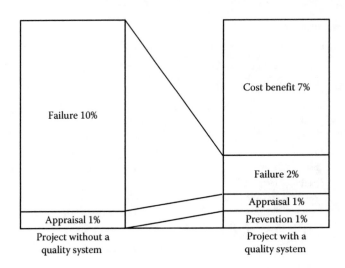

Cost benefit 7%

Failure 10%

Failure 2%

Appraisal 1%

Appraisal 1%

Prevention 1%

Project without a
quality system

Project with a
quality system

FIGURE 10.4
Implementation of quality management. (From H. W. Chung, *Understanding Quality: Assurance in Construction*, 1999. Reprinted with permission from Cengage Learning Services Limited.)

costs may represent an increase of cost in one area and a reduction of cost in another.

Quality Cost in Construction

Quality of construction is defined as

1. Scope of work
2. Time
3. Budget

Cost of quality refers to the total cost incurred during the entire life cycle of construction project in preventing nonconformance to owner requirements (defined scope). There are certain hidden costs that may not directly affect the overall cost of the project; however, it may cost the consultant/designer to complete the design within the stipulated schedule to meet owner requirements and conformance to all the regulatory codes/standards, and for the contractor to construct the project within the stipulated schedule meeting all the contract requirements. Rejection/nonapproval of executed/installed works by the supervisor due to noncompliance with specifications will cause the contractor loss in terms of

- Material
- Manpower
- Time

The contractor shall have to rework or rectify the work, which will need additional resources and will need extra time to do the work as specified. This may disturb the contractor's work schedule and affect execution of other activities. The contractor has to emphasize the "zero defect" policy, particularly for concrete works. To avoid rejection of works, the contractor has to take the following measures:

1. Execution of works as per approved shop drawings using approved material
2. Following approved method of statement or manufacturer's recommended method of installation
3. Conduct continuous inspection during construction/installation process
4. Employ properly trained workforce
5. Maintain good workmanship
6. Identify and correct deficiencies before submitting the checklist for inspection and approval of work
7. Coordinate requirements of other trades, for example, if any opening is required in the concrete beam for crossing of services pipe

Timely completion of a project is one of the objectives to be achieved. To avoid delay, proper planning and scheduling of construction activities are necessary. Since construction projects have the involvement of many participants, it is essential that the requirements of all the participants are fully coordinated. This will ensure execution of activities as planned resulting in timely completion of the project.

Normally, the construction budget is fixed at the inception of the project; therefore, it is necessary to avoid variations during the construction process as it may take time to get approval of an additional budget resulting in time extension to the project. Quality costs related to construction projects can be summarized as follows:

Internal failure costs

- Rework
- Rectification
- Rejection of checklist
- Corrective action

External failure costs

- Breakdown of installed system
- Repairs
- Maintenance
- Warranty

Appraisal costs

- Design review/preparation of shop drawings
- Preparation of composite/coordination drawings
- On-site material inspection/test
- Off-site material inspection/test
- Pre-checklist inspection

Prevention costs

- Preventive action
- Training
- Work procedures
- Method statement
- Calibration of instruments/equipment

Quality costs during the design phases are different from those of the construction phase. Costs of quality during design phases are mainly to ensure development of project design and documents to ensure conformance to the client's requirements/TOR (terms of reference)/matrix of owner's requirements. Quality costs related to design development/contract documents of construction projects can be summarized as follows:

Internal failure costs

- Redesign/redraw to meet requirements of other trades
- Redesign/redraw to meet fully coordinated design
- Rewrite specifications/documents to meet requirements of all other trades

External failure costs

- Incorporate design review comments by client/project manager
- Incorporate specifications/documents review comments by client/project manager
- Incorporate comments by regulatory authority(ies)
- Resolve RFI (request for information) during construction

Appraisal costs

- Review of design drawings
- Review of specifications
- Review of contract documents to ensure meeting owner's needs, quality standards, constructability, and functionality
- Review for regulatory requirements, codes

Prevention costs

- Conduct technical meetings for proper coordination
- Follow quality system
- Meeting submission schedule
- Training of project team members
- Update of software used for design

Quality Performance Management System

Quality performance management system (QPMS) is a product of the CII. QPMS is one of the tools available for a total quality management (TQM) project and is a good implementation tool for a project to utilize in a TQM environment. It is a management tool developed by CII to give management the information necessary to identify quality improvement opportunities.

QPMS focuses on reducing the cost of quality in four ways:

1. It provides a process that facilitates awareness of individual and group quality performance (how well we do things right) by measuring these costs in dollars.
2. It arms managers with information on quality costs and activities that enable proactive decisions affecting quality outcome.
3. It provides a database for estimating quality performance on future projects.
4. If and when widely accepted, the data should provide benchmarking information throughout the industry. (Benchmarking is a point of reference by which the performance is judged or measured.)

According to CII, the QPMS has been developed as a management tool to meet the following criteria. It must

1. Be capable of tracking quality-related costs that are involved in the design and construction of engineered projects and answer the following four questions:

   What quality management activities and deviation costs are involved?

   When were the quality management activities and deviation costs incurred?

   Why did the deviations occur (i.e., their root causes)?

   How did the rework relate to the quality management?

2. Provide valuable cost-of-quality information to establish baseline and identify opportunities for improvement, without providing either too much or too little detail

3. Be adaptable to various types and aspects of design and construction projects
4. Be easily implementable by owners, designers, and contractors
5. Be cost effective
6. Be compatible with existing cost systems used by management

Thus, it can be summarized that with implementation of quality management system, the cost of quality is reduced and ultimately results in savings.

Systems Engineering

Introduction

Systems are pervasive throughout the universe in which we live. This world can be divided into the natural world and the human-made world. Systems appeared first in natural forms and subsequently with the appearance of human beings. Systems were created based on components, attributes, and relationships.

Systems engineering and analysis, when coupled with new emerging technologies, reveal unexpected opportunities for bringing new improved systems and products into being that will be more competitive in the world economy. Product competitiveness is desired by both commercial and public-sector producers worldwide to meet consumer expectations. These technologies and processes can be applied to construction projects. The systems engineering approach to construction projects help us understand the entire process of project management in order to manage its activities at different levels of various phases to achieve economical and competitive results. The cost effectiveness of the resulting technical activities can be enhanced by giving more attention to what they are to do, before addressing what they are composed of. To ensure economic competitiveness regarding the product, engineering must become more closely associated with economics and economic facilities. This is best accomplished through the life cycle approach to engineering.

Experience in recent decades indicates that properly coordinated and functioning human-made systems will result in a minimum of undesirable side effects through the application of this integrated, life cycle-oriented "systems" approach. The consequences of not applying systems engineering in the design and development and/or reengineering of systems have been disruptive and costly.

The systems approach is a technique that represents a broad-based systematic approach to problems that may be interdisciplinary. It is particularly useful when problems are affected by many factors, and it entails the creation of a problem model that corresponds as closely as possible to reality. The systems approach stresses the need for the engineer to look for all the relevant factors, influences, and components of the environment that surround the problem. The systems approach corresponds to a comprehensive attack on a problem and to an interest in, and commitment to, formulating a problem in the widest and fullest manner that can be professionally handled.

System Definition

There are many definitions of *system*. One dictionary definition calls it "a group or combination of interrelated, independent or interacting elements forming a collective entity." A system is an assembly of components or elements having a functional relationship to achieve a common objective for useful purpose. A system is composed of components, attributes, and relationships. These are described as follows:

1. Components are the operating parts of the system consisting of input, process, and output. Each system component may assume a variety of values to describe a system state, as set by some control action and one or more restrictions.
2. Attributes are the properties or discernible manifestations of the components of a system. These attributes characterize the system.
3. Relationships are the links between components and attributes.

The properties and behavior of each component of the set have an effect on the properties and behavior of the set as a whole and depend on the properties and behavior of at least one other component on the list. The components of the system cannot be divided into independent subsets. A system is more than the sum of its components and parts. Not every set of items, facts, methods, or procedures is a system. To qualify the system, it should have a functional relationship, interaction between many components, and useful purpose. The purposeful action performed by a system is its function. A basic behavioral concept of a system is that it is a device that accepts one or more inputs and generates from them one or more outputs. This simple behavioral approach to systems is generally known as the Black Box and is represented schematically in Figure 10.5. The Black Box system phenomenon establishes the functional relationship between system inputs and outputs.

Every system is made up of components and components that can be broken down into similar components. If two hierarchical levels are involved in a given system, the lower one is conveniently called a subsystem. The designation of

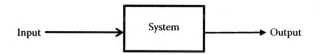

Input   ⟶   System   ⟶ Output

FIGURE 10.5
Black Box.

system, subsystem, and components are relative because the system at one level in the hierarchy is the component at another level. Everything that remains outside the boundaries of the system is considered to be environmental. Material, energy, and/or information that pass through the boundaries are called "inputs" to the system. In reverse, material, energy, and/or information that pass from the system to the environment is called output.

Accordingly, a system is an assembly of components or elements having a functional relationship to achieve a common objective for a useful purpose.

Systems Engineering

INCOSE (International Council on Systems Engineering) defines systems engineering as follows:

> An interdisciplinary approach and means to enable the realization of successful system. It focuses on defining customer needs and required functionality early in the development cycle, documenting requirements, the proceeding with design synthesis and system validation while considering the complete problem:
>
> Operations
> Test
> Cost and schedule
> Disposal
> Performance
> Manufacturing
> Training and support
>
> Systems engineering integrates all the disciples and specialty groups into a team effort forming a structural development process that proceeds from concept to production to operation. Systems engineering considers both the business and the technical needs of all customers with the goal of providing a product that meets the user needs.

The system life cycle process is illustrated in Figure 10.6 and is fundamental to the application of system engineering.

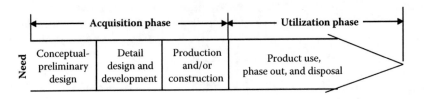

FIGURE 10.6
The product life cycle. (From B. S. Blanchard, W. J. Fabrycky, and J. Wolter, *Systems Engineering and Analysis*, 1998. Reprinted with permission from Pearson Education, Inc.)

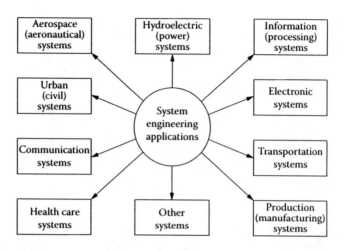

FIGURE 10.7
Application areas of systems engineering. (From B. S. Blanchard, W. J. Fabrycky, and J. Wolter, *Systems Engineering and Analysis*, 1998. Reprinted with permission from Pearson Education, Inc.)

The life cycle begins with the identification of need and extends through conceptual and preliminary design, detailed design, and development, production and/or construction, product use, phase-out, and disposal. The program phases are classified as acquisition and utilization to recognize procedure and customer activities. This classification represents a generic approach. Sometimes the acquiring process may involve both the customer and the producer (or contractor), whereas acquiring may include a combination of contractor and consumer (or ultimate user) activities.

In general, engineering has focused mainly on product performance as the main objective rather than on the development of overall system of which the product is a part. Application of a systems engineering process leads to reduction in the cost of design development, production/construction, and operation, and hence results in reduction in life cycle cost of the product; thus, the product becomes more competitive and economical. Systems engineering provides the basis for a structural and logical approach. The need for systems engineering increases with the size of projects. Application areas of systems engineering are illustrated in Figure 10.7.

Construction Project Life Cycle

Most construction projects are custom-oriented, having a specific need and a customized design. It is always the owner's desire that his project should be unique and better. Further, it is the owner's goal and objective that the facility is completed on time. Expected time schedule is important from both financial and acquisition of the facility by the owner/user.

The system life cycle is fundamental to the application of systems engineering. Detailed presentations of the elaborate technological activities and interaction that must be integrated over the system life cycle are shown in Figure 10.8. This figure summarizes major technical functions performed during the acquisition and utilization process of the system life cycle.

A systems engineering approach to construction projects helps to understand the entire process of project management and to manage and control its activities at different levels of various phases to ensure timely completion of the project with economical use of resources to make the construction project most qualitative, competitive, and economical.

Systems engineering starts from the complexity of the large-scale problem as a whole and moves toward the structural analysis and partitioning process until the questions of interest are answered. This process of decomposition is called a work breakdown structure. The WBS is a hierarchical representation of system levels. Being a family tree, the WBS consists of a number of levels, starting with the complete system at level 1 at the top and progressing downward through as many levels as necessary to obtain elements that can be conveniently managed.

Benefits of systems engineering applications are

- Reduction in the cost of system design and development, production/construction, system operation and support, system retirement, and material disposal
- Reduction in system acquisition time
- More visibility and reduction in the risks associated with the design decision-making process

Shtub, Bard, and Globerson (1994) have divided the project into five phases as illustrated in Figure 10.9.

Representative construction project life cycle, as per Morris, has four stages (phases) for construction project and is illustrated in Figure 10.10.

Though it is difficult to generalize project life cycle to system life cycle, considering that there are innumerable processes that make up the construction process, the technologies and processes as applied to systems engineering can also be applied to construction projects. The number of phases shall depend on the complexity of the project. Duration of each phase may vary from project to project. Based on the concept of project life cycle shown in Figures 10.8 through 10.10, it is possible to evolve a comprehensive life cycle for construction projects, which may have five of the most common phases.

These are as follows:

1. Conceptual design
2. Preliminary design

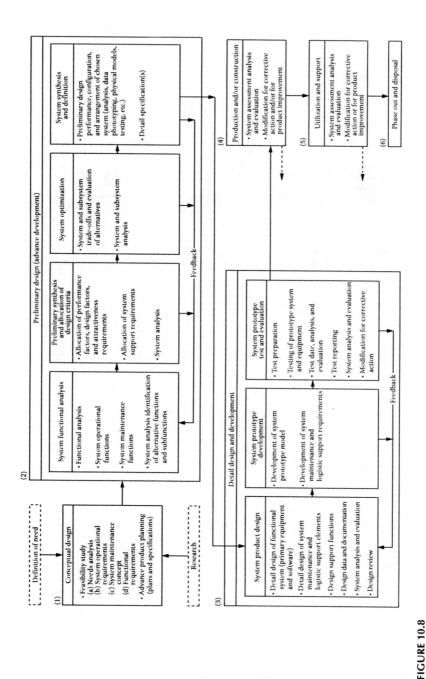

FIGURE 10.8

The system life cycle process. (From B. S. Blanchard, W. J. Fabrycky, and J. Wolter, *Systems Engineering and Analysis*, 1998. Reprinted with permission from Pearson Education, Inc.)

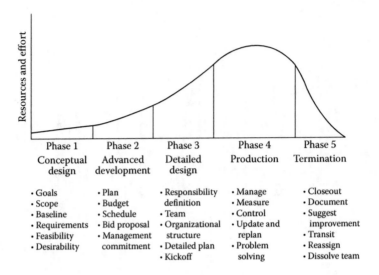

FIGURE 10.9
Project life cycle. (From A. Shtlub, J. F. Bard, and S. Globerson, *Project Management*, 1994. Reprinted with permission from Pearson Education, Inc.)

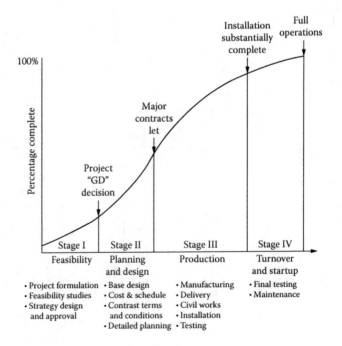

FIGURE 10.10
Representative construction project life cycle. (From Project Management Institute, *PBOK® Guide*, 2004. Reprinted with permission from PMI.)

3. Detailed design

4. Construction

5. Testing, commissioning, and handover

Each phase can be further subdivided into the WBS principle to reach a level of complexity where each element/activity can be treated as a single unit that can be conveniently managed. WBS represents a systematic and logical breakdown of the project phase into its components (activities). It is constructed by dividing the project into major elements with each of these being divided into subelements. This is done until a breakdown is done in terms of manageable units of work for which responsibility can be defined. WBS involves envisioning the project as a hierarchy of goal, objectives, activities, subactivities, and work packages. The hierarchical decomposition of activities continue until the entire project is displayed as a network of separately identified and nonoverlapping activities. Each activity will be single purposed, of a specific time duration, and manageable; its time and cost estimates easily derived, deliverables clearly understood, and responsibility for its completion clearly assigned. The WBS helps in

- Effective planning by dividing the work into manageable elements, which can be planned, budgeted, and controlled
- Assignment of responsibility for work elements to project personnel and outside agencies
- Development of control and information system

WBS facilitates the planning, budgeting, scheduling, and control activities for the project manager and its team. By application of WBS phenomenon, the construction phases are further divided into various activities. Division of these phases will improve the control and planning of the construction project at every stage before a new phase starts. The components/activities of construction project life cycle phases divided on WBS principle are listed as follows:

1. Conceptual design
   Identification of need
   Feasibility
   Identification of project team
   Identification of alternatives
   Financial implications/resources
   Time schedule
   Development of concept design

2. Preliminary design
   General scope of works/basic design
   Regulatory/authorities' approval
   Budget
   Schedule
   Contract terms and conditions
   Value engineering study
3. Detailed design
   Detailed design of the works
   Regulatory/authorities' approval
   Contract documents and specifications
   Detailed plan
   Budget
   Estimated cash flow
   Tender/bid documents
4. Construction
   Mobilization
   Execution of works
   Planning and scheduling
   Management of resources/procurement
   Monitoring and control
   Quality
   Inspection
5. Testing, commissioning, and handover
   Testing
   Commissioning
   Regulatory/authorities' approval
   As-built drawings/records
   Technical manuals and documents
   Training of user's personnel
   Hand over facility to owner/end user
   Move-in-plan
   Substantial completion

Table 10.3 illustrates the subdivided activities/components of the construction project life cycle.

TABLE 10.3

Construction Project Life Cycle

Conceptual Design	Preliminary Design	Detailed Design	Construction	Testing, Commissioning, and Handover
Identification of need	General scope of work/basic design	Detailed design of the works	Mobilization	Testing
Feasibility	Regulatory approval	Regulatory/ authorities' approval	Execution of works	Commissioning
Identification of project team	Budget	Contract documents and specifications	Planning and scheduling	Regulatory/ authorities' approval
Identification of alternatives	Schedule	Detailed plan	Management of resources/ procurement	As-built drawings/ records
Financial implications/ resources	Contract terms and conditions	Budget	Monitoring and control	Technical manuals and documents
Time schedule	Value engineering study	Estimated cash flow	Quality	Training of user's personnel
Development of concept design		Tender/ bidding	Inspection	Handover of facility to owner/end user
				Move-in plan
				Substantial completion

These activities may not be strictly sequential; however, the breakdown allows implementation of project management functions more effectively at different stages.

Quality in Construction Projects

Construction projects are mainly capital investment projects. They are customized and nonrepetitive in nature. Construction projects have become more complex and technical, and the relationships and the contractual grouping of those who are involved are also more complex and contractually varied. The products used in construction projects are expensive, complex, immovable, and long-lived. Generally, a construction project comprises building materials (civil), electromechanical items, finishing items, and equipment. These are normally produced by other construction-related industries/

manufacturers. These industries produce products as per their own quality management practices complying with certain quality standards or against specific requirements for a particular project. Owners of construction projects or their representatives have no direct control over these companies unless they themselves, their representatives, or appointed contractors commit to buying their product for use in their facility. These organizations may have their own quality management program. In manufacturing or service industries quality management of all in-house manufactured products is performed by the manufacturer's own team or is under the control of the same organization that has jurisdiction over its manufacturing plants at different locations. Quality management of vendor-supplied items/products is carried out as stipulated in the purchasing contract as per the quality control specification of the buyer.

Construction projects are constantly increasing in technological complexity. Electromechanical services constitute between 25% and 35% of the total cost of a building project, depending on what type of technologically advanced services are required for the project. Figure 10.11 illustrates typical values of various trades of a major building construction project. In this project, the electromechanical work constitutes approximately 36% of the total project value, which shows the increasing technological complexity of building construction projects.

In addition, the requirements of construction clients are on the increase and, as a result, construction products (buildings) must meet varied performance standards (climate, rate of deterioration, maintenance, etc.). Therefore, to ensure the adequacy of client brief, which addresses the numerous complex client/user needs, it is now necessary to evaluate the requirements in terms of activities and their interrelationships.

Quality management in construction projects is different from that in manufacturing. Quality in construction projects encompasses not only the quality of products and equipment used in the construction but also the total management approach to completing the facility as per the scope of works to customer/owner satisfaction within the budget and in accordance with the specified schedule to meet the owner's defined purpose. The nature of the contracts between the parties plays a dominant part in the quality system required from the project, and the responsibility for fulfilling them must therefore be specified in the project documents. The documents include plans, specifications, schedules, bill of quantities, and so on. Quality control in construction typically involves ensuring compliance with minimum standards of material and workmanship in order to ensure the performance of the facility according to the design. These minimum standards are contained in the specification documents. For the purpose of ensuring compliance, random samples and statistical methods are commonly used as the basis for accepting or rejecting work completed and batches of materials. Rejection of a batch is based on nonconformance or violation of the relevant design specifications.

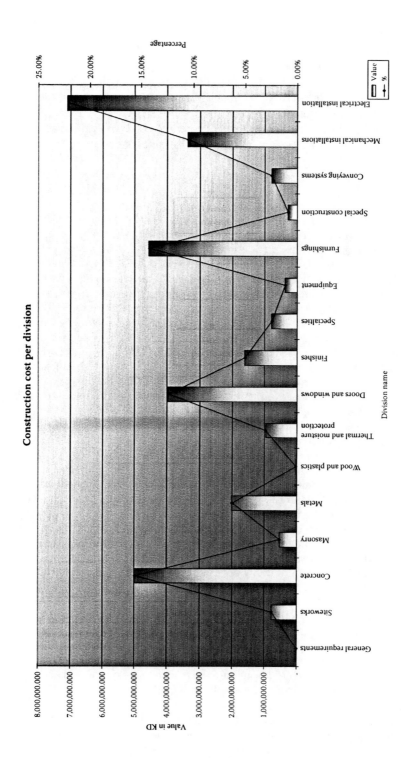

FIGURE 10.11
Division values of a construction project.

The survey of Quality of Construction by FIDIC (Federation Internationale des Ingenieurs-Conseils, the international association of consulting engineers) confirmed that failure to achieve appropriate quality of construction is a problem worldwide. Lack of quality in construction is manifested in poor or nonsustainable workmanship, unsafe structure, delays, cost overruns, and disputes in construction contracts.

Defects or failure in construction facilities can result in very large costs. Even with minor defects, reconstruction may be required and facility operation impaired.

Chung (1999) has described the quality of construction as follows:

> The quality of building work is difficult, and often impossible, to quantify since a lot of construction practices cannot be assessed in numerical terms. The framework of reference is commonly the appearance of final product. "How good is good enough?" is often a matter of personal judgment and consequently a subject of contention. In fact, a building is of good quality if it will function as intended for its design life. As the true quality of the building will not be revealed until many years after completion, the notion of quality can only be interpreted in terms of the design attributes. So far as the builder is concerned, it is fair to judge the quality of his work by the degree of compliance with the stipulations in the contract, not only the technical specifications but also the contract sum and the contract period. His client cannot but be satisfied if the contract is executed as specified, within budget and on time. Therefore, a quality product of building construction is one that meets all contractual requirements (including statutory regulations) at optimum cost and time. (p. 4)

About quality in construction, CII Source Document 79 (1992) describes that

> Quality has many meanings; however, for projects, conformance to established requirements has relevance and clarity. While simple, this definition cannot stand alone. Another term is needed for the term *requirements*. Requirements are contractually established characteristics of a product, process, or service. A characteristic is a physical or chemical property, a dimension, a temperature, a pressure, or any other specification used to define the nature of product, process or service.
>
> The requirements are initially set by client/customer (ordinarily the user/operator of the facility) and are then translated during the preplanning phase into a conceptual design and estimate developed into a project scope and more fully defined. During the Design phase, the requirements are translated into specific design documents (drawings, plans, specification, purchase orders, and the like). Procurement of fabricated items often proceeds concurrently with design. The products of design and procurement reach the construction site for erection and installation during the construction phase. (p. 5)

An implicit assumption in the traditional quality control practices is the notion of an acceptable quality level, which is an allowable fraction of defective items. Materials obtained from suppliers or work performed percentage should be within the acceptable quality level. Problems with materials or goods are corrected after delivery of the product. In contrast to this traditional approach of quality control is the goal of total quality control. In this system, no defective items are allowed anywhere in the construction process. While the zero defects goal can never be permanently obtained, it provides a goal so that an organization is never satisfied with its quality control program even if defects are reduced by substantial amounts year after year. This concept and approach to quality control was first developed in manufacturing firms in Japan and Europe, but has since spread to many construction companies. Total quality control is a commitment to quality expressed in all parts of an organization and typically involves many elements. Design reviews to ensure safe and effective construction procedures are a major element. Other elements include extensive training for personnel, shifting the responsibility for detecting defects from quality control inspectors to workers, and continually maintaining equipment. Worker involvement in improved quality control is often formalized in quality circles in which groups of workers meet regularly to make suggestions for quality improvement. Material suppliers are also required to ensure zero defects in delivered goods. Initially, all materials from a supplier are inspected and batches of goods with any defective items are returned. Suppliers with good records can be certified, and such suppliers will not be subject to complete inspection subsequently.

Total quality management is an organization-wide effort centered on quality to improve performance that involves everyone and permeates every aspect of an organization to make quality a primary strategic objective. It is a way of managing an organization to ensure the satisfaction at every stage of the needs and expectations of both internal and external customers.

In case of construction projects, an organizational framework is established and implemented mainly by three parties: owner, designer/consultant, and contractor. Project quality is the result of aggressive and systematic application of quality control and quality assurance. Figure 10.12 illustrates Juran's triple concept applied to construction.

Construction projects being unique and nonrepetitive in nature need specified attention to maintain quality. Each project has to be designed and built to serve a specific need. TQM in construction projects typically involves ensuring compliance with minimum standards of material and workmanship in order to ensure the performance of the facility according to the design. TQM in a construction project is a cooperative form of doing the business that relies on the talents and capabilities of both labor and management to continually improve quality. The important factor in construction projects is to complete the facility as per the scope of works to customer/owner satisfaction within the budget and to complete the work within the specified schedule to meet the owner's defined purpose. Figure 10.13 shows various elements influencing the quality of construction.

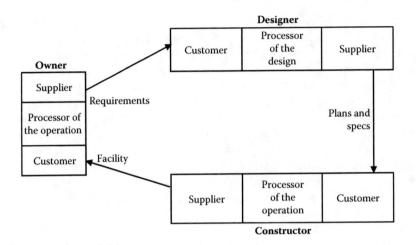

FIGURE 10.12
Triple role concept applied to construction. (From CII Publication 10-10. Reprinted with permission of CII, University of Texas.)

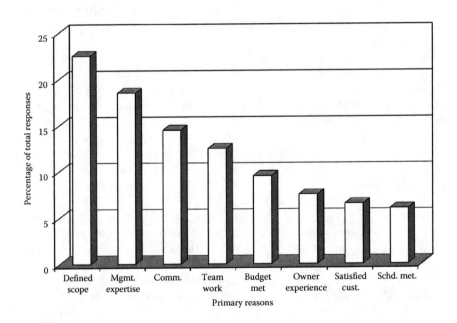

FIGURE 10.13
Primary reasons for quality. (From CII Publication 79. Reprinted with permission of CII, University of Texas.)

Oberlender (2000) has observed:

> Quality in construction is achieved by the people who take pride in their work and have the necessary skills and experience to do the work. The actual quality of construction depends largely upon the control of construction itself, which is the principle responsibility of the contractor. ... What is referred to today as "quality control," which is a part of a quality assurance program, is a function that has for years been recognized as the inspection and testing of materials and workmanship to see that the work meets the requirements of the drawings and specifications. (p. 278)

Crosby's quality definition is "conformance to requirements" and that of Oakland is "meeting the requirements." Juran's philosophy of quality is "fitness for use or purpose."

Based on the philosophies of quality gurus, quality of construction projects can be evolved as follows:

1. Properly defined scope of work
2. Owner, project manager, design team leader, consultant, and constructor's manager are responsible to implement quality
3. Continuous improvement can be achieved at different levels as follows:
   a. Owner—Specify the latest needs
   b. Designer—Specification should include the latest quality materials, products, and equipment
   c. Constructor—Use the latest construction equipment to build the facility
4. Establishment of performance measures
   a. Owner
      i. To review and ensure that designer has prepared the contract documents that satisfy his needs
      ii. To check the progress of work to ensure compliance with the contract documents
   b. Consultant
      i. As a consultant designer, to include the owner's requirements explicitly and clearly define them in the contract documents
      ii. As a supervision consultant, supervise contractor's work as per contract documents and the specified standards
   c. Contractor—To construct the facility as specified and use the materials, products, and equipment that satisfy the specified requirements
5. Team approach—Every member of the project team should know that TQM is a collaborative effort, and everybody should participate

in all the functional areas to improve the quality of the project work. They should know that it is a collective effort by all the participants.

6. Training and education—Both consultant and contractor should have customized training plans for their management, engineers, supervisors, office staff, technicians, and laborers.

7. Establish leadership—Organizational leadership should be established to achieve the specified quality. Encourage and help the staff and laborers to understand the quality to be achieved for the project.

These definitions when applied to construction projects relate to the contract specifications or owner/end user requirements to be formulated in such a way that construction of the facility is suitable for the owner's use or meets the owner's requirements. Quality in construction is achieved through the complex interaction of many participants in the facilities development process.

The quality plan for construction projects is part of the overall project documentation consisting of the following:

1. Well-defined specification for all the materials, products, components, and equipment to be used to construct the facility

2. Detailed construction drawings

3. Detailed work procedure

4. Details of the quality standards and codes to be compiled

5. Cost of the project

6. Manpower and other resources to be used for the project

7. Project completion schedule

Figure 10.14 illustrates functional relationships between various participants.

Table 10.4 identifies the key quality assurance activities that would take place during the life cycle of a typical construction project. These activities are performed by all the participants of the construction projects at various phases/stages of the project.

Ishikawa (1985) has set forth 10 principles to improve quality assurance and to eliminate unsatisfactory relations between the purchaser (vendee) and the supplier (vendor). Table 10.5 summarizes the quality control principles and their application in construction projects.

In order to process the construction project in an effective and efficient manner and to improve control and planning, construction projects are divided into various phases. In traditional thinking, there are five phases of a construction project life cycle, which are further broken down into various activities. These are conceptual design, preliminary design, detailed engineering, construction, and commissioning and handing over.

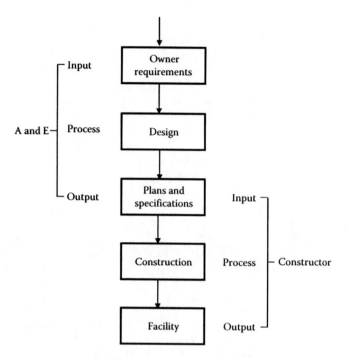

FIGURE 10.14
Juran's triple role-functional relationship. (From CII Publication 51. Reprinted with permission of CII, University of Texas.)

TABLE 10.4

Key Quality Assurance Activities on a Typical Construction Project

Client	Design Consultant	Contractor
Establish project brief/objectives/specification (include QA conditions)	Carry out tender review, prepare outline PQP, and submit	
Accept outline PQP		
Place contract	Set up project team, carry out contract review, prepare PQP	
Approve PQP	Submit for approval	
Approve DQP	Prepare DQP, if appropriate	
Approve key drawings	Prepare drawings	
Approve calculations	Prepare calculations, carry out design reviews, prepare detailed specifications	
Monitor design consultant's activities by audit	Issue enquires for construction work, including QA conditions	Carry out tender review, prepare QA submission
	Carry out bid appraisal assess QA submission	

continued

TABLE 10.4 **(continued)**

Key Quality Assurance Activities on a Typical Construction Project

Client	Design Consultant	Contractor
	Place contract with QA conditions	Carry out contract review setup site team
	Approve PQP	Prepare PQP and submit for approval
	Approve shop drawings and other documents	Place subcontracts, including QA condition when appropriate to work package. Include requirement for documentation submissions, approvals, and records
		Receive DQPs from subcontractors for approval before commencement of work
	Approve DQPs	Prepare DQPs for own work if required
	Project management contract. Conduct progress meeting. Monitor work by inspection test and review of documentation	Place "hold points," and so on, on DQPs to monitor work packages. Approve DQPs
		Monitor off-site work against DQPs
	Carry out audits per agreed-upon schedules	Carry out goods inwards inspection per agreed procedure
		Control work on site against PQP. DQPs, inspection checklist, and so on
		Carry out audits on- and off-site per agreed audit schedule
		Generate records as construction proceeds
		Mark up drawing to as-built state
		Prepare handover packages and submit
Accept documentation package	Check handover package and submit with design records	

Source: Thorpe, B., Sumner, P., and Duncan, J. 1996. *Quality Assurance in Construction.* Reprinted with permission from Ashgate Publishing Limited.

Note: This table identifies the key quality assurance activities during the life cycle of a typical construction project, from project initiation to handover.

TABLE 10.5

Ten Quality Control Principles to Improve Vendee–Vendor Relations

Principle	Ishikawa Principle	Application in Construction Projects
Principle 1	Both vendee and vendor are fully responsible for the quality control application and harmonizing their quality control systems	The owner, consultant, and contractor are fully responsible for application of quality control to meet the defined scope of work in the contract documents
Principle 2	Both vendee and vendor should be independent of each other and respect the independence of the other party	The contractor shall follow an agreed-upon quality control plan, and the consultant shall be responsible for overseeing compliance with contract documents
Principle 3	The vendee is responsible for supplying clear and adequate information and requirements to the vendor so that the vendor can know precisely how and what should be manufactured	The consultant is responsible for providing the owner's requirements explicitly and clearly defining them in the contract documents
Principle 4	Both vendee and vendor, before entering into business transactions, should conclude a rational contract between them in respect to quality, quantity, price, delivery systems, and method of payment	The contractor should study all the documents during the tendering/bidding stage and submit his proposal taking into consideration all the requirements specified in the contract documents
Principle 5	The vendor is responsible for the assurance of quality that will give satisfaction to the vendee, and he is also responsible for submitting necessary and actual data upon the vendee's request	The contractor is responsible for constructing the facility as specified and using materials, products, equipment, and methods that satisfy the specified requirements. The contractor shall follow the submittal procedure specified in the contract documents
Principle 6	Both vendee and vendor should decide the evaluation method of various items beforehand, which will be accepted as satisfactory to both parties	Method of payment (work progress, material, equipment, etc.) to be clearly defined in the contract documents. Rate analysis of BOQ or BOM item to be agreed upon before signing of the contract
Principle 7	Both vendee and vendor should establish in their contract the systems and procedures through which they can amicably settle disputes whenever any problems occur	Contract documents should include a clause to settle disputes arising during the construction stage itself
Principle 8	Both vendee and vendor, taking into consideration the other party's standing, should exchange the information necessary to carry out better quality control	Each member of the project team should participate in all the functional areas to improve the quality of the project

continued

TABLE 10.5   (continued)

Ten Quality Control Principles to Improve Vendee–Vendor Relations

Principle	Ishikawa Principle	Application in Construction Projects
Principle 9	Both vendee and vendor should always perform control business activities sufficiently, such as ordering, production and inventory planning, clerical work, and systems, so that their relationship is maintained and remains amicable and satisfactory	The contractor should perform the work per the agreed-upon construction program and hand over the project per the contracted schedule. The contractor is responsible for providing all the resources, manpower, material, equipment, and so on, to build the facility per specifications
Principle 10	Both vendee and vendor, when dealing with business transactions, should always take full account of the consumer's interests	The contractor should build the facility as stipulated in the contract documents, plan, and specifications within budget and on schedule to meet the owner's objectives

Source: K. Ishikawa, 1985. *What Is Total Quality Control? The Japanese Way.* David J. Lu, tr. Reprinted with permission from Pearson Education, Inc.

Participation of all three parties at different levels during construction is required to develop a quality system and apply quality tools and techniques. With the application of various quality principles, tools, and methods by all the participants at different stages of construction, rework can be reduced, resulting in savings in the project cost and making the project qualitative and economical. This will ensure completion of a construction project and make the project qualitative, competitive, and economical in a way that will meet the owner's needs and specification requirements.

There are several types of contracting system that these parties are involved in at different levels of contracts. The following are the types of contracting systems that are normally used in building and civil engineering construction:

1. Design/bid/build-type contracting system (traditional contracting system)
2. Design/build-type contracting system
3. Project manager-type contracting system
4. Construction manager-type contracting system
5. Guaranteed maximum price
6. Build–own–operate–transfer
7. Turnkey contract

Design/Bid/Build

In this method, the owner contracts design professionals to prepare detailed design and contract documents. These are used to receive competitive bids

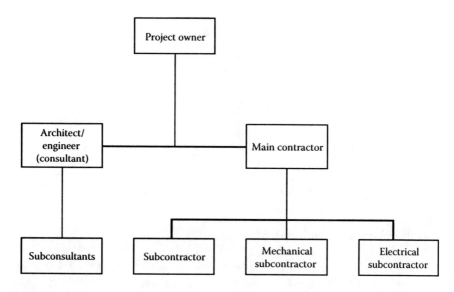

FIGURE 10.15
Design/bid/build-type contracting system.

from the contractors. A design/bid/build contract has a well-defined scope of work. This method involves three steps:

1. Preparation of complete detailed design and contract documents for tendering
2. Receiving bids from prequalified contractors
3. Award of contract to successful bidder

In this method, two separate contracts are awarded, one to the designer/consultant and one to the contractor. In this type of contract structure, design responsibility is primarily that of the architect or engineer employed by the client and the contractor is primarily responsible for construction only. In most cases, the owner contracts the designer/consultant to supervise the construction process. These types of contracts are lump-sum, fixed-priced contracts. Any variation, or change, during the construction needs prior approval from the owner. Since a complete design is prepared before construction, the owner knows the cost of the project, time of completion of the project, and the configuration of the project. The client, through the architect or engineer, retains control of design during construction. This type of contracting system requires considerable time; each step must be completed before starting the next step. Figure 10.15 illustrates the design/bid/build type of contract relationship.

Design/Build

In a design/build contract, the owner contracts a single firm to design and build the project. In this type of contracting system, the contractor is

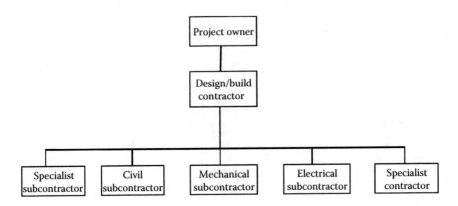

FIGURE 10.16
Design/build-type contracting system.

appointed based on an outline design or design brief to understand the project owner's intent. The owner has to clearly define his or her needs and scope of work before the signing of the contract. It is imperative that the project definition be understood by the contractor to avoid any conflict, as the contractor is responsible for detailed design and construction of the project. A design/build type of contract is often used to shorten the time required to complete a project. Since the contract with the design/build firm is awarded before starting any design or construction, a cost plus contract or reimbursable arrangement is normally used instead of lump-sump, fixed-cost arrangement. This type of contract requires extensive involvement on the part of the owner during the entire life cycle of the project. He or she has to be involved in making decisions during the selection of design alternatives and the monitoring of costs, schedules, and quality during construction and, therefore, the owner has to maintain/hire a team of qualified professionals to perform these activities. Design/build contracts are used for relatively straightforward work, where no significant risk or change is anticipated and when the owner is able to specify precisely what is required. Figure 10.16 illustrates the design/build type of contract relationship.

Project Manager

A project manager contract is used when the owner decides to turn over the entire project management to a professional project manager. In the project manager type of contract, the project manager is the owner's representative and is directly responsible to the owner. The project manager is responsible for planning, monitoring, and managing the project. In its broadest sense, the project manager has responsibility for all the phases of the project from inception of the project until the completion and handing over of the project to the owner/end user. The project manager is involved in giving advice to the owner and is responsible for the appointment of design professionals, consultants,

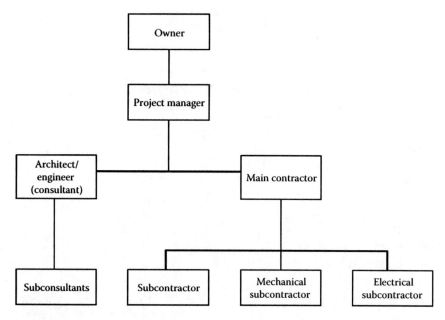

FIGURE 10.17
Project manager-type contracting system.

supervision firm, and selection of the contractor who will implement the project. Figure 10.17 illustrates the project manager type of contract relationship.

Construction Manager

In this method, the owner contracts a construction management firm to coordinate the project and to provide construction management services. The construction management type of contract system is a four-party arrangement involving the owner, designer, construction management firm, and contractor. The construction manager provides advice to the owner regarding cost, time, safety, and about the quality of materials/products/systems to be used on the project. The architect/engineer or supervisor is responsible for maintaining the construction quality and supervising the construction process. The basic prerequisite for the construction manager type of contract is that the firm be knowledgeable and capable of coordinating all aspects of the project to meet the intended use of the project by the owner. There are two general types of construction manager type of contracts. These are

1. Agency construction manager
2. At-risk construction manager

The agency construction manager acts as an advisor to the owner/client, whereas the at-risk construction manager is responsible for on-site

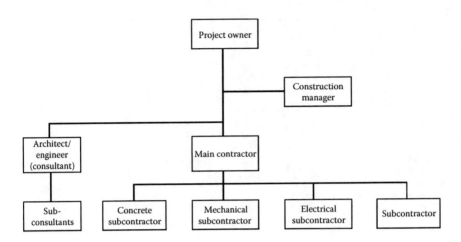

FIGURE 10.18
Construction manager-type contracting system.

performance and actually performs some of the project work. The agency construction manager firm performs no design or construction, but assists the owner in selecting design firms and contractors to implement the project. Figure 10.18 illustrates the agency construction manager type of relationship.

Guaranteed Maximum Price

In this method, the contractor is compensated for the actual costs incurred in connection with design and construction of the project, plus a fixed fee—all subject, however, to a ceiling above which the client is not obligated to pay. A guaranteed maximum price contract specifies a target profit (or fees), a price ceiling (but not for a profit ceiling or floor), and a profit (or fee) adjustment formula. These elements are all negotiated at the outset. The guaranteed maximum price contract combines construction management with design/build. With a guaranteed maximum price contract, amounts below the maximum are typically shared between the client and the contractor, while the contractor is responsible for absorbing the cost above the maximum. Any changes due to specific instructions issued by the owner fall outside the scope of the guaranteed price. A cost plus guaranteed maximum price, as it is also known, type of contract is typically used

- When time pressure requires letting of the contract before design development is sufficiently advanced to allow a conventional lump-sum type of contract to be fixed
- If this type of contract is likely to be less costly than other types
- When financing or other constraints preclude the use of alternatives such as two-stage contract of construction management

- If it is impractical to obtain certain types of services with improved delivery or technical performance, or quality without the use of this type of contract

In this method, the contractor and owner know that the drawings and specifications are not complete, and the contractor and owner agree to work together to complete the drawings and specifications as provided in the contract agreement. This type of contract is weighted heavily in favor of the owner. The contractor takes on all the risk in this type of contracting system. Value engineering studies are conducted to identify design alternatives to help the project contractor maintain the budget and schedule. This type of contract needs

- The contractor's tendering/bidding department should have adequate information to provide the necessary data to support negotiation of final cost and incentive price revision
- Adequate cost pricing information for establishing a reasonable firm target price at the time of initial contract negotiation
- High administration cost from the owner side to monitor what the contractor is actually spending to get the benefit of underspending
- Evaluation of a minimum of two or three proposals for any major subcontract work

In certain guaranteed maximum price (GMP) contracts, the owner monitors and controls the contractor's expenses toward the project resources such as construction equipment, machinery, manpower, and staff on a monthly basis by fixing the basic price and profit percentage agreed upon at the initial stage.

Build–Own–Operate–Transfer

This type of method is generally used by governments to develop public infrastructure by involving the private sector in financing, designing, operating, and managing the facility for a specified period and then transferring the facility to the same government free of charge. The terms BOOT (Build-Own-Operate-Transfer) and BOT (Build-Own-Transfer) are used synonymously.

Examples of BOT projects include

- Airports
- Bridges
- Motorways/toll roads
- Parking facilities
- Tunnels

Certain countries allow the private sector to develop commercial and recreational facilities on government land through the BOT scheme.

The Turnkey Contract

As the name suggests, these are the types of contracts where, upon completion, one turns a key in the door and finds everything working to full operating standards. In this type of method, the owner employs a single firm to undertake design, procurement, construction, and commissioning of the entire work. The firm is also involved in management of the project during the entire process of the contract. The client is responsible for preparation of their statement of requirements, which becomes the strict responsibility of the contractor to deliver. This type of contract is used mainly for the process type of project and is sometimes called engineering, procurement, and construction (EPC).

There are two general types of owners: single-builder owners and multiple-builder owners. Single-builder owners are organizations that do not have a need for projects on a repetitive basis, normally have a limited project staff, and contract all design and construction activities to outside organizations. They usually handle projects with a design/bid/build or construction management contract. Multiple-builder owners are generally large organizations that have a continual need for projects and generally have a staff assigned to project work. They typically handle small-sized, short-duration projects by design/bid/build. For a project in which they desire extensive involvement, a design/build, construction management, or an owner/agent contract arrangement is often used.

All the foregoing contract deliverable systems follow generic life cycle phases of a construction project; however, the involvement/participation of various parties differs depending on the type of deliverable system adapted for a particular project.

In case of the design/build type of deliverable system, the contractor is contracted right from the early stage of the construction project and is responsible for design development of the project. Figure 10.19 shows the typical logic flow diagram for the design/bid/build type of construction project, and Figure 10.20 shows the diagram for the design/build type of contracting system. Details of activities performed during the various phases of the design/bid/build type of contract delivery system are discussed in related sections.

Conceptual Design

Conceptual design is the first phase of the construction project life cycle. The conceptual design is initiated once the need is recognized. In this phase, the idea is conceived and given an initial assessment. Conceptual design, or the design development phase, is often viewed as most critical to achieving outstanding project performance. During the conceptual phase, the environment is examined, forecasts are prepared, objectives and alternatives are evaluated, and the first examination of the technical performance, cost, and

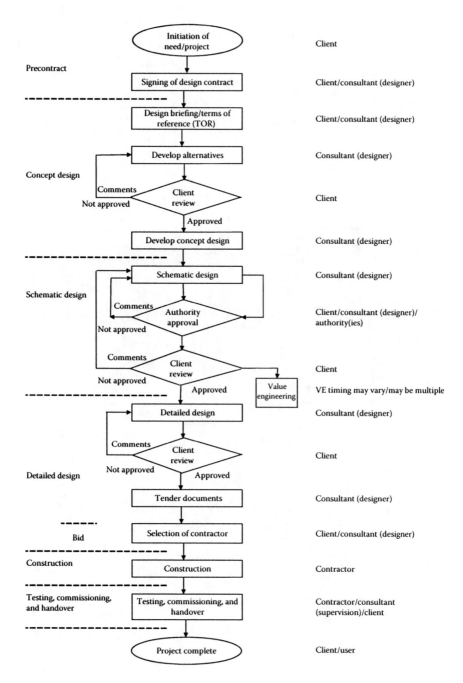

FIGURE 10.19
Logic flow diagram for construction projects: design/bid/build.

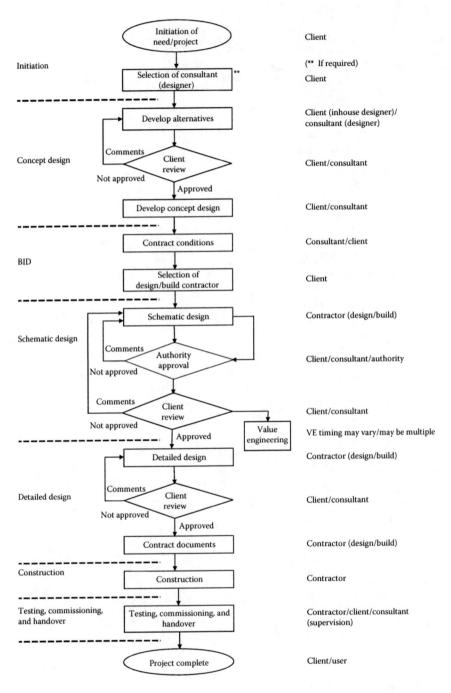

FIGURE 10.20
Logic flow diagram for construction projects: design/build.

time objectives of the project is made. The conceptual phase includes

- Identification of need by the owner, and establishment of main goals
- Feasibility study, which is based on owner's objectives
- Identification of project team by selecting other members and allocation of responsibilities
- Identification of alternatives
- Financial implications, resources, based on estimation of life cycle cost of the favorable alternative
- Time schedule
- Development of concept design

The most significant impacts on the quality in a project occur during the conceptual phase. This is the time when specifications, statement of work, contractual agreements, and initial design are developed. Initial planning has the greatest impact on a project because it requires the commitment of processes, resources schedules, and budgets. A small error that is allowed to stay in the plan is magnified several times through subsequent documents that are second or third in the hierarchy.

Figure 10.21 shows major activities in the conceptual design phase.

Identification of Need

Most construction projects begin with recognition of the new facility. The owner of the facility could be an individual, a public/private sector company, or a governmental agency. The need for the project is created by the owner and is linked to the financial resources available to develop the facility. The owner's needs are quite simple and are based on the following:

- To have best use of the money, that is, to have maximum profit or services at a reasonable cost
- On-time completion, that is, to meet the owner's/user's schedule
- Completion within budget, that is, to meet the investment plan for the facility

The conceptual design is initiated once the owner's need is recognized. Therefore, it is important that the owner defines the requirements and objectives of the potential project clearly at the start of the formulation of design. The need statement is an expression of an unfulfilled requirement. It provides a specifically focused requirement that can be addressed as a way of providing a solution.

The owner's need must be well defined, indicating the minimum requirements of quality and performance, an approved main budget, and required

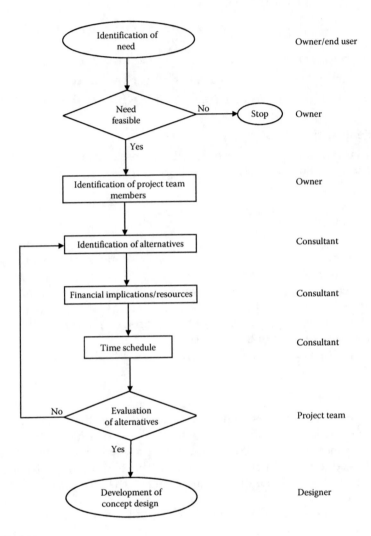

FIGURE 10.21
Major activities in the conceptual design phase.

completion date. Sometimes, the project budget is fixed and, therefore, the quality of the building system, materials, and completion of the project needs to be balanced with the budget.

Figure 10.22 illustrates a preliminary appraisal and the steps in the project identification.

Feasibility

Once the owner's need is identified, the traditional approach is pursued through a feasibility study or an economical appraisal of owner needs or

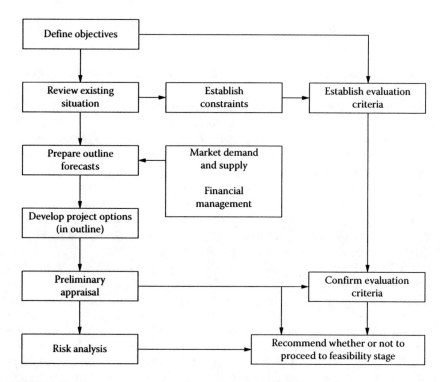

FIGURE 10.22
Steps in project identification. (From R. K. Corrie, *Engineering Management Project Evaluation*, 1991. Reprinted with permission from Thomas Telford Publishing, UK.)

benefits, also taking into account the many relevant moral, social, environmental, and technical constraints. The feasibility study takes its starting point from the output of the project identification need.

ICE (1996) listed some of the investigations to be carried out for a major project. These are as follows:

- Outline design
- Studies of novel requirements and risks
- Public consultation
- Geotechnical study of site, sources of materials, storage areas, and access routes
- Environmental impact analysis
- Health and safety studies
- Testing for contaminated land and requirements for the disposal of waste
- Estimates of capital and operating costs

- A master program of work, expenditure, and financing
- Assessment of funding

Depending on the circumstances, the feasibility study may be short or lengthy, simple or complex. In any case, it is the principle requirement in project development as it gives the owner an early assessment of the viability of the project and the degree of risk involved. The outcome of the feasibility study helps select a defined project that meets the stated project objectives, together with a broad plan of implementation. Figure 10.23 illustrates various stages for a feasibility study.

Technical studies are performed to analyze that the proposed facility is suitable for intended use by the owner/user. The proposed project is economically feasible if the total value of the benefits that result from the project exceeds the cost that results from the project. Economic feasibility depends on technical feasibility because the facility must be suitable for intended use. Financial feasibility may or may not be related to economic feasibility.

The project study is usually performed by the owner through his own team or by engaging individuals/organizations involved in preparation of economic and financial studies. Once the project definition has been ascertained, the owner selects other team members of the project and finalizes the contract delivery system for the project. If the feasibility study shows that the objectives of the owner are best met through the ideas generated, then the owner will select and engage a project team based on the project delivery system to develop his notional ideas into a more workable form.

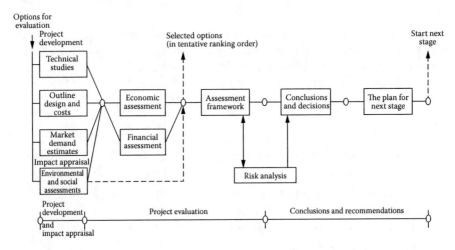

FIGURE 10.23
Project feasibility stages. (From R. K. Corrie, *Engineering Management Project Evaluation*, 1991. Reprinted with permission from Thomas Telford Publishing, UK.)

Identification of Project Team

Most construction projects involve three major groups or parties:

1. Owner: A person or an organization that articulated the need for the facility and is responsible for arranging the financial resources for the creation of the facility.
2. Designer/consultant: This consists of architects, engineers, or consultants. They are the owner's appointed entity accountable for converting the owner's conception and need into a specific facility with detailed directions through drawings and specifications, within the economic objectives and schedule. They are responsible for the design process and assist the owner in preparation of tender and contract documents. The owner may engage the designer to supervise construction.
3. Contractor: A construction firm engaged by the owner to complete the specified facility by providing the necessary staff, workforce, materials, equipment, tools, and other accessories to complete the project to the satisfaction of the owner/end user in compliance with the contract documents.

The owner is the first member of the project team. The owner's relationship with other team members and his or her responsibilities depend on the type of deliverable system the owner would prefer to go with. There are many types of contract delivery systems; however, design/bid/build is the most predominantly used contracting system preferred by the owners.

For the design/bid/build type of contract system, the first thing the owner has to do is select design professionals/consultants. Generally, the owner selects a designer/consultant with whom he or she has worked before with satisfactory results. The owner can use his or her preferred designer/consultant or select one by obtaining proposals from several design professionals/consultants. The owner may contract a designer or A&E to provide site supervision during the construction process. Thus, the designer or A&E firm acts as the prime professional to design the project and also supervises the construction.

Once the project delivery system is finalized and the designer/consultant is selected and contracted by the owner to proceed with the project design, a TOR is issued to the designer/consultant to prepare a design proposal and contract documents. A TOR is a document that describes the purpose and structure of a project. It gives the project team a clear understanding of the development of the project.

Table 10.6 illustrates the typical requirements of project team members. Table 10.7 illustrates the typical responsibilities of project team members. Figure 10.24 illustrates the responsibilities of different parties.

TABLE 10.6

Typical Requirements of Project Team Members

Owner/Project Manager	Design Professional or A&E	Constructor
Adequate function and appearance of the new facility	An adequate project scope definition An adequate budget	A well-defined set of contract documents
Project completion on time and within budget	A reasonable schedule Timely decisions from the owner	Timely decisions from the owner and design professional
Desirable balance of life cycle and initial capital costs	Realistic and fair sharing of project risks	Realistic and fair sharing of project risks
Addressing of environmental, health, permitting, safety, user impacts, and sustainable development considerations	Adequate communication with the owner regarding performance A fair and reasonable process for resolving disputes	Adequate communication with the owner regarding performance A fair and reasonable process for resolving disputes
A fair and reasonable process for resolving disputes	Timely payment and a reasonable profit	Timely payment and a reasonable profit

Source: American Society of Civil Engineers, *Quality in the Constructed Project*, 2000. Reprinted with permission from ASCE.

The TOR generally requires the designer (consultant) to perform the following:

- Predevelopment studies, which includes data collection and analysis related to the project
- Development of conceptual alternatives
- Evaluation of conceptual alternatives and selection of preferred alternatives in consultation with the owner
- Preparation of preliminary design, budget, and schedule and obtaining authorities' approvals
- Preparation of detailed design and contract documents for tendering purpose

Table 10.8 illustrates the contribution of various participants during all the phases of the construction project life cycle for the design/bid/build type of contracting system.

Identification of Alternatives

Table 10.9 shows a quantitative comparison of functional alternatives to cost alternatives that may be analyzed to select the preferred alternative.

Each alternative is based on a predetermined set of performance measures to meet the owner's requirements. In case of construction projects, it is mainly

TABLE 10.7

Typical Responsibilities of Project Team Members

Owner/Project Manager	Design Professional or A&E	Constructor
Fulfillment of contractual obligations to other team members, including furnishing site and related information, and timely payment	Fulfillment of contractual obligations to other team members	Fulfillment of contractual obligations to other team members
Compliance with applicable laws, regulations, codes, standards, and practices	Compliance with applicable laws, regulations, codes, standards, and practices	Compliance with applicable laws, regulations, codes, standards, and practices
Provision of adequate funding	Fulfillment of professional standards	Interpretation of plans and specifications
Provision of necessary real estate or rights of way	Development and drafting of well-defined contract documents	Construction of facility as described in contract documents
Provision of project goals and objectives	Responsiveness to project schedule, budget, and program	Management of construction site activities and safety program
Fulfillment of insurance and legal requirements	Provision of construction-phase design services	Management, quality control, and payment of subcontractors and vendors
Assignment of site safety responsibility		
Acceptance of completed facility		

Source: American Society of Civil Engineers, *Quality in the Constructed Project,* 2000. Reprinted with permission from ASCE.

the extensive review of development options that are discussed between the owner and the designer/consultant. The consultant engineer provides engineering advice to the owner to enable him to assess its feasibility and the relative merits of various alternative schemes to meet his requirements. Social, economical, and environmental impacts, functional capability, safety, and reliability should be taken into account while considering the development of alternatives. Each alternative is evaluated based on the predetermined set of performance measures to meet the owner's requirements. Figure 10.25 summarizes the general steps in the systematic process of studying project alternatives and evaluating associated impacts.

Quantitative comparison and evaluation of conceptual alternatives are carried out by considering the advantages and disadvantages of each item systematically. The designer makes a brief presentation to the owner, and the project is selected based on preferred conceptual alternatives. Various possibilities are considered during this stage, and the technological and economical feasibility is assessed and compared to select the best possible alternative.

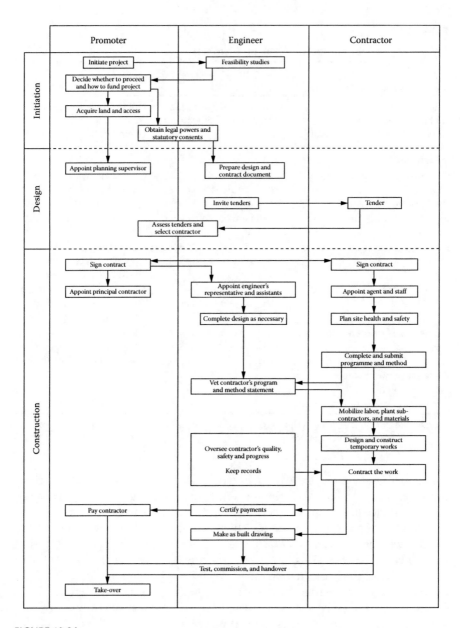

FIGURE 10.24
Division of responsibility. (From *Civil Engineering Procedure* by ICE. Reprinted with permission from Thomas Telford Publishing, UK.)

TABLE 10.8

Contribution of Various Participants (Design/Bid/Build Type of Contracts)

| Phase | Example of Contribution | | |
	Owner	Designer	Contractor
Conceptual design	Identification of need Selection of alternative Selection of team members Approval of time schedule Approval of budget TOR	Feasibility Development of alternatives Cost estimates Schedule Development of concept design	
Preliminary design	Approval of preliminary (schematic) design	Develop general layout/scope of facility/project Regulatory approval Budget Schedule Contract terms and conditions	
Detailed design	Approval of budget Approval of time schedule Approval of design Contract negotiation Signing of contract	Development of detailed design Authorities' approval Detailed plan Budget Schedule BOQ Tender documents Evaluation of bids	Collection of tender documents Preparation of proposal Submission of bid
Construction	Approve subcontractors Approve contractor's core staff Legal/regulatory clearance Site works instruction Variation order Payments	Supervision Approve plan Monitor work progress Approve shop drawings Approve material Recommend payment	Execution of work Contract management Selection of subcontractors Selection of core staff Planning Resources Procurement Quality Safety
Testing, commissioning, and handover	Training Acceptance of project Substantial completion certificate Payments	Witness tests Check closeout requirements Recommend takeover Recommend issuance of substantial completion certificate	Testing Commissioning Authorities' approvals Documents Training Handover

TABLE 10.9

Conceptual Alternatives

Functional Alternatives	Cost Alternatives
Materials handling methods	Design cost
Traffic flow arrangements (patterns in air, water, land, people, or products)	Capital cost of construction
	Operation and maintenance costs
Types of travel modes (vehicle type, size, style)	Life expectancy or design-life periods
Method to provide fish passage at barriers in waterways	
Space allocations	Return on investment
Clear-span requirements in buildings	Project phasing (initial opening or operating segments)
Public/private (joint development) options	
Methods to avoid or minimize impacts to the natural environment	Extra cost for aesthetics
	Cost/benefit ratios

Source: American Society of Civil Engineers, *Quality in the Constructed Project*, 2000. Reprinted with permission from ASCE.

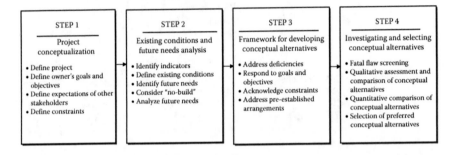

FIGURE 10.25
Alternative study and impact analysis process. (From American Society of Civil Engineers, *Quality in the Constructed Project*, 2000. Reprinted with permission from ASCE.)

Financial Implications/Resources

The next step is to refine cost estimates for the conceptual alternatives as this is required by the owner to determine the capital cost of construction, so that he or she can arrange the finances. It is the owner's responsibility to provide an approved maximum finance to complete the facility. It is required that the owner formulate his or her thoughts on project financing, as the financial conditions will affect the possible options from the beginning. Normally the following points should be considered:

1. What are the sources of funding?
2. What criteria or rules apply?
3. How could the project best respond to those rules?

In case any funding agency is involved in financing the project, it may impose certain conditions that affect the project feasibility and

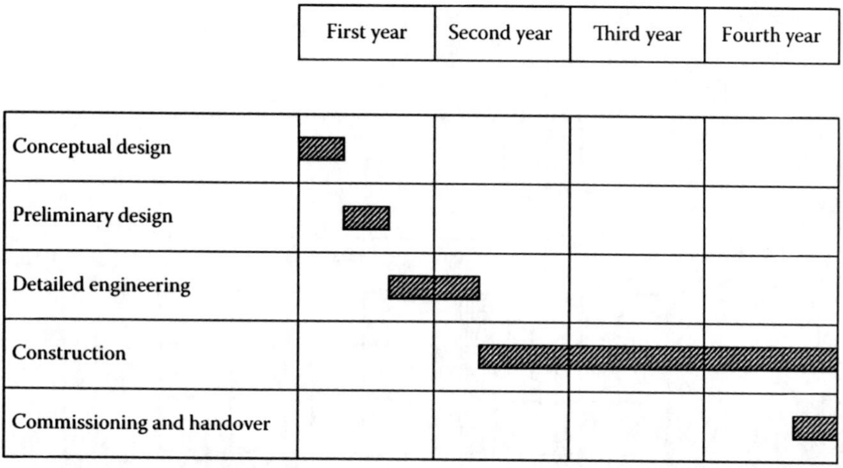

	First year	Second year	Third year	Fourth year
Conceptual design	▨			
Preliminary design	▨			
Detailed engineering	▨▨			
Construction			▨▨▨▨▨▨	
Commissioning and handover				▨

FIGURE 10.26
Typical time schedule.

implementation. It is likely that such funding agencies may also insist on the adoption of a particular contract strategy.

Time Schedule

The duration of a construction project is finite and has a definite beginning and a definite end; therefore, during the conceptual phase, the expected time schedule for the completion of the project/facility is worked out. The expected time schedule is important from both financial and facility acquisition points of view of the owner/end user. It is the owner's goal and objective that the facility is completed in time. Figure 10.26 illustrates a time schedule for a typical construction project.

Development of Concept Design

The selected preferred alternative is the base for development of the concept design. The designer can use techniques such as QFD to translate the owner's need into technical specifications. Figure 10.27 illustrates the house of quality concept for an office building project based on certain specific requirements by the customer.

While developing the concept design, the designer must consider the following:

1. Project goals
2. Usage
3. Technical and functional capability

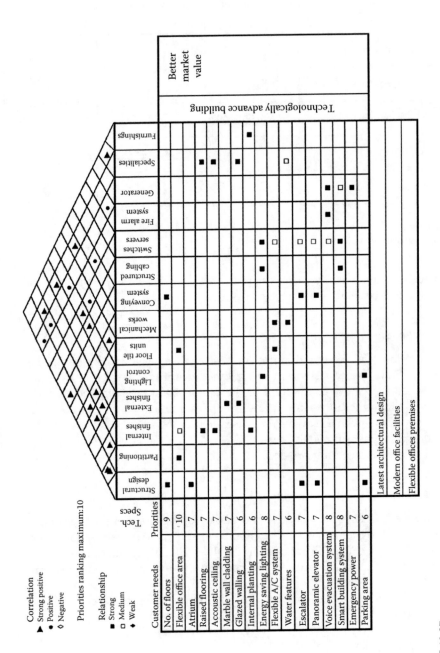

FIGURE 10.27
House of quality for office building project.

4. Aesthetics

5. Constructability

6. Sustainability (environmental, social, and economical)

7. Health and safety

8. Reliability

9. Environmental compatibility

10. Fire protection measures

11. Supportability during maintenance/maintainability

12. Cost effectiveness over the entire life cycle (economy)

It is the designer's responsibility to pay greater attention to improving the environment and to achieve sustainable development. Numerous UN meetings (such as the first United Nations conference on Human Development held in Stockholm in 1972; the 1992 Earth Summit in Rio de Janeiro; the 2002 Earth Summit in Johannesburg; the 2005 World Summit; and the Brundtland Commission on Environment and Development in 1987) have emphasized "sustainability," whether it be sustainable environment, sustainable economic development, sustainable agricultural and rural development, and so on. Accordingly, the designer has to address environmental and social issues and comply with local environmental protection codes. A number of tools and rating systems have been created by LEED (the United States), BREEAM (the United Kingdom), and HQE (France) in order to assess and compare the environmental performance of the buildings. These initiatives have a great impact on how buildings are designed, constructed, and maintained. Therefore, during the implementation of building projects, the following need to be considered:

1. Accretion with the environment by using natural resources such as sunlight, solar energy, and appropriate ventilation configuration

2. Energy conservation by energy-efficient measures to diminish energy consumption

3. Environmental protection to reduce environmental impact

4. Use of materials harmonizing with the environment

5. Aesthetic match between a structure and its surrounding natural and built environment

6. Good air quality

7. Comfortable temperature

8. Comfortable lighting

9. Comfortable sound

10. Clean water

11. Less water consumption
12. Integration with social and cultural environment

During the design stage, the designer must work jointly with the owner to develop details regarding the owner's needs and give due consideration to each part of the requirements. The owner on his part should ensure that the project objectives are

- Specific
- Realistic
- Measurable
- Agreed upon by all the team members
- Possible to complete within the stipulated time
- Within the budget

The following are the requirements for a building construction project, normally mentioned in the TOR, to be prepared by the designer during the conceptual phase for submission to the owner:

1. Site plan
   a. Civil
   b. Services
   c. Landscaping
   d. Irrigation
2. Architectural design
3. Building and engineering systems
4. Structural
   a. Mechanical (HVAC)
   b. Public health
   c. Fire suppression systems
   d. Electrical
   e. Low-voltage systems
   f. Others
5. Cost estimates
6. Schedules

The designer is required to submit all the above requirements in the form of

- Report
- Drawings

- Models
- Presentation

Preliminary Design

Preliminary design is mainly a refinement of the elements in the conceptual design phase. Preliminary design is also known as *schematic design*. During this phase, the project is planned to a level where sufficient details are available for the initial cost and schedule. This phase also includes the initial preparation of all documents necessary to implement the facility/construction project. The central activity of preliminary design is the architect's design concept of the owner's objective, which can help make the detailed engineering and design for the required facility. Preliminary design is a subjective process transforming ideas and information into plans, drawings, and specifications of the facility to be built. Component/equipment configurations, material specifications, and functional performance are decided during this stage. At this stage, the owner can alter the scope and consider alternatives. The owner seeks to optimize certain facility features within the constraints of other factors such as cost, schedule, vendor capabilities, and so on.

Design is a complex process. Before design is begun, the scope must adequately define deliverables, that is, what will be furnished. These deliverables are design drawings, contract specifications, type of contracts, construction inspection record drawings, and reimbursable expenses.

Preliminary design is the basic responsibility of the architect (designer/consultant or A&E). In the case of building construction projects, a preliminary design determines

1. General layout of the facility/building/project
2. Required number of buildings/number of floors in each building/area of each floor
3. Different types of functional facilities required such as offices, stores, workshops, recreation, training centers, parking, and so on
4. Type of construction such as reinforcement concrete or steel structure, precast, or cast *in situ*
5. Type of electromechanical services required
6. Type of infrastructure facilities inside the facilities area
7. Type of landscape

The designer has to consider the following points while preparing the preliminary design:

1. Concept design deliverables
2. Calculations to support the design

3. System schematics for electromechanical system
4. Coordination with other members of the project team
5. Authorities' requirements
6. Availability of resources
7. Constructability
8. Health and safety
9. Reliability
10. Energy conservation issues
11. Environmental issues
12. Selection of systems and products that support the functional goals of the entire facility
13. Sustainability
14. Requirements of all stakeholders
15. Optimized life cycle cost (value engineering)

General Scope of Works/Basic Design

The purpose of a general scope of works is to provide sufficient information to identify the works to be performed and to allow detailed design to proceed without significant changes that may adversely affect the project budget and schedule.

At the preliminary design stage, the scope must define deliverables, that is, what will be furnished. It should include a schedule of dates for delivering drawings, specifications, calculations and other information, forecasts, estimates, contracts, materials, and construction. The designer develops a design concept with the plan, elevation, and other related information that meet the owner's requirements. The designer also develops a concept of how various systems such as heating and cooling systems, communication systems, and so on will fit into the system.

Bennet (2003) has given a list of preliminary design drawings required for preliminary approval quoting one building agency of a U.S. state. These are

- The basic design approach drawn at an agreed-upon scale
- Site location in relation to the existing environment
- Relation to master plans
- Circulation
- Organization of building functions
- Functional/aesthetic aspects of the design concepts under study
- Graphic description of critical details
- Visual and functional relationship
- Compatibility with the surrounding environment

Bennet (2003) further states that the same agency requires the design professional to prepare a narrative description of the following building systems upon completion of schematic design:

- Structure
- Floor grade and systems
- Roof
- Exterior/interior walls and partitions
- Interior finishes
- Sight lines
- Stairs and elevators
- Specialty items
- Electrical systems
- Mechanical systems
- Built-in equipment
- Site construction

Regulatory Approval

Once the preliminary design is approved, it should be submitted to regulatory bodies for their review and approval for compliance with the regulations, codes, and licensing procedure.

Budget

Based on the preliminary design, the budget is prepared by estimating the cost of activities and resources. The preparation of the budget is an important activity that results in a timed phased plan summarizing the expected expenses toward the contract and also the income or the generation of funds necessary to achieve the milestone. The budget for a construction project is the maximum amount the owner is willing to spend for design and construction of the facility that meets the owner's need. The budget is determined by estimating the cost of activities and resources and is related to the schedule of the project. If the cash flow or resulting budget is not acceptable, the project schedule should be modified. It is required that while preparing the budget, the risk assessment of the project is also performed.

Schedule

After the preliminary scope of works, the preliminary design and budget for the facility/project are finalized; the logic of the construction program is set. On the basis of logic, a CPM schedule (bar chart) is prepared to determine the critical path and set the contract milestones.

Contract Terms and Conditions

Normally it is the consultant/designer team that is responsible for developing a set of contract documents that meets the owner's needs, and specifies the required level of quality, budget, and schedule. At this stage, a contract exists between the consultant and the client for the development of the project, and any good management test will demand that the contract be clearly understood by all parties associated with it. There are numerous combinations of contract arrangements for handling the construction projects; however, design/bid/build is predominantly used in most construction project contracts. This delivery system has been chosen by owners for many centuries and is called the traditional contracting system. In the traditional contracting system, the detailed design for the project is completed before tenders for construction are invited. The detailed engineering is carried out by the consultant/design professional to make the project qualitative and economical.

Based on the type of contracting arrangements with which the owner would like to handle the project, necessary documents are prepared by establishing a framework for execution of the project. Generally, FIDIC's model conditions for international civil engineering contracts are used as a guide to prepare these contract documents. Preliminary specifications and documents are prepared in line with model contract documents.

Value Engineering Study

Value engineering (VE) studies can be conducted at various phases of a construction project; however, the studies conducted in the early stage of a project tend to provide the greatest benefit. In most projects, VE studies are performed during the schematic phase of the project. At this stage the design professionals have considerable flexibility to implement the recommendations made by the VE team, without significant impacts on the project's schedule or design budget. In certain countries, for a project over US$5 million, a VE study must be conducted as part of the schematic design process. The team members who perform the VE study depend on the client's/ owner's requirement. It is advisable that a SAVE international registered certified value specialist be assigned to lead this study.

Detailed Design

Detailed design is the third phase of the construction project life cycle. It follows the preliminary design phase and takes into consideration the configuration and the allocated baseline derived during the preliminary phase. Detailed design is also known as *design development/detailed engineering*. During this phase, all suggested changes are reevaluated to ensure that the changes will not detract from meeting the project design goals/objectives. Detailed design involves the process of successively breaking down, analyzing, and designing the structure and its components so that it complies with the recognized codes and standards of safety and performance while rendering the design in the form of drawings and specifications that will tell the contractors exactly how to build the facility to meet the owner's need. During this phase, detailed design of the work, contract documents, detailed plan, budget, estimated cash flow, regulatory approval, and tender/bidding documents are prepared. Depending on the type of contract the owner would like to have for completing the facility, the designer (consultant) can start preparing the detailed design. The success of a project is highly correlated with the quality and depth of the engineering plans prepared during this phase.

Figure 10.28 illustrates major activities in the detailed design phase.

Detailed Design of the Works

The detailed design process starts once the preliminary design is approved by the owner. Detailed design is enhancement of work carried out during the preliminary stage. During this phase, a comprehensive design of the works with a detailed WBS and work packages are prepared. In general, specific and detailed scopes of works lead to better-quality projects. The detailed design phase is the traditional realm of design professionals, including architects, interior designers, landscape architects, and several other disciplines such as civil, electrical, mechanical, and other engineers as needed.

Accuracy in the project design is a key consideration of the life cycle of the project; therefore, it is required that the designer/consultant be not only an expert in the technical field but also should have a broad understanding of engineering principles, construction methods, and value engineering. The designer must know the availability of the latest products in the market and to use proven technology, methods, and materials to meet the owner's objectives. He or she must refrain from using a monopolistic product, unless its use is important or critical for the proper functioning of the system. He or she must ensure that at least two or three sources are available in the market producing the same type of product that complies with all its required features and intent of use. This will help the owner get competitive bidding during the tender stage.

The authors of *Quality in the Constructed Project* (ASCE, 2000) have listed the general functions and responsibilities of the design professional as follows:

- Being fully qualified and licensed to offer and provide the services contractually undertaken and provided
- Applying appropriate skills to the design
- Being proactive and clear in communication
- Being responsive to the established budget schedule and program
- Making timely interpretation, evaluation, and decisions
- Disclosing fully related external interests

- Avoiding conflicts of interest
- Complying with applicable codes, regulations, and laws
- Interpreting contract documents impartially
- Representing the owner's interests as required by contract
- Performing project-specific duties outlined in the contract between the design professional and owner

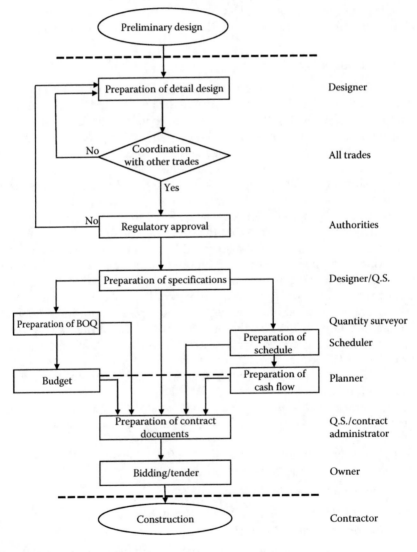

FIGURE 10.28
Major activities in the detailed design phase.

ASCE (2000) further describe that the design professional (or in-house design team) can help ensure project quality through several activities, including

- Developing a scope of services that meet the owner's requirements and the project goals and objectives
- Developing a design activity plan for the project
- Defining project design guidelines
- Estimating accurately the hours of effort and costs involved in achieving a quality design
- Building flexibility into the design activity plan to allow for changes and future project development, as well as associated budget and schedule revisions
- Developing a realistic schedule with appropriate milestones to confirm progress
- Monitoring design progress constantly

Detailed design activities are similar, although more in-depth than the design activities in the preliminary design stage. The size, shape, levels, performance characteristics, technical details, and requirements of all the individual components are established and integrated into the design. Design engineers of different trades have to take into consideration all these at a minimum while preparing the scope of works. The range of design work is determined by the nature of the construction project.

The following are the aspects of works to be considered by design professionals while preparing the detailed design. These can be considered as a base for the development of design to meet customer requirements and will help achieve the qualitative project.

Architectural Design

- Intent/use of building/facility
- Property limits
- Aesthetic look of the building
- Environmental conditions
- Elevations
- Plans
- Axis, grids, levels
- Room size to suit the occupancy and purpose
- Zoning as per usage/authorities requirements
- Identification of zones, areas, rooms

- Modules to match with structural layout/plan
- Number of floors
- Ventilation
- Thermal insulation details
- Stairs, elevators (horizontal and vertical transportation)
- Fire exits
- Ceiling height and details
- Reflected ceiling plan
- Internal finishes
- Internal cladding
- Partition details
- Masonry details
- Joinery details
- Schedule of doors and windows
- Utility services
- Toilet details
- Required electromechanical services
- External finishes
- External cladding
- Glazing details
- Finishes schedule
- Special equipment
- Fabrication of items, such as space frame, steel construction, retaining wall, having special importance for appearance/finishes
- Special material/product to be considered, if any
- Any new material/product to be introduced
- Conveying system core details
- Ramp details
- Hard and soft landscape
- Parking areas

Concrete Structure

- Type of foundation
- Energy-efficient foundation
- Design of foundation based on field and laboratory tests of soil investigation that give the following information:

a. Subsurface profiles, subsurface conditions, and subsurface drainage
b. Allowable bearing pressure, and immediate and long-term settlement of footing
c. Coefficient of sliding on foundation soil
d. Degree of difficulty for excavation
e. Required depth of stripping and wasting
f. Methods for protecting below-grade concrete members against impact of soil and groundwater (water and moisture problems, termite control, and radon where appropriate)
g. Geotechnical design parameters such as angle of shear resistant, cohesion, soil density, modulus of deformation, modulus of subgrade reaction, and predominant soil type
h. Design loads such as dead load, live load, wind load, and seismic load

- Grade and type of concrete
- Size of bars for reinforcement and the characteristic strength of bars
- Clear cover for reinforcement for
    a. Raft foundation
    b. Underground structure
    c. Exposed to weather structure such as columns, beams, slabs, walls, and joists
    d. Not exposed to weather columns, beams, slabs, walls, and joists
- Reinforcement bar schedule, stirrup spacing
- Location of columns in coordination with architectural requirements
- Number of floors
- Height of each floor
- Beam size and height of beam
- Superstructure:
    a. Columns
    b. Stairs
    c. Walls
    d. Beams
    e. Slabs
- Deflection that may cause fatigue of structural elements; cracks or failure of fixtures, fittings, or partitions; or discomfort to occupants
- Movement and forces due to temperature

- Equipment vibration criteria
- Reinforcement bar schedule, stirrup spacing
- Shaft and pit for conveying system
- Building services to fit in the building
- Environmental compatibility
- Excavation
- Dewatering
- Shoring
- Backfilling
- Property limits/surrounding areas

Elevator Works

- Type of elevator
- Loading capacity
- Speed
- Number of stops
- Travel height
- Cabin, cabin accessories, cabin finishes, and car operating system
- Door, door finishes, and door system
- Safety features
- Drive, size, and type of motor
- Floor indicators, call button
- Control system
- Cab overhead dimensions
- Pit depth
- Hoist way
- Machine room
- Operating system

HVAC Works

- Environmental conditions
- Air-conditioning calculations
- Room pressurizing and leakage calculations
- Energy consumption calculations
- Air-conditioning calculations for IT equipment rooms
- Selection of chillers, cooling towers

- Selection of the duct work systems plant and equipment, for example, air-handling units, fan coil units, filters, coils, fans, humidifiers, duct heaters, and so on
- Selection of pumps
- Smoke extract ventilation calculations
- Exhaust ventilation calculations (toilets, chemicals storage, basement parking)
- Selection of fans
- Preparation of the plan and section layouts and plant room drawings
- Electrical load calculations
- Comparison of electrical consumption with electrical conservation code
- Ductwork sizing calculations
- Selection of the ductwork components such as balancing dampers, constant volume boxes, variable air volume boxes, attenuators, grills and diffusers, fire dampers, pressure relief dampers, and so on
- Pipework sizing calculations
- Selection of the inline pipe work components, for example, valves, strainers, air vents, commissioning sets, flexible connections, sensors, and so on
- Selection of boilers, pressurization units, air-conditioning calculations
- Pipework and duct work insulation selection
- Details of grills and diffusers, control valves, and so on
- Preparation of equipment schedules
- Control details
- HVAC-related electrical works
- Starter panels, MCC panels, schematic diagram of MCC
- Preparation of point schedule for building management system (BMS)
- Schematic diagram for BMS

Fire Protection System

The fire protection system provides protection against fire to life and property. The system is designed taking into consideration the local fire code and National Fire Protection Association (NFPA) standards. The system includes the following:

- Sprinkler system for fire suppression in all the areas of the building
- Hydrants (landing valve) for professional fighting
- Hose reel for public use throughout the building
- Gaseous fire protection system for communication rooms

- Fire protection system for diesel generator room
- Size of fire pumps and controls
- Water storage facility
- Interface with other related systems

Plumbing Works

- Maximum working pressure to have adequate pressure and flow of water supply
- Maximum design velocity
- Maximum probable demand
- Demand weight of fixture in fixture units for public uses
- Friction loss calculation
- Maximum hot water temperature at fixture outlet
- Water heater outlet hot temperature
- Providing isolating valves to ensure that the system is easily maintainable
- Hot water system
- Central water storage capacities
- Size of pumps and controls
- Location of storage tank
- Schematic diagram for water distribution system

Drainage System

While designing the drainage system, the schedule of foul drainage demand units and frequency factors for the following items should be considered for sizing the piping system, number of manholes, capacity of sump pump, and capacity of sump pit:

- Washbasins
- Showers
- Urinals
- Restrooms
- Kitchen sinks
- Other equipment such as dishwashers and washing machines

Electrical System

- Lighting calculations for different areas based on illumination level recommended by Commission Internationale de l'Eclairage (CIE)/ Committee de Europeen Normalisation (CEN)/Chartered Institution of Building Services (CIBSE) and Isolux diagrams

- Selection of light fittings, type of lamps
- Exit/emergency lighting system
- Circuiting references, normal as well as emergency
- Power for wiring devices
- Power supply for equipment (HVAC, PH and FF, conveying system, and others)
- Sizing of conduits
- Sizing of cable tray
- Sizing of cable trunking
- Selection of wires and cables
- Voltage drop calculations for wires and cables
- Selection of upstream and downstream breakers
- Derating factor
- Sensitive of breakers (degree of protection)
- Selection of isolators
- Schedule of distribution boards, switchboards, and main low-tension boards
- Location of distribution boards, switch boards and low-tension panels
- Short circuit calculations
- Sizing of diesel generator set for emergency supply
- Sizing of transformers
- Schematic diagrams
- Substation layout
- IP ratings (degree of protection)
- Calculations for grounding (earthing) system
- Calculations for lightning protection system
- Interface with other trades

Fire Alarm System

A fire alarm system is designed taking into consideration the local fire code and NFPA standards. The system includes the following:

- Conduiting and raceways
- Type of system: analog/digital/addressable
- Type of detectors based on the area and spacing between the detectors and the walls
- Break glass/pull station

- Type of horns/bells
- Voice evacuation system, if required
- Type of wires and cables
- Mimic panel, if required
- Repeater panel, if required
- Main control panel
- Interface with other systems such as HVAC, elevator
- Riser diagram

Telephone/Communication System

- Structured cabling considering type and size of cable: copper, fiber optic
- Type and size of the cables
- Racks
- Wiring accessories/devices
- Access/distribution switches
- Internet switches
- Core switch
- Access gateway
- Router
- Network management system
- Servers
- Telephone handsets

Public Address System

- Conduiting and raceways
- Type of system: analog/digital/IP based
- Types of wires and cables
- Types of speakers
- Distribution of speakers
- Required noise level in different areas
- Calculations for sound pressure level
- Zoning of system, if required
- Size and type of premixer
- Size and type of amplifier
- Microphones
- Paging system

- Message recorder/player
- Interface with other systems

Audiovisual System

- Conduiting and raceways
- Type of system: analog/digital/IP based
- Types of wires and cables
- Racks
- Type, size, and brightness of projectors
- Type and size of speakers and sound pressure level
- Type and size of screens
- Microphones
- Cameras (visualizers)
- CD/DVD players–recorders
- Control processors
- Video switch matrix
- Mounting details of equipment

Security System/CCTV

- Type of system: digital/IP based
- Conduiting and raceways
- Wires and cabling network
- Level of security required
- Type and size of cameras
- Types of monitors/screens
- Video/event recording
- Video servers
- Database server
- System software
- Schematic diagram
- System console

Security System Access Control

- Conduiting and raceways
- Wires and cabling network
- Proximity radio frequency identification (RFID) reader
- Fingerprint and proximity combine reader

- Magnetic lock
- Release button
- Door contact
- RFID card
- Reader control panel
- Server
- Multiplexer
- Monitors
- Workstation
- Metal detector

Landscape

As a landscape architect, the following points are to be considered while designing the landscape system:

- Property boundaries
- Size and shape of the plot
- Shape and type of dwelling
- Integration with surrounding areas
- Orientation to the sun and wind
- Climatic/environmental conditions
- Ecological constraints (soil, vegetation, etc.)
- Location of pedestrian paths and walkways
- Pavement
- Garage and driveway
- Vehicular circulation
- Location of sidewalk
- Play areas and other social/community requirements
- Outdoor seating
- Location of services, positions of both under- and aboveground utilities and their levels
- Location of existing plants, rocks, or other features
- Site clearance requirements
- Foundation for paving, including front drive
- Top soiling or top soil replacement
- Soil for planting
- Planting of trees, shrubs, and ground covers
- Grass area

- Sowing grass or turfing
- Lighting poles/bollard
- Special features, if required
- Signage, if required
- Surveillance, if required
- Installation of irrigation system
- Marking out the borders
- Storage for landscape maintenance material

External Works (Infrastructure and Road)

External works are part of the contract requirements of a project that involves construction of a service road and other infrastructure facilities to be connected to the building and also includes care of existing services passing through the project boundary line. The designer has to consider the following while designing external works:

- Grading material
- Asphalt paving for road or street
- Pavement
- Pavement marking
- Precast concrete curbs
- Curbstones
- External lighting
- Cable routes
- Piping routes for water, drainage, storm water system
- Trenches or tunnels
- Bollards
- Manholes and hand holes
- Traffic marking
- Traffic signals
- Boundary wall/retaining wall, if required

Bridges

Designer should use relevant authorities' design manual and standards and consider the following points while designing bridges.

- Soil stability
- Alignment with road width, property lines
- Speed

- Intersections/interchanges
- Number of lanes, width
- Right-of-way lines
- Exits, approaches, and access
- Elevation datum
- Superelevation
- Clearance with respect to railroad, roadway, navigation (if applicable)
- High and low levels of water (if applicable)
- Utilities passing through the bridge length
- Slopes
- Number and length of span
- Live loads, bearing capacity
- Water load, wind load, earthquake effect (seismic effect)
- Bridge rails, protecting screening, guard rails, barriers
- Shoulder width
- Footings, columns, and piles
- Abutment
- Beams
- Substructure
- Superstructure, deck slab
- Girders
- Slab thickness
- Reinforcement
- Supporting components, deck hanger, tied arch
- Expansion and fixed joints
- Retaining walls, crash wall
- Drainage
- Lighting
- Aesthetic
- Sidewalk, pedestrian and bike facilities
- Signage, signals
- Durability
- Sustainability

Highways

Designer should use relevant authorities' design manual and standards and consider the following point while designing highways:

- Type of highway
- Soil stability
- Speed
- Number of lanes, width
- Shoulder width
- Gradation
- Type of pavement and thickness
- Right-of-way lines
- Exits, approaches, access, and ramp
- Superelevation
- Slopes, curvature, turning
- Median, barriers, curb
- Sidewalks, driveways
- Pedestrian accommodation
- Bridge roadway width
- Drainage
- Gutter
- Special conditions, such as snow and rain
- Lighting
- Signage, signals
- Durability
- Sustainability

Furnishings/Furniture (Loose)

In building construction projects, loose furnishings/furniture is tendered as a special package and is normally not the part of the main contract. In order to give sufficient information about the product, the descriptive features and specifications of the furnishing/furniture products are accompanied by a pictorial view/cutout sheet/photo of the product and the furniture layout. Figures 10.29 through 10.34 illustrate the detailed specifications for the furnishing of the director's and manager's room of an office building.

It is unlikely that the design of a construction project will be right in every detail the first time. Effective management and design professionals who are experienced and knowledgeable in the assigned task will greatly reduce the chances of error and oversight. However, so many aspects must be considered, especially for designs involving multiple disciplines and enfaces, and changes will be inevitable. The design should be reviewed taking into consideration requirements of all the disciplines before release of design drawings for a construction contract. Engineering design has significant

Director and office manager and advisor room layouts

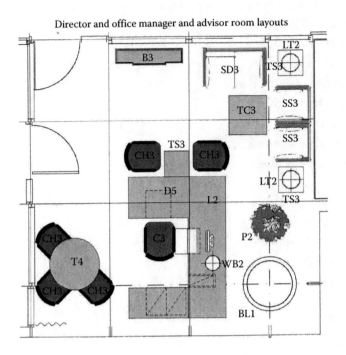

FIGURE 10.29
Room layout.

importance to the construction projects and must meet the customer's requirement at the start of project implementation. Engineering design has significant importance for construction projects. Engineering weakness can adversely impact the quality of design to such an extent that marginal changes can easily increase costs beyond the budget, which may affect schedule. Some areas are deemed critical to the proper design of a product; therefore, explicit design, material specification, and grades of the material specified in documentation have great importance. Most of the products used in construction projects are produced by other construction-related industries/manufacturers; therefore, the designer, while specifying the products, must specify related codes, standards, and technical compliance of these products.

CII Publication 10-1 (1989) has summarized that deviation costs averaged 12.4% of the total installed project cost; design deviation averaged 79% of the total deviation costs and 9.5% of the total project cost. Furthermore, design changes accounted for two-thirds of the design deviations. It has also given construction deviation averages, which are

- 17% of the total deviation costs
- 2.5% of the total installed project cost

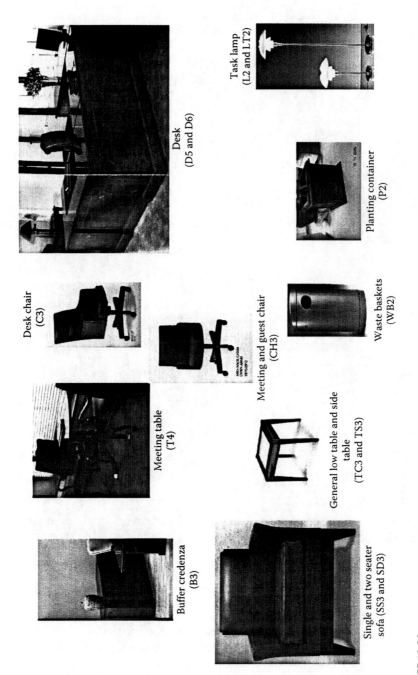

FIGURE 10.30
Furniture index.

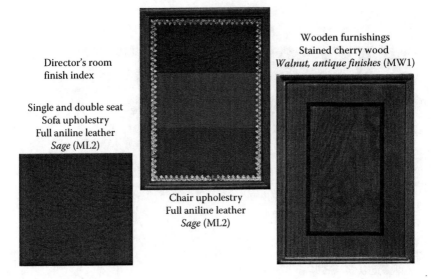

Director's room
finish index

Single and double seat
Sofa upholstery
Full aniline leather
Sage (ML2)

Chair upholstery
Full aniline leather
Sage (ML2)

Wooden furnishings
Stained cherry wood
Walnut, antique finishes (MW1)

FIGURE 10.31
Finishes index.

DESCRIPTION: Desk units			CODE: D5 & D6		QUANTITY:
LOCATION	RECOMMENDED MANUFACTURERS	OPTION ONE	OPTION TWO	OPTION THREE	OR APPROVED EQUAL
		HAWORTH	STEELCASE	KIMBALL	

DIMENSIONS: (as shown on the drawing)
D6: H 760 mm x D 914mm x W 2030mm, Credenza H760 mm x D 508mm x W 2030mm and Bridge: H 760mm x D 610mm x L 1200mm ;
D6: H 760 mm x D 914mm x W 1830mm, Credenza H760 mm x D 508mm x W 1830mm and Bridge: H 760mm x D 610mm x L 1200mm ;

SPECIFICATIONS / CONSTRUCTION: The contractor to check the number of right-handed and left-handed desk units prior to order. High grade executive desk with full front modesty panel made of kiln dried cherry wood with walnut antique finish as specified as per engineer's control sample with solid traditional moulding polished in walnut finish. The table top is constructed of 22mm high density marine grade MDF with stained walnut with three pencil drawers as shown in figure. The pattern along the top edge should match the wall paneling pattern and is to be approved prior to order and manufacture. The desk with bridge and credenza's top to have tanned aniline dyed leather inlay of color same as desk chair.

Two full fixed pedestal to be with box/ box/ box and the credenza to have four full fixed pedestals with box/ Be fixed below the pencil drawer and should also have back finished panel. The central drawer to have pencil tray. For detailed specifications, refer appendix C.

FINISHES:
Top to have leather inlay and all hardware and other accessories to be in chrome finish (brass finish for level 8 only). The contractor to submit different ranges of the specified color to get the approval and match engineer's control sample and prior to order. All exposed wood machine sanded; Pieces pre-stained to check for scratches; Pre-stain completely sanded off; lacquered on spray - catalyzed urethane or catalyzed conversion varnish topcoat. Wooden parts must be stained walnut, antique finish; to match engineer's sample prior to order.

REMARKS: This specification work formats is just design intent, must be read in conjunction with other schedules and all relevant drawings, all dimensions must be verified on site prior to order or manufacture of materials or production of shop drawings. The Contractor must fully coordinate with all other engineering works, especially the audio & visual works, to ensure proper installation. All necessary accessories, optional features, hardware including locks with keys etc , all to be included in order to make each item complete fully functional with maximum utility. The finishes required must be verified with respect to architectural interiors' wall paneling and coordinate its approval with respect to the engineer's control sample prior to order or manufacture

FIGURE 10.32
Specification for desk units.

It further states that design deviation related to construction projects are results of design errors and design omissions. Design errors are the result of mistakes or errors made in the project design. Design omissions result when a necessary item or component is omitted from the design. Design changes occur when changes are made in the project design or requirements. Table 10.10 shows the major causes of rework.

DESCRIPTION: Desk Lamp			CODE:		QUANTITY:
LOCATION	RECOMMENDED MANUFACTURERS	OPTION ONE	OPTION TWO	OPTION THREE	OR APPROVED EQUAL
		Louis Poulsen	W & D	PORTA ROMANA	

DIMENSIONS: (shape and size to be verified as per architectural interiors' drawings and prepared shop drawings at site)
H 472 mm x W 290mm

SPECIFICATIONS / CONSTRUCTION:
The shade is of blown glass with a top plate and base of flow formed high polished chrome. The stem is high luster steel plated brass. The desk lamp is fitted with a 2.6m black plastic cable with plug and toggle switch in the base plate. It should be of IP20 protection code and electric shock protection Class II. Light source 150 W .

FINISHES:
Transparent blown glass

All hardwares and other accessories to be in polished chrome finish (from level 3-7) & brass finish (for level 8).

The contractor to submit different ranges of colors to get the approval and match engineer's control sample and prior to order.

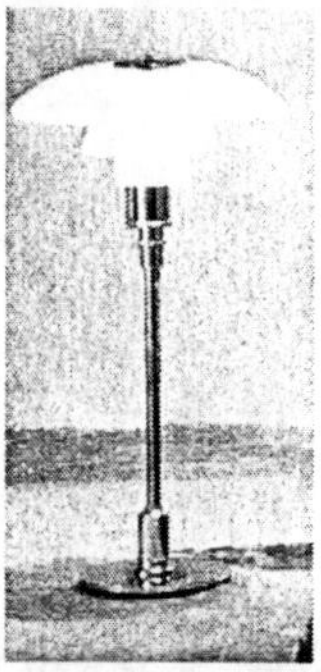

REMARKS: This specification work formats is just design intent, must be read in conjunction with other schedules and all relevant drawings, all dimensions must be verified on site prior to order or manufacture of materials or production of shop drawings. The Contractor must fully coordinate with all other engineering works, especially the audio & visual works, to ensure proper installation. All necessary accessories, optional features, hardwares including locks with keys etc., all to be included in order to make each item complete fully functional with maximum utility. The finishes required must be verified with respect to architectural interiors, wall paneling and coordinate its approval with respect to the engineer's control sample prior to order or manufacture.

FIGURE 10.33
Specification for desk lamp.

DESCRIPTION: Single and Two Seater Sofa			CODE:		QUANTITY:
LOCATION	RECOMMENDED MANUFACTURERS	OPTION ONE	OPTION TWO	OPTION THREE	OR APPROVED EQUAL
		David Edward			

DIMENSIONS:
SO3 : W 1350 X D 750 X H 810 MM
SS3: W 720 X D 750 X H 810 MM

SPECIFICATIONS / CONSTRUCTION: Executive hand craved family sofas fully upholstered seat and back, with closed wooden topped arms upholstered in full aniline leather. The frame is constructed of kiln dried hard cherry, Mortise / tenon on main joinery; in solid parts but stained in walnut, antique finish. The chair with carved legs. The seat and back are upholstered with leather approved by the engineer and are trim nailed with brass pins. Ultraflex seat construction stretched an additional 20% for firm seat; Back: high strength nylon weave, 2430 Propex, stretched & stapled to frame ; Polyurethane foam; All construction materials to meet CAL 117, Pre-upholstered muslin or bonded Dacron. Eight-way bed seat springs, pocket sprung back with frames all glued, screwed and doweled together. For additional specifications for wood, refer appendix C and for upholstery specification, refer appendix G. It consists of spring down seat cushion that is made of : an insulating box of high quality density polyurethane foam filled with coil springs individually wrapped and held in place by fabric pockets. The unit is inserted into a down-proof envelope of fabric ticking that is channeled into sections on the top and bottom. Each ticking section is filled with a special blend of down, feathers and polyester fiber to achieve luxurious softness and resilience. The down content is 25% and 75% feathers.

FINISHES: Seat and both sides of back are upholstered in fully in full aniline leather. The contractor to submit different ranges of colors to get the approval and match engineer's control sample and prior to order. All exposed wood machine sanded, Pieces pre-stained to check for scratches, Pre-stain completely sanded off , lacquered on spray - catalyzed urethane or catalyzed conversion varnish top coat. Wooden parts must be stained walnut, antique finish; to match engineer's sample prior to order.

GRAPHICAL REPRESENTATION

REMARKS: This specification work formats is just design intent, must be read in conjunction with other schedules and all relevant drawings, all dimensions must be verified on site prior to order or manufacture of materials or production of shop drawings. The Contractor must fully coordinate with all other engineering works, especially the audio & visual works, to ensure proper installation. All necessary accessories, optional features, hardwares including locks with keys etc., all to be included in order to make each item complete fully functional with maximum utility. The finishes required must be verified with respect to architectural interiors, wall paneling and coordinate its approval with respect to the engineer's control sample prior to order or manufacture.

FIGURE 10.34
Specification for sofa.

In order to reduce the rework resulting from quality deviation in design, CII Publication 10-1 (1989) has made the following recommendations:

1. Reduce the number of design changes by
   - Establishing definitive project scope
   - Performing periodic reviews with participation of all parties
   - Establishing procedures to limit scope modifications

TABLE 10.10

Major Causes of Rework, by Phase

Primary Cause		When Detected (Phase)			
Party and Type	Design	Procurement	Construction	Startup
Owner change	X	X	X	X
Designer error/omission	X	X	X	X
Designer change	X	X	X	X
Vendor error/omission	X	X	X	X
Vendor change	X	X	X	X
Constructor error/omission			X	
Constructor change			X	
Transporter error		X	X	X

Source: CII Publication 10-3. Reprinted with permission of CII, University of Texas.

2. Implement a quality management program that has total commitment at all levels of the firm

3. Adopt the standard set of quality-related terminology

4. Develop and implement a system that incorporates a database to identify deviation costs and quality problem areas

5. Implement a quality performance management system to identify costs associated with both quality management and correcting deviation costs

It is, therefore, necessary to have quality control personnel from the project team review and check the design for quality assurance using thorough itemized review checklists to ensure that design drawings fully meet the owner's objectives/goals. It is also necessary to review the design with the owner prior to initiation of work to ensure a mutual understanding of the build process. The design drawings should be fully coordinated with all the trades. The installation specification details are comprehensively and correctly described, and the installation quality requirements for systems are specified in detail.

Figure 10.35 illustrates the design data review cycle, which can be applied to review construction project design drawings. This process can be termed as continuous improvement of design.

Regulatory/Authorities' Approval

Government agency regulatory requirements have considerable impact on precontract planning. Some agencies require that the design drawings be submitted for their preliminary review and approval to ensure that the designs are compatible with local codes and regulations. These include

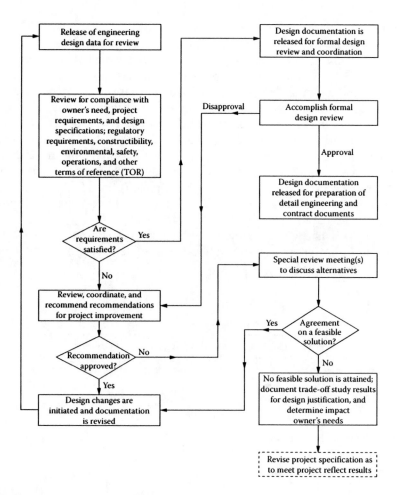

FIGURE 10.35

Design data review cycle. (Modified from B.S. Blanchard, W.J. Fabrycky, and J. Wolter. *Systems Engineering and Analysis*, 1998. Reprinted with permission from Pearson Education, Inc.)

submission of drawings to electrical authorities showing the antici-pated electrical load required for the facility, approval of fire alarm and fire fighting system drawings, and approval of drawings for water supply and drainage system. Technical details of the conveying system are also required to be submitted for approval from the concerned authorities.

Contract Documents and Specifications

Preparation of detailed documents and specifications as per master format is one of the activities performed during this phase of the construction project. The contract documents must specify the scope of works, location, quality, and duration for completion of the facility. As regards the technical specifications of the construction project, master format specifications are included in the contract documents. The master format is a master list of section titles and numbers for organizing information about construction requirements, products, and activities into a standard sequence. It is a uniform system for organizing information in project manuals, for organizing cost data, for filling product information and other technical data, for identifying drawing objects, and for presenting construction market data. MasterFormat™ (1995 edition) consisted of 16 divisions; however, MasterFormat (2004 edition) consists of 48 divisions (49 is reserved). MasterFormat contract documents produced jointly by the Construction Specifications Institute (CSI) and Construction Specifications Canada (CSC) are widely accepted as standard practice for preparation of contract documents.

Table 10.11 lists division numbers and titles of MasterFormat 2004 published by the Construction Specifications Institute and Construction Specifications Canada.

TABLE 10.11

MasterFormat 2004 Division Numbers and Titles

Procurement and Contracting Requirements Group Division 00 Procurement and Contracting Requirements	
Specifications Group	
General Requirements Subgroup	
Division 01	General requirements
Facility Construction Subgroup	
Division 02	Existing conditions
Division 03	Concrete
Division 04	Masonry
Division 05	Metals
Division 06	Wood, plastics, and composites
Division 07	Thermal and moisture protection
Division 08	Openings
Division 09	Finishes
Division 10	Specialties
Division 11	Equipment
Division 12	Furnishings
Division 13	Special construction
Division 14	Conveying equipment
Division 15	Reserved
Division 16	Reserved
Division 17	Reserved
Division 18	Reserved
Division 19	Reserved

TABLE 10.11 (continued)

MasterFormat 2004 Division Numbers and Titles

	Procurement and Contracting Requirements Group Division 00 Procurement and Contracting Requirements

Facility Services Subgroup

Division 20	Reserved
Division 21	Fire suppression
Division 22	Plumbing
Division 23	Heating, ventilation, and air conditioning
Division 24	Reserved
Division 25	Integrated automation
Division 26	Electrical
Division 27	Communication
Division 28	Electronic safety and security
Division 29	Reserved

Site and Infrastructure Subgroup

Division 30	Reserved
Division 31	Earthwork
Division 32	Exterior improvements

Site and Infrastructure Subgroup

Division 33	Utilities
Division 34	Transportation
Division 35	Waterway and marine

Construction

Division 36	Reserved
Division 37	Reserved
Division 38	Reserved
Division 39	Reserved

Process Equipment Subgroup

Division 40	Process integration
Division 41	Material processing and handling equipment
Division 42	Process heating, cooling, and drying equipment
Division 43	Process gas and liquid handling, purification, and storage equipment
Division 44	Pollution control equipment
Division 45	Industry-specific manufacturing equipment
Division 46	Reserved
Division 47	Reserved
Division 48	Electric power generation
Division 49	Reserved

Source: The Construction Specifications Institute and Construction Specifications Canada. Reprinted with permission from CSI.

Particular specifications consist of many sections related to a specific topic. Detailed requirements are written in these sections to enable the contractor understand the product or system to be installed in the construction project. The designer has to interact with the project team members and owner while preparing the contract documents.

Typical sections are as follows:

Section No.
Title

Part 1—General

    1.01—General reference/related sections

    1.02—Description of work

    1.03—Related work specified elsewhere in other sections

    1.04—Submittals

    1.05—Delivery, handling, and storage

    1.06—Spare parts

    1.07—Warranties

In addition to the foregoing, a reference is made for items such as preparation of mock-up, quality control plan, and any other specific requirements related to the product or system specified herein.

Part 2—Product

    2.01—Materials

    2.02—List of recommended manufacturers

Part 3—Execution

    3.01—Installation

    3.02—Site quality control

Shop Drawing and Materials Submittals

The detailed procedure for submitting shop drawings, materials, and samples is specified under the section titled "Submittal" of contract specifications. The contractor has to submit the specifications to the owner/consultant for review and approval. The following are the details of preparation of shop drawings and materials.

A—Shop Drawings

The contractor is required to prepare shop drawings taking into account the following partial list of considerations:

1. Reference to contract drawings. This helps the A&E (consultant) to compare and review the shop drawing with the contract drawing
2. Detailed plans and information based on the contract drawings
3. Notes of changes or alterations from the contract documents
4. Detailed information about fabrication or installation of works
5. Verification of all dimensions at the job site
6. Identification of product
7. Installation information about the materials to be used
8. Type of finishes, color, and textures
9. Installation details relating to the axis or grid of the project
10. Roughing in and setting diagram
11. Coordination certification from all other related trades (subcontractors)

The shop drawings are to be drawn accurately to scale and shall have project-specific information in it. They should not be reproductions of contract drawings.

Immediately after approval of individual trade shop drawings, the contractor has to submit builder's workshop drawings, composite/coordinated shop drawings taking into consideration the following at a minimum.

A1—Builder's Workshop Drawings

Builder's workshop drawings indicate the openings required in the civil or architectural work for services and other trades. These drawings indicate the size of openings, sleeves, and level references with the help of detailed elevation and plans.

A2—Composite/Coordination Shop Drawings

The composite drawings indicate the relationship of components shown on the related shop drawings and indicate the required installation sequence. Composite drawings should show the interrelationship of all services with one another and with the surrounding civil and architectural work. Composite drawings should also show the detailed coordinated cross sections, elevations, reflected plans, and so on, resolving all conflicts in levels, alignment, access, space, and so on. These drawings are to be prepared taking into consideration the actual physical dimensions required for installation within the available space.

B—Materials

Similarly, the contractor has to submit the following, at a minimum, to the owners/consultants to get their review and approval of materials, products, equipment, and systems. The contractor cannot use these items unless they are approved for use in the project.

B1—Product Data

The contractor has to submit the following details:

- Manufacturer's technical specifications related to the proposed product
- Installation methods recommended by the manufacturer
- Relevant sheets of manufacturer's catalogs
- Confirmation of compliance with recognized international quality standards
- Mill reports (if applicable)
- Performance characteristic and curves (if applicable)
- Manufacturer's standard schematic drawings and diagrams to supplement standard information related to project requirements and configuration of the same to indicate product application for the specified works (if applicable)
- Compatibility certificate (if applicable)
- Single-source liability (this is normally required for systems approval when different manufacturers' items are used)

B2—Compliance Statement

The contractor has to submit a specification comparison statement along with the material transmittal.

The consultant reviews the transmittals and action as follows:

a. Approved
b. Approved as noted
c. Approved as noted, resubmit
d. Not approved

In certain projects, the owner is involved in the approval of materials.

In case of any deviation from specifications, the contractor has to submit a schedule of such deviations listing all the points not conforming to the specification.

B3—Samples

The contractor has to submit (if required) the samples from the approved material to be used for the work. The samples are mainly required to

- Verify color, texture, and pattern
- Verify that the product is physically identical to the proposed and approved material
- Comparison with products and materials used in the works

At times it may be specified to install the samples in such a manner as to facilitate review of qualities indicated in the specifications.

Contractor's Quality Control Plan

The contract documents specify the details of the contents of the quality control plan (QCP) to be prepared by the contractor for the construction project; the plan has to be submitted to the consultant for approval. The following is the outline for the preparation of a QCP:

1. Purpose of the QCP
2. Project description
3. Site staff organization chart for quality control
4. Quality control staff and their responsibilities
5. Construction program and subprograms
6. Schedule for submission of subcontractors, manufacturer of materials, and shop drawings
7. QC procedure for all the main activities such as
   a. Procurement (direct bought out items)
   b. Off-site manufacturing, inspection, and testing
   c. Inspection of site activities (checklists)
   d. Inspection and testing procedure for systems
   e. Procedure for laboratory testing of material
   f. Inspection of material received at site
   g. Protection of works
8. Method statement for various installation activities
9. Project-specific procedures for site work instructions, and remedial notes
10. Quality control records
11. List of quality procedures applicable to project in reference to the company's quality manual and procedure
12. Periodical testing procedure for construction equipment and tools
13. Quality updating program
14. Quality auditing program
15. Testing

16. Commissioning

17. Handover

18. Site safety

Specifications of work quality are an important feature of construction project design. Specifications of required quality and components represent part of the contract documents and are detailed under various sections of particular specifications. Generally, the contract documents include all the details as well as references to generally accepted quality standards published by international standards organizations. Proper specifications and contract documentations are extremely important as these are used by the contractor as a measure of quality compliance during the construction process.

A contract for construction commits the contractor to construct the facility and the owner to pay. Once the contract is signed, it commits all the parties to obligations and liabilities and is enforceable by law. A breach of contract by either party may make that party liable for payment of damages to the other.

There are standard sets of conditions of contract published by engineering institutes/societies and other bodies. Depending on the need for the construction project and the type of contract arrangements, an appropriate set of condition of contracts is selected. The contract document must include health and safety programs to be followed by the contractor during the construction process.

Detailed Plan

As per PMBOK, a project plan is used to

- Guide project execution
- Document project planning assumptions
- Document project planning decisions regarding alternatives chosen
- Define key management views regarding content, scope, and timing
- Provide a baseline for progress measurement and project control

A project plan is a formal, approved document used to manage project execution. It is the evaluation of time and effort to complete the project. Based on the detailed engineering and design drawings and contract documents, the design team (consultant) prepares a detailed plan for construction. The plan is based on the following:

- Assessment of owner's capabilities and final estimated cost (budget)
- Scheduling information
- Resource management, which includes availability of financial resources, expected cash flow statement, supplies, and human resources

A typical preliminary work program prepared based on the contracted construction documents is illustrated in Figure 10.36.

Preliminary work program

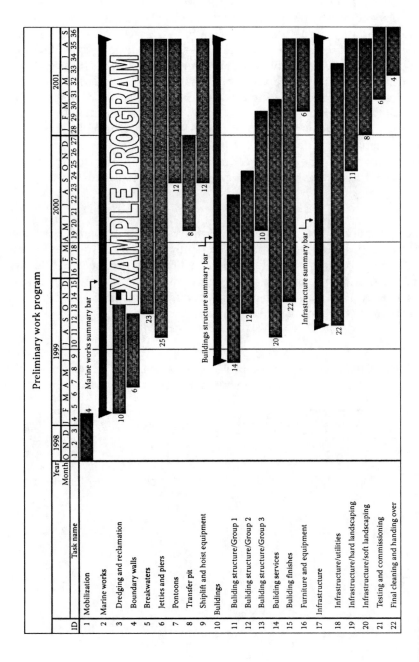

FIGURE 10.36
Preliminary work program.

Budget

The budget for a project is the maximum amount of money the owner is willing to spend for its design and construction. The preparation of a budget is an important activity that results in a time-phased plan summarizing expected expenditure, income, and milestones. Normally, project budgeting starts with the identification of need; however, the detailed cost estimate is done during the engineering phase. On the basis of work packages, the consultant/designer starts computing the project budget. A bill of material or bill of quantities is prepared based on the approved design drawings. The BOQ is considered as a base for computing the budget. If the budget exceeds the owner's capability of financing the project, then the designs are reviewed to ensure that it meets the owner's estimated cost to build the facility.

Figure 10.37 illustrates a project S-curve for a building construction project.

Cash Flow

The estimate of cash flow requirement for the project is prepared from the preliminary estimate and preliminary work program. An accurate cash flow projection helps owners plan the payments on time as per the schedule for the project. A simple cash flow projection based on prior planning helps owners make available all the resources required from their side. Cash flow is used as part of the control package during construction.

Tender/Bidding

Most of the cost of the construction project is expended during the construction phase. In most cases, the contractor is responsible for procurement of

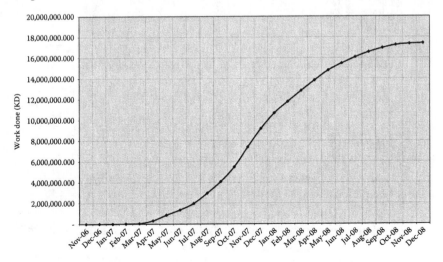

FIGURE 10.37
Project S-curve.

TABLE 10.12

Contract Documents

Document No. (I) tendering procedures consisting of the following:
I.1 Tendering invitation
I.2 Instructions for tenderers
I.3 Form of tender and appendix
I.4 Initial bond (form of bank guarantee)
I.5 Performance bond (form of bank guarantee)
I.6 Form of agreement
I.7 List of tender documents
I.8 Declaration No. (1)
Document No. (II) contract conditions consisting of the following:
II.1 General conditions (legal clauses and conditions 1971 [May 1985 edition] and amendments until closing date of tender)
II.2 Particular conditions
II.3 Kuwait tender law (currently valid)
Document No. (III) technical conditions and amendments, consisting of the following:
III.1 General specifications for building and engineering works, specific to ministry of public works. 1990 edition and all amendments
III.2 Particular specifications
III.3 Drawings
III.4 Bills of quantities
III.5 Price analysis schedule
III.6 Addenda (if any)
III.7 Technical requirements (if any), and any other instructions issued by the employer

all the material, providing construction equipment and tools, and supplying the manpower to complete the project in compliance with the contract documents.

Table 10.12 lists contract documents consisting of tendering procedures, contract conditions, and technical conditions of major construction projects in Kuwait.

In many countries, it is a legal requirement that government-funded projects employ the competitive bidding method. This requirement gives an opportunity to all qualified contractors to participate in the tender, and normally the contract is awarded to the lowest bidder. Private-funded projects have more flexibility in evaluating the tender proposal. Private owners may adopt the competitive bidding system, or the owner may select a specific contractor and negotiate the contract terms. Negotiated contract systems have flexibility of pricing arrangement as well as the selection of the contractor based on his expertise or the owner's past experience with the contractor successfully completing one of his or her projects. The negotiated contract systems are based on following forms of payment:

1. Cost plus contracts: It is a type of contract in which the contractor agrees to do the work for the cost of time and material, plus an

agreed-upon amount of profit. The following are the different types of cost plus contracts:

   a. Cost plus percentage fee contract

   b. Cost plus fixed fee contract

   c. Cost plus incentive fee contract

2. Reimbursement contracts: It is a type of contract in which the contractor agrees to do the work for the cost as per schedule of rates, or bill of quantities, or bill of material.

3. Fixed price contracts: With this type of contract, the contractor agrees to work with a fixed price (it is also called *lump sum*) for the specified and contracted work. Any extra work is executed only upon receipt of instruction from the owner. Fixed price contracts are generally inappropriate for work involving major uncertainties, such as work involving new technologies.

4. Target cost contracts: A target cost contract is based on the concept of a top-down approach, which provides a fixed price for an agreed range of out-turn costs around the target. In this type of contract, overrun or underspend is shared by the owner and the contractor at predetermined agreed-upon percentages.

5. Guaranteed maximum price contracts (cost plus guaranteed maximum price): With this type of contract, the owner and contractor agree to a project cost guaranteed by the contractor as maximum.

It is the owner's desire that his or her facility be of good quality and the price reasonable. In order to achieve this, the owner has to share risks and/or provide incentives and safeguards to enhance the quality of construction. The risk involved in various types of contracts based on forms of payment is as follows:

1. Cost plus—high risk

2. Reimbursement—intermediate

3. Fixed price—low risk

In order to maintain a climate of mutual cooperation during construction, the owner has to develop an understanding with the contractor. The contract needs to be adapted through mutual agreement with the contractor. The contract strategy needs to provide incentives and safeguards to deal with the risks.

Turner (2003) has suggested a twofold methodology in contract selection. The aim is

1. To develop a cooperative project organization

2. To appropriately allocate resources

Turner (2003) has further described that, as per Oliver Williamson (1995, 1996), there are two schemas or vectors to describe the ability of contracts to provide ex-ante incentivization and flexible, farsighted ex-post governance. The first schema has three parameters:

- The reward it provides the contractor to share the owner's objectives and perform
- The associated risk
- The safeguard provided by the owner through contract to shield the contractor from the risk

Figure 10.38 illustrates a sample contractual schema for ex-ante incentivization. Although the schema in the figure assumes a safeguard risk, it can deal with only a risk that is foreseen.

Williamson has further proposed four parameters to describe the ability of a contract form to provide flexible, farsighted, ex-post governance:

1. The incentive intensity
2. The ease of making uncontested, bilateral adaptations to contract
3. The reliance on monitoring and related administrative controls (transaction costs)
4. The reliance on court ordering

Risk	Present	High incentive	Medium incentive
	Absent	Low incentive	No safeguard or incentive necessary
		Absent	Present
		Safeguard	

FIGURE 10.38
A sample contractual schema for ex-ante incentivization. (From J. R. Turner, *Contracting for Project Management*, 2003. Reprinted with permission from Ashgate Publishing Limited, Farnham, UK.)

Contract form	Reward	Risk	Safeguard
Cost plus			
Cost +%fee	High but misaligned	High	High
Cost+ fixed fee	Medium but misaligned	High	High
Cost + incentive fee	Medium	High	High
Alliance	Medium	High	Medium
Remeasurement			
r-sor	Low and misaligned	Low	High
r-boq	Low	Medium	Medium
r-bom	Low	Medium	Low
Fixed price			
Build only	Low	Low	Low
Specification a	Low	Low	Low
Specification b	Medium	Medium	Low
Cardinal points	High	High	Low (insurance)
Other			
Target price	Medium	Medium	Medium
Time and materials	High	High	Low
Budget or gmp			
Routine contracts			
Market	High	Low	Low
Hierarchy	Low	High	High

FIGURE 10.39

Contract forms and ex-ante incentivization. r, remeasurement based; sor, schedule of rates; boq, bill of quantities; bom, bill of materials. (From J.R. Turner, *Contracting for Project Management*, 2003. Reprinted with permission from Ashgate Publishing Limited, Farnham, UK.)

The incentive profiles of the contract types are summarized in Figure 10.39, and the governance profiles in Figure 10.40.

In the case of a competitive bidding system, it is necessary that the detailed design and specifications for the project be prepared by the designer for bidding purposes. Under the competitive bidding system, normally there are four stages in tendering of a construction project:

1. Selection of tenderer (prequalification)
2. Invitation to bid
3. Tender preparation and submission
4. Appraisal of tenders, negotiation, and decision

For most construction projects, selection of a tenderer is based on the lowest tender price. Tenders received are opened and evaluated by the owner/owner's representative. Normally, tender results are declared in the official gazette or by some sort of notifications. The successful tenderer is informed

Contract form	Incentive intensity	Adaptiveness	Transaction cost	Safeguard
Cost plus				
Cost + %fee	Misaligned	High	High	Low
Cost + fixed fee	Low	High	High	Low
Cost + incentive fee	Medium	High	High	Low
Alliance	High	High	High	Low
Remeasurement				
r-sor	Misaligned	Medium	High	Low
r-boq	Low	Medium	Medium	Low
r-bom	Low	Medium	Low	Low
Fixed price				
Build only	Low	Low	High	High
Specification a	Low	Low	High	High
Specification b	Medium	Medium	Medium	Medium
Cardinal points	High	High	Low	Low
Other				
Target price	Medium	Low	High	Medium
Time and materials	Low	Low	High	High
Budget or gmp				
Routine contracts				
Market	High	Low	Low	High
Hierarchy	Low	High	High	Low

FIGURE 10.40

Contract forms and flexible, farsighted, ex-post governance. r, remeasurement based; sor, schedule of rates; boq, bill of quantities; bom, bill of materials. (From J.R. Turner, *Contracting for Project Management*, 2003. Reprinted with permission from Ashgate Publishing Limited, Farnham, UK.)

of the acceptance of the proposal and is invited to sign the contract. The tenderer has to submit the performance bond before the formal contract agreement is signed. If a successful tenderer fails to submit the performance bond within the specified period or withdraws his tender, then the contractor loses the initial bond and may be subjected to other regulatory applicable conditions.

The signing of contract agreement between the owner/owner's representative and the contractor binds both parties to fulfill their contractual obligations.

Construction

Construction is the translation of the owner's goals and objectives into a facility built by the contractor as stipulated in the contract documents, plans, and specifications within budget and on schedule. The construction phase is an important phase in construction projects. A majority of total project budget and schedule is expended during construction. Similar to costs, the time required for the construction phase of the project is much higher than the time required for the preceding phases. Construction usually requires a large workforce and a variety of activities. Construction activities involve erection, installation, or construction of any part of the project. Construction activities are actually carried out by the contractor's own workforce or by subcontractors. Construction therefore requires more detailed attention of its planning, organizations, monitoring and control of project schedule, budget, quality, safety, and environment concerns.

Oberlender (2000) has described the importance of construction in the following words:

> The construction phase is important because the quality of the completed project is highly dependent on the workmanship and management of construction. The quality of construction depends on the completeness and quality of the contract documents that are prepared by the designer and three other factors: laborers who have the skills necessary to produce the work, field supervisors who have the ability to coordinate the numerous activities that are required to construct the project in the field, and the quality of materials that are used for construction of project. Skilled laborers and effective management of the skilled laborers are both required to achieve a quality project. (p. 258)

It is usual to invite contractors to compete for a contract for construction work, in the expectation that they will plan to do the work efficiently and therefore at a minimum cost. Once the contract is awarded to the successful bidder (contractor), it is the responsibility of the contractor to respond to the needs of the client (owner) by building the facility as specified in the contract documents, drawings, and specifications within the budget and on time.

The owner also appoints an engineer to supervise the construction process. It is a normal practice for the designer/consultant of the project to be contracted by the owner to supervise the construction process. The engineer is responsible for achieving project quality goals and is also responsible for implementing the procedures specified in the contract documents. Table 10.13 lists the responsibilities the owner delegates to the engineer.

Sometimes the owner engages a construction manager or project manager during the construction process to act as the owner's representative and delegates the following activities, thus leaving the engineer to perform project quality-related work only:

1. Review of contract documents
2. Approval of contractor's construction schedule
3. Cost control
4. Time control
5. Project methodology

The engineer appoints an engineer's representative to supervise the project construction process. The engineer's representative is supported by a supervision team consisting of professionals having experience and expertise in

TABLE 10.13

Responsibilities of Supervision Consultant

Serial Number	Description
1	Achieving the quality goal as specified
2	Review contract drawings and resolve technical discrepancies/errors in the contract documents
3	Review construction methodology
4	Approval of contractor's construction schedule
5	Regular inspection and checking of executed works
6	Review and approval of construction materials
7	Review and approval of shop drawings
8	Inspection of construction material
9	Monitoring and controlling construction expenditure
10	Monitoring and controlling construction time
11	Maintaining project record
12	Conduct progress and technical coordination meetings
13	Coordination of owner's requirements and comments related to site activities
14	Project-related communication with contractor
15	Coordination with regulatory authorities
16	Processing of site work instruction for the owner's action
17	Evaluation and processing of variation order/change order
18	Recommendation of contractor's payment to owner
19	Evaluating and making decisions related to unforeseen conditions
20	Monitor safety at site
21	Supervise testing, commissioning, and handover of the project
22	Issue substantial completion certificate

supervision and administration of similar construction projects. The engineer's representative is also called the resident engineer. Depending on the type and size of the project, the supervision team usually consists of the following personnel:

1. Resident engineer
2. Contract administrator/quantity surveyor
3. Planning/scheduling engineer
4. Engineers from different trades such as architectural, structural, mechanical, HVAC, electrical, low-voltage system, landscape, infrastructure
5. Inspectors from different trades
6. Interior designer
7. Document controller
8. Office secretary

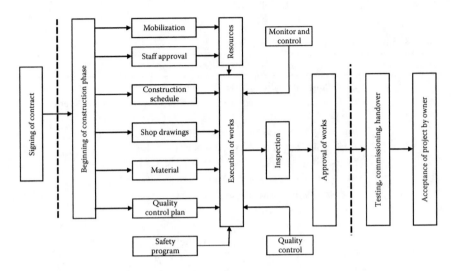

FIGURE 10.41
Major activities during the construction phase.

The construction phase consists of various activities such as mobilization, execution of work, planning and scheduling, control and monitoring, management of resources/procurement, quality, and inspection. Figure 10.41 illustrates major activities to be performed during the construction phase.

These activities are performed by various parties having contractual responsibilities to complete the specified work. Coordination among these parties is essential to ensure that the constructed facility meets the owner's objectives.

Mobilization

The contractor is given a few weeks to start the construction work after the signing of the contract. The activities to be performed during the mobilization period are defined in the contract documents. During this period, the contractor is required to perform many of the activities before the beginning of actual construction work at the site. Necessary permits are obtained from the relevant authorities to start the construction work. After being granted access to the construction site by the owner, the contractor starts mobilization work, which consists of preparation of site offices/field offices for the owner, the supervision team (consultant), and the contractor himself. This includes all the necessary on-site facilities and services necessary to carry out specific tasks. Mobilization activities usually occur at the beginning of a project but can occur anytime during a project when specific on-site facilities are required.

Examples of mobilization activities include

- Set up site offices and storage
- Construct temporary access roads, lay down areas and perimeter fences
- Install the necessary utilities for construction
- Set up a temporary firefighting system
- Perform site survey and testing
- Satisfy health and site safety requirements
- Submit preliminary construction program
- Selection of core staff as mentioned in the contract documents
- Insurance policies
- Selection of subcontractor (this may be an ongoing activity per the approved schedule)

In anticipation of the award of contract, the contractor begins the following activities much in advance, but these are part of contract documents, and the contractor's action is required immediately after signing of the contract in order to start construction:

- Mobilization of construction equipment and tools
- Workforce to execute the project

For a smooth flow of construction process activities, proper communication and submittal procedure need to be established among all concerned parties at the beginning of the construction activities. Table 10.14 illustrates an example matrix for site administration of a building construction project.

Proper adherence to these duties helps a smooth implementation of the project. Correspondence between consultant and contractor is normally through letters or job site instructions. Figure 10.42 is a job site instruction form used by the consultant to communicate with the contractor.

Execution of Works

According to ICE (1996):

> The contractor is responsible for construction and maintaining the works in accordance with the contract drawings, specifications and other documents and also further information and instruction issued in accordance with the contract. The contractor should be as free as possible to plan and execute the works in the way he wishes within the terms of his contract, so should the sub-contractors. Any requirements for part of a project to be finished before the rest and all limits of contractor's freedom should therefore have been stated in the tender document. (p. 77)

TABLE 10.14

Matrix for Site Administration and Communication

Serial Number	Description of Activities	Contractor	Consultant	Owner
1	Communication			
	1.1 General correspondence	P	P	P
	1.2 Job site instruction	D	P	C
	1.3 Site works instruction	D	P/B	A
	1.4 Request for information	P	A	C
	1.5 Request for modification	P	B	A
2	Submittals			
	2.1 Subcontractor	P	B/R	A
	2.2 Materials	P	A	C
	2.3 Shop drawings	P	A	C
	2.4 Staff approval	P	B	A
	2.5 Premeeting submittals	P	D	C
3	Plans and programs			
	3.1 Construction schedule	P	R	C
	3.2 Submittal logs	P	R	C
	3.3 Procurement logs	P	R	C
	3.4 Schedule update	P	R	C
4	Monitor and control			
	4.1 Progress	D	P	C
	4.2 Time	D	P	C
	4.3 Payments	P	R/B	A
	4.4 Variations	P	R/B	A
	4.5 Claims	P	R/B	A
5	Quality			
	5.1 Quality control plan	P	R	C
	5.2 Checklists	P	D	C
	5.3 Method statements	P	A	C
	5.4 Mock up	P	A	B
	5.5 Samples	P	A	B
	5.6 Remedial notes	D	P	C
	5.7 Nonconformance report	D	P	C
	5.8 Inspections	P	D	C
	5.9 Testing	P	A	B
6	Site safety			
	6.1 Safety program	P	A	C
	6.2 Accident report	P	R	C
7	Meetings			
	7.1 Progress	E	P	E
	7.2 Coordination	E	P	C
	7.3 Technical	E	P	C
	7.4 Quality	P	C	C

TABLE 10.14 (continued)

Matrix for Site Administration and Communication

Serial Number	Description of Activities	Contractor	Consultant	Owner
	7.5 Safety	P	C	C
	7.6 Closeout		P	
8	Reports			
	8.1 Daily report	P	R	C
	8.2 Monthly report	P	R	C
	8.3 Progress report		P	A
	8.4 Progress photographs		P	A
9	Close out			
	9.1 Snag list	P	P	C
	9.2 Authorities' approvals	P	C	C
	9.3 As-built drawings	P	D/A	C
	9.4 Spare parts	P	A	C
	9.5 Manuals and documents	P	R/B	A
	9.6 Warranties	P	R/B	A
	9.7 Training	P	C	A
	9.8 Handover	P	B	A
	9.9 Substantial completion certificate	P	B/P	A

P, prepare/initiate; B, advise/assist; R, review/comment; A, approve; D, action; E, attend; C, information.

Construction activities mainly consist of the following:

- Site work such as cleaning and excavation of project site
- Construction of foundations, including footings and grade beams
- Construction of columns and beams
- Forming, reinforcing, and placing of the floor slab
- Laying up masonry walls and partitions
- Installation of roofing system
- Finishes
- Furnishings
- Conveying system
- Installation of firefighting system
- Installation of water supply, plumbing, and public health system
- Installation of heating, ventilation, and air-conditioning system
- Integrated automation system
- Installation of electrical lighting and power system
- Emergency power supply system

Project Name
Consultant Name

JOB SITE INSTRUCTION (JSI)

CONTRACTOR: _____   JSI No. : _____

CONTRACT No.: _____   DATE : _____

The work shall be carried out in accordance with the Contract Documents without change in Contract Sum or Contract Time. Proceeding with the work in accordance with these instructions indicates your acknowledgement that there will be no change in the Contract Sum or Contract Time.

Subject:

_____

SAMPLE FORM

_____
_____
_____
_____
_____

ATTACHMENTS: (List attached documents that support description.)

_____
_____
_____
_____
_____
_____

Signed:	Received by Contractor :
Resident Engineer	Date:

Distribution: ☐ Owner   ☐ A/E   ☐ Contractor

FIGURE 10.42
Job site instruction.

- Fire alarm system
- Communication system
- Electronic security and access control system
- Landscape works
- External works

Planning and Scheduling

Project planning is a logical process to ensure that the work of the project is carried out

- In an organized and structured manner
- Reducing uncertainties to a minimum
- Reducing risk to a minimum
- Establishing quality standards
- Achieving results within budget and scheduled time

Prior to the start of execution of a project or immediately after the actual project starts, the contractor prepares the project construction plans based on the contracted time schedule of the project. Detailed planning is needed at the start of construction to decide how to use resources such as laborers, plant, materials, finance, and subcontractors economically and safely to achieve the specified objectives. The plan shows the periods for all sections of the works and activities, indicating that everything can be completed by the date specified in the contract and ready for use or for installation of equipment by other contractors.

According to Oberlender (2000), "Project Planning is the heart of good project management because it provides the central communication that coordinates the work of all parties. Planning also establishes the benchmark for the project control system to track the quantity, cost, and timing of work required to successfully complete the project. Although the most common desired result of planning is to finish the project on time, there are other benefits that can be derived from good project planning" (p. 140).

Effective project management requires planning, measuring, evaluating, forecasting, and controlling all aspects of project quality and quality of work, cost, and schedules. The purpose of the project plan is to successfully control the project to ensure completion within the budget and schedule constraints. Project planning is the evolution of the time and efforts to complete the project. Table 10.15 lists the benefits of project planning and scheduling.

Planning is a mechanism that conveys or communicates to project participants what activity is to be done, how, and in what order to meet the project objectives by scheduling the activities. Project planning is required to bring the project to completion on schedule, within budget, and in accordance with the owner's needs as specified in the contract. The planning process considers all the individual tasks, activities, or jobs that make up the project and must be performed. It takes into account all the resources available, such as human resources, finances, materials, plant, equipment, and so on. It also considers the works to be executed by the subcontractors.

TABLE 10.15

Benefits of Project Planning and Scheduling

1. Finish the project on time
2. Continuous (uninterrupted) flow of work (no delays)
3. Reduced amount of rework (least amount of changes)
4. Minimize confusion and misunderstandings
5. Increased knowledge of status of project by everyone
6. Meaningful and timely reports to management
7. You run the project instead of the project running you
8. Knowledge of scheduled times of key parts of the project
9. Knowledge of distribution of costs of the project
10. Accountability of people, defined responsibility/authority
11. Clear understanding of who does what, and how much
12. Integration of all work to ensure a quality project for the owner

Source: G. D. Oberlender, 2000. *Project Management for Engineering.* Reprinted with permission of The McGraw-Hill Companies.

The following is the list of activities of construction projects normally included in the construction program/plan:

A. General activities

1. Mobilization

B. Engineering

1. Subcontractor submittal and approval

2. Materials submittal and approval

3. Shop drawing submittal and approval

4. Procurement

C. Site activities

1. Site earthworks

2. Dewatering and shoring

3. Excavation and backfilling

4. Raft works

5. Retaining wall works

6. Concrete foundation and grade beams

7. Waterproofing

8. Concrete columns and beams

9. Casting of slabs

10. Wall partitioning

11. Interior finishes

12. Furnishings

13. External finishes
14. Equipment
15. Conveying systems works
16. Plumbing and public health works
17. Firefighting works
18. HVAC works
19. Electrical works
20. Fire alarm system works
21. Communication system works
22. Low-voltage systems works
23. Landscape works
24. External works

D. Close out
1. Testing and commissioning
2. Completion and handover

The contractor also submits the following:

1. Resources (equipment and manpower) schedule
2. Cost loading (schedule of item's pricing based on bill of quantities)

Planning and scheduling are often used synonymously for preparing a construction program because both are performed interactively. Planning is the process of identifying the activities necessary to complete the project, while scheduling is the process of determining the sequential order of the planned activities and the time required to complete the activity. Scheduling is the mechanical process of formalizing the planned functions, assigning the starting and completion dates to each part or activity of the work in such a manner that the whole work proceeds in a logical sequence and in an orderly and systematic manner.

The first step in preparation of a construction program is to establish the activities, and the next step is to establish the estimated time duration of each activity. The deadline for each activity is fixed, but it is often possible to reschedule by changing the sequence in which the tasks are performed, while retaining the original estimated time. Figure 10.43 illustrates the steps in project planning.

Construction projects are unique and nonrepetitive in nature. Construction projects consist of many activities aimed at the accomplishment of a desired objective. In order to achieve the quality objectives of the project, each activity has to be completed within the specified limit, using the specified product and approved method of installation. A construction project consists of

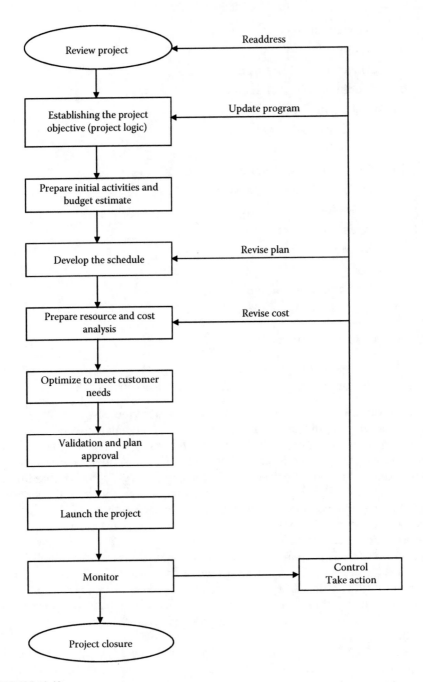

FIGURE 10.43
Project planning steps.

TABLE 10.16

Key Principles for Planning and Scheduling

1. Begin planning before starting work, rather than after starting work
2. Involve people who will actually do the work in the planning and scheduling process
3. Include all aspects of the project: scope, budget, schedule, and quality
4. Build flexibility into the plan, include allowance for changes and time for reviews and approvals
5. Remember the schedule is the plan for doing the work, and it will never be precisely correct
6. Keep the plan simple, eliminate irrelevant details that prevent the plan from being readable
7. Communicate the plan to parties; any plan is worthless unless it is known

Source: G. D. Oberlender, 2000. *Project Management for Engineering.* Reprinted with permission of The McGraw-Hill Companies.

a number of related activities that are dependent on other activities and cannot be started until others are completed, and some that can run in parallel. The most important point while starting the planning is to establish all the activities that constitute the project. Table 10.16 lists key principles for planning and scheduling.

Planning involves defining the objectives of the project; listing of tasks or jobs that must be performed; determining gross requirements for material, equipment, and manpower; and preparing costs and durations for the various jobs or activities needed for the satisfactory completion of the project. The techniques for planning vary depending on the project's size, complexity, duration, personnel, and owner's requirements. Techniques used during the construction phase of the project should make possible the evaluation of the project's progress against the plan. There are many different analytical and graphical techniques that are commonly used for planning of the project. These are

1. The bar chart
2. CPM
3. Progress curves
4. Matrix schedule

The most widely used forms of program are bar charts and network diagrams. The bar chart is the oldest planning method used in project management. It is a graphical representation of the estimated duration of each activity and the planned sequence of activities. The horizontal axis represents the time schedule, whereas the project activities are shown along the vertical axis.

Network diagrams such as PERT and CPM are used for scheduling of complex projects. PERT/CPM diagrams consist of nodes and links and represent the entire project as a network of arrows (activities) and nodes (events). In order to draw a network diagram, work activities have to be identified, the relationships among the activities need to be specified, and a precedence relationship between the activities in a particular sequence needs to be established.

The most widely used scheduling technique is CPM. CPM analysis represents the set of sequence of predecessor/successor activities that will take the longest time to complete. The duration of the critical path is the sum of all the activity durations along the path. Thus, the critical path is the longest possible path of the project activities network. The duration of the critical path represents the minimum time required to complete the project.

There are many computer-based programs available for preparing the network and critical path of activities for construction projects. These programs can be used to analyze the use of resources, review project progress, and forecast the effects of changes in the schedule of works or other resources. Most computer programs automate preparation and presentation of various planning tools such as the bar chart, PERT, and CPM analysis. These programs are capable of storing enormous quantities of data and help process and update the program quickly. They manipulate data for multiple usages from the planning and scheduling perspectives.

In order to manage and control the project at different levels in the most effective manner, it is broken down into a group of smaller subprojects/subsystems and then into small, well-defined activities. This breakdown is necessary because of the size and complexity of construction projects, and is referred to as work breakdown structure. To begin the preparation of the detailed construction program, the contractor prepares a WBS. Its purpose is to define various activities that must be executed to complete the project. WBS helps the construction project planner to

1. Plan and schedule the work
2. Estimate costs and budget
3. Control schedule, cost, and quality

Activities are those operations of the plan that take time to carry out and on which resources are expended. Depending on the size of the project, the project is divided into multiple zones, and relevant activities are considered for each zone to prepare the construction program. Electromechanical activities are further divided into first fix, second fix, and final fix, depending on their relationship with civil and architectural works. The construction program is prepared by selecting appropriate activities relevant to a particular floor/zone. These activities are also considered for preparation of cost and resource loading schedule for the project. While preparing the program, the relationships between project activities and their dependency and precedence are considered by the planner. These activities are connected to their predecessor and successor activity based on the way the task is planned to be executed. There are four possible relationships that exist between various activities. These are (1) the finish-to-start relationship, (2) the start-to-start relationship, (3) the finish-to-finish relationship, and (4) the start-to-finish relationship.

In order to prepare a project plan, the logic is reviewed for correctness and ascertained that all activities are shown, the scope of the project has been interpreted correctly, and the resources that are required for performing each job are applied. Figure 10.44 illustrates a logic flowchart for firefighting works. Similarly, Figures 10.45 through 10.47 illustrate logic flowcharts for plumbing, HVAC, and electrical works, respectively.

Once all the activities are established by the planner and the estimated duration of each activity has been assigned, the planner prepares a detailed program fully coordinating all the construction activities.

CPM calculates the minimum completion time for a project along with the possible start and finish times for the project activities. The critical path is the longest in the network, whereas the other paths may be equal or shorter than that path. Therefore, there is a possibility that some of the events and activities can be completed before they are actually needed and, accordingly, it is possible to develop a number of activity schedules from the CPM analysis to delay the start of each activity as long as possible but still finish the project with minimum possible time without extending the completion date of the project. To develop such a schedule, it is required to find out when each activity needs to start and when it needs to be finished. There may be some activities in the project with some leeway for when they can start and finish. This is called slack time, or float, in an activity. For each activity in a project, there are four points in time: early start, early finish, late start, and late finish. Early start and early finish are the earliest times an activity can start and finish, respectively. Similarly, late start and late finish are the latest times an activity can start and finish, respectively. The difference between late start time and early start time is the slack time, or float.

With the advent of powerful computer programs such as Primavera® and Microsoft Project™, it is possible for the details of the work breakdown to be fed to these software programs. The software is capable of producing network diagrams and schedules and a limitless number of different reports, which also help in the efficient monitoring the project schedule by comparing actual with planned progress. The software can be used to analyze the project for use of resources, forecasting the effects of changes in the schedule, and cost control.

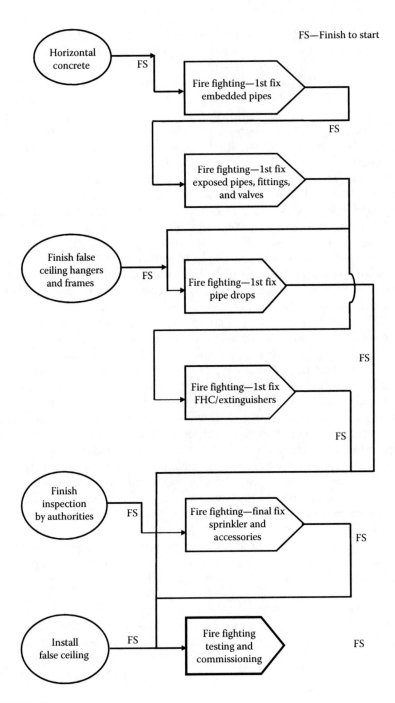

FIGURE 10.44
Logic flowchart for firefighting work.

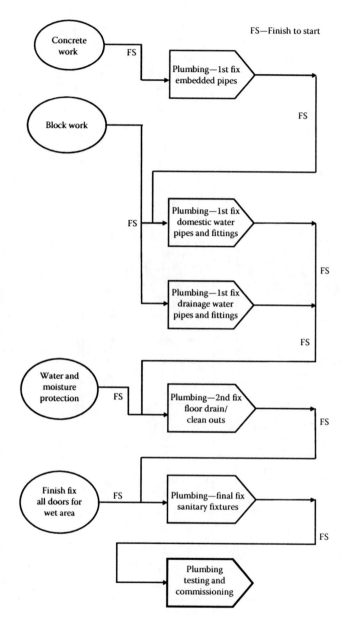

FIGURE 10.45
Logic flowchart for plumbing work.

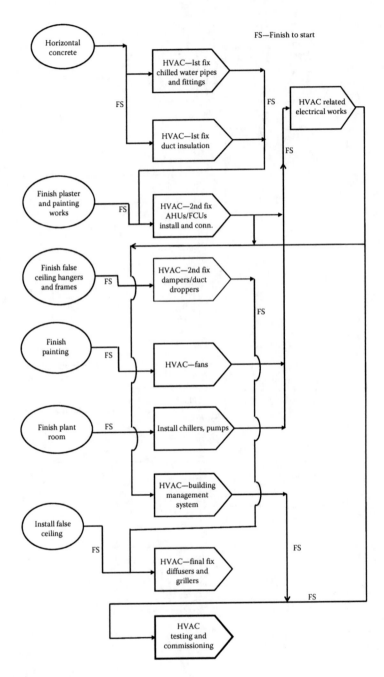

FIGURE 10.46
Logic flowchart for HVAC work.

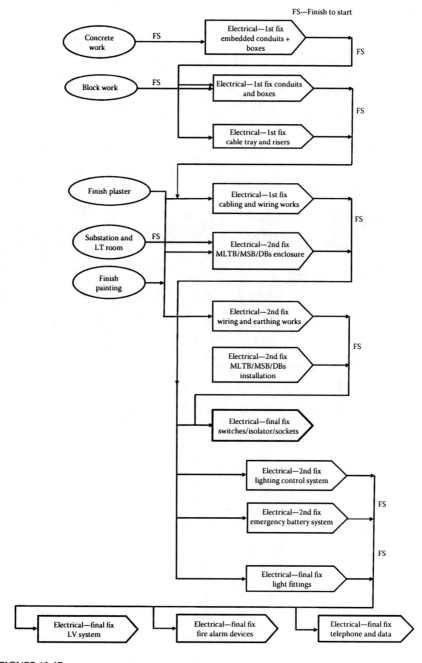

FIGURE 10.47
Logic flowchart for electrical work.

References

R. Abdul Razzak, 2011, *Quality Management in Construction Projects*, Taylor & Francis/ CRC Press, Boca Raton, FL.

ASCE. 2000. *Quality in the Constructed Project: A Guide for Owners, Designers, and Constructors*, American Society of Civil Engineers, Reston, VA.

B. S. Blanchard, W. J. Fabrycky, and J. Wolter, *Systems Engineering and Analysis*, 1998. Pearson Education, Inc., Upper Saddle River, NJ.

H. W. Chung, *Understanding Quality: Assurance in Construction*, 1999. Cengage Learning Services Limited, Hampshire, UK.

R. K. Corrie, *Engineering Management Project Evaluation*, 1991. Thomas Telford Publishing, London, UK.

ICE. 1996. *Civil Engineering Procedure*, Institution of Civil Engineers (Great Britain), London, UK.

L. R. Ireland, *Quality Management for Projects & Programs*, 1991. PMI.

K. Ishikawa, *What Is Total Quality Control? The Japanese Way*, 1985. David J. Lu, tr. Pearson Education, Inc.

H. Kerzner, *Project Management*, 2001. John Wiley & Sons.

G. D. Oberlender, *Project Management for Engineering*, 2000. The McGraw-Hill Companies, New York, NY.

PMI. 2004. *A Guide to the Project Management Body of Knowledge (PMBOK)*, 3rd Edition, Project Management Institute (PMI), Newtown Square, PA.

A. Shtlub, J. F. Bard, and S. Globerson, *Project Management*, 1994. Pearson Education, Inc., Upper Saddle River, NJ.

B. Thorpe, P. Sumner, and J. Duncan, *Quality Assurance in Construction*, 1996. Ashgate Publishing Limited, Farnham, UK.

J. R. Turner, *Contracting for Project Management*, 2003. Ashgate Publishing Limited, Farnham, UK.

11

Engineering Economics for Oil and Gas

> To be gas wise, but oil foolish is to make room for a bad economy.

In this chapter, the basic computational techniques of engineering economic analysis are presented for application to energy economics. The contents of this chapter are based on techniques, models, and examples from Badiru (1993, 1996, 2009), Badiru and Pulat (1995), Badiru and Omitaomu (2007), and Badiru et al. (2008). Cost management and economic analysis are two of the primary functions of energy project management. Cost is a vital criterion for assessing project performance. Cost management involves having an effective control over project costs through the use of quantitative techniques of estimation, forecasting, budgeting, and reporting. Cost estimation requires collecting relevant data needed to estimate elemental costs during the life cycle of a project. This could be particularly dicey in the oil and gas business where global volatility may instantaneously dictate the direction (up or down) of the economic outlook of a project. Cost planning involves developing an adequate budget for the planned work. Cost control involves continual process of monitoring, collecting, analyzing, and reporting cost data. Oil and gas project cost management is impacted by the state of technology and several concomitant cost factors. The primary components of cost management within any project undertaking are:

- Cost estimating
- Cost budgeting
- Cost control

Cost control must be exercised across the other elements of the project management knowledge areas according to PMI's PMBOK. The technique of earned value management plays a major and direct role in cost management.

Cost Management: Step-by-Step Implementation

The cost management component of the PMBOK consists of the elements shown in the block diagram in Figure 11.1. The three elements in the block diagram are carried out across the process groups presented earlier in Chapter 1. The overlay of the elements and the process groups are shown in

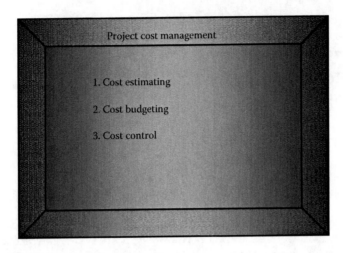

FIGURE 11.1
Block diagram of project cost management.

Table 11.1. Thus, under the knowledge area of cost management in PMBOK, the required steps are

Step 1: Cost estimation
Step 2: Cost budgeting
Step 3: Cost control

Tables 11.2 through 11.4 present the inputs, tools, techniques, and outputs of each step.

Project Portfolio Management

Project portfolio management is the systematic application of the tools and techniques of management to the collection of cost-based element of a project. Examples of project portfolios would be planned initiatives, ongoing projects and ongoing support services, and investment in emerging technology. A formal project portfolio management strategy enables measurement and objective evaluation of investment scenarios. Some of the key aspects of an effective project portfolio management are

1. Define the project, supporting program, and enabling system as well as the required portfolio.
2. Define business value and desired return on investment (ROI) and prioritize projects.

TABLE 11.1

Implementation of Project Cost Management across Process Groups

Initiating	Planning	Executing	Monitoring and Controlling	Closing
Project cost management	1. Cost estimating 2. Cost budgeting		3. Cost control	

TABLE 11.2

Tools and Techniques for Cost Estimating within Project Cost Management

Step 1: Cost Estimation		
Inputs	**Tools and Techniques**	**Output(s)**
Enterprise environmental factors	Analogous estimating	Activity cost estimates
Organizational process assets	Resource cost rates	Activity cost supporting detail
Project scope statement	Goal programming	Requested changes
WBS	Return on investment analysis	Cost management plan (updates)
WBS dictionary	Bottom-up estimating	Other in-house outputs, reports, and data inferences of interest to the organization
Project management plan	Parametric estimating	
Other in-house (custom) factors of relevance and interest	Project management cost software	
	Vendor bid analysis	
	Reserve analysis	
	Cost of quality	
	CMMI (capability maturity model integration)	
	Other in-house (custom) tools and techniques	

TABLE 11.3

Tools and Techniques for Cost Budgeting within Project Cost Management

Step 2: Cost Budgeting		
Inputs	**Tools and Techniques**	**Output(s)**
Project scope statement	Cost aggregation	Cost baseline
Work breakdown structure	Portfolio management	Project funding requirements
WBS dictionary	Reserve analysis	Cost management plan (updates)
Activity cost estimates	Parametric estimating	Requested changes
Activity cost estimate supporting detail	Funding limit reconciliation	Other in-house outputs, reports, and data inferences of interest to the organization
Project schedule	Balance scorecard	
Resource calendars	Critical chain elements budgeting	
Contract	Other in-house (custom) tools and techniques	
Cost management plan		
Other in-house (custom) factors of relevance and interest		

TABLE 11.4

Tools and Techniques for Cost Control within Project Cost Management

Step 3: Cost Control		
Inputs	**Tools and Techniques**	**Output(s)**
Cost baseline	Process control charts	Cost estimates (updates)
Project funding requirements	Cost change control system	Cost baseline (estimates)
Performance reports	Performance measurement	Performance measurements
Work performance	analysis	Forecasted completion
information	Forecasting	Requested changes
Approved change requests	Trend analysis	Recommended corrective
Project management plan	Project performance	actions
Other in-house (custom)	reviews	Organizational process assets
factors of relevance and	Project management	(updates)
interest	software	Project management plan
	Variance analysis	(updates)
	Variance management	Other in-house outputs,
	Earned value management	reports, and data inferences
	Other in-house (custom)	of interest to the
	tools and techniques	organization

3. Define an overall project portfolio management methodology.

4. Delineate an overall project portfolio in translating strategy into results.

5. Introduce a balanced scorecard that synthesizes and integrates the numerous and complex metrics related to different portfolio management processes into one framework.

6. Clarify projects that will provide effective allocation and management of limited resources.

7. Introduce progressive project assessment approach, including initial project assessment, mid-cycle project assessment, and closing project assessment.

8. Employ quantitative techniques to objectively assess a project for its absolute merit and relative merit against other projects.

9. Utilize weighted scoring models to quantify intangible benefits of the project.

10. Evaluate project decision techniques that clarify choices involving both risks and opportunities.

11. Build a business case for each project and rank order projects based on strategic fit, risks, opportunities, and the changing nature of science and technology.

12. Establish criteria for phasing out a project when it is no longer serving the desired purpose.

Project Cost Elements

Cost management in a project environment refers to the functions required to maintain effective financial control of the project throughout its life cycle. There are several cost concepts that influence the economic aspects of managing industrial projects. Within a given scope of analysis, there will be a combination of different types of cost factors as defined below:

Actual cost of work performed   The cost actually incurred and recorded in accomplishing the work performed within a given time period.

Applied direct cost   The amounts recognized in the time period associated with the consumption of labor, material, and other direct resources, without regard to the date of commitment or the date of payment. These amounts are to be charged to work-in-process (WIP) when resources are actually consumed, material resources are withdrawn from inventory for use, or material resources are received and scheduled for use within 60 days.

Budgeted cost for work performed   The sum of the budgets for completed work plus the appropriate portion of the budgets for level of effort and apportioned effort. Apportioned effort is effort that by itself is not readily divisible into short-span work packages but is related in direct proportion to measured effort.

Budgeted cost for work scheduled   The sum of budgets for all work packages and planning packages scheduled to be accomplished (including work in process), plus the amount of level of effort and apportioned effort scheduled to be accomplished within a given period of time.

Burdened costs   Burdened costs are cost components that are fully loaded with overhead charges as well as other pertinent charges. This includes cost of management and other costs associated with running the business.

Cost baseline   The cost baseline is used to measure and monitor project cost and schedule performance. It presents a summation of costs by period. It is used to measure cost and schedule performance and sometimes called performance measurement baseline (PMB).

Diminishing returns   The law of diminishing returns refers to the phenomenon of successively less output for each incremental resource input.

Direct cost   Cost that is directly associated with actual operations of a project. Typical sources of direct costs are direct material costs and direct labor costs. Direct costs are those that can be reasonably measured and allocated to a specific component of a project.

Economies of scale   This is a term referring to the reduction of the relative weight of the fixed cost in total cost, achieved by increasing the quantity of output. Economies of scale help to reduce the final unit cost of a product and are often simply referred to as the savings due to mass production.

Estimated cost at completion   This refers to the sum of actual direct costs, plus indirect costs that can be allocated to a contract, plus the estimate of costs (direct and indirect) for authorized work remaining to be done.

First cost   The total initial investment required to initiate a project or the total initial cost of the equipment needed to start the project.

Fixed cost   Costs incurred regardless of the level of operation of a project. Fixed costs do not vary in proportion to the quantity of output. Examples of costs that make up the fixed cost of a project are administrative expenses, certain types of taxes, insurance cost, depreciation cost, and debt servicing cost. These costs usually do not vary in proportion to quantity of output.

Incremental cost   The additional cost of changing the production output from one level to another. Incremental costs are normally variable costs.

Indirect cost   This is a cost that is indirectly associated with project operations. Indirect costs are those that are difficult to assign to specific components of a project. An example of an indirect cost is the cost of computer hardware and software needed to manage project operations. Indirect costs are usually calculated as a percentage of a component of direct costs. For example, the direct costs in an organization may be computed as 10% of direct labor costs.

Life cycle cost   This is the sum of all costs, recurring and nonrecurring, associated with a project during its entire life cycle.

Maintenance cost   This is a cost that occurs intermittently or periodically for the purpose of keeping project equipment in good operating condition.

Marginal cost   Marginal cost is the additional cost of increasing production output by one additional unit. The marginal cost is equal to the slope of the total cost curve or line at the current operating level.

Operating cost   This is a recurring cost needed to keep a project in operation during its life cycle. Operating costs may consist of items such as labor, material, and energy costs.

Opportunity cost   This refers to the cost of forgoing the opportunity to invest in a venture that, if it had been pursued, would have produced an economic advantage. Opportunity costs are usually incurred due to limited resources that make it impossible to take advantage of all investment opportunities. It is often defined as the cost of the best-rejected opportunity. Opportunity costs can also be incurred due to a missed opportunity rather than due to an intentional rejection. In many cases, opportunity costs are hidden or implied because they typically relate to future events that cannot be accurately predicted.

Overhead cost   These are costs incurred for activities performed in support of the operations of a project. The activities that generate overhead costs support the project efforts rather than contributing directly to the project goal. The handling of overhead costs varies widely from company to company. Typical overhead items are electric power cost, insurance premiums, cost of security, and inventory carrying cost.

Standard cost   This is a cost that represents the normal or expected cost of a unit of the output of an operation. Standard costs are established in advance. They are developed as a composite of several component costs, such as direct labor cost per unit, material cost per unit, and allowable overhead charge per unit.

Sunk cost   Sunk cost is a cost that occurred in the past and cannot be recovered under the present analysis. Sunk costs should have no bearing on the prevailing economic analysis and project decisions. Ignoring sunk costs can be a difficult task for analysts. For example, if $950,000 was spent 4 years ago to buy a piece of equipment for a technology-based project, a decision on whether or not to replace the equipment now should not consider that initial cost. But uncompromising analysts might find it difficult to ignore that much money. Similarly, an individual making a decision on selling a personal automobile would typically try to relate the asking price to what was paid for the automobile when it was acquired. This is wrong under the strict concept of sunk costs.

Total cost   This is the sum of all the variable and fixed costs associated with a project.

Variable cost   This cost varies in direct proportion to the level of operation or quantity of output. For example, the costs of material and labor required to make an item will be classified as variable costs since they vary with changes in level of output.

Basic Cash-Flow Analysis

Economic analysis is performed when a choice must be made between mutually exclusive projects that compete for limited resources. The cost performance of each project will depend on the timing and levels of its expenditures. The techniques of computing cash-flow equivalence permit us to bring competing project cash flows to a common basis for comparison. The common basis depends on the prevailing interest rate. Two cash flows that are equivalent at a given interest rate will not be equivalent at a different interest rate. The basic techniques for converting cash flows from one point in time to another are presented in the following sections.

Time Value of Money Calculations

Cash-flow conversion involves the transfer of project funds from one point in time to another. The following notations are used for the variables involved in the conversion process:

i = interest rate per period

n = number of interest periods

P = a present sum of money

F = a future sum of money

A = a uniform end-of-period cash receipt or disbursement

G = a uniform arithmetic gradient increase in period-by-period payments or disbursements

In many cases, the interest rate used in performing economic analysis is set equal to the minimum attractive rate of return (MARR) of the decision maker. The MARR is also sometimes referred to as *hurdle rate, required internal rate of return (IRR), ROI,* or *discount rate.* The value of MARR is chosen for a project based on the objective of maximizing the economic performance of the project.

Calculations with Compound Amount Factor

The procedure for the single payment compound amount factor finds a future amount, F, that is equivalent to a present amount, P, at a specified interest rate, i, after n periods. This is calculated by the following formula:

$$F = P(1 + i)^n$$

A graphic representation of the relationship between P and F is shown in Figure 11.2.

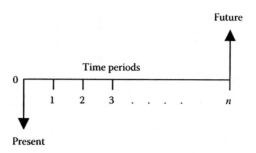

FIGURE 11.2
Single payment compound amount cash flow.

Example

A sum of $5000 is deposited in a project account and left there to earn interest for 15 years. If the interest rate per year is 12%, the compound amount after 15 years can be calculated as follows:

$$F = \$5000(1 + 0.12)^{15} = \$27,367.85$$

Calculations with Present Value Factor

Present value (PV or P), also called present worth, is the present-day at-hand value of a cash flow. The present value factor computes PV when F is given. The present value factor is obtained by solving for P in the equation for the compound amount factor. That is

$$P = F(1 + i)^{-n}$$

Supposing it is estimated that $15,000 would be needed to complete the implementation of a project 5 years from now, how much should be deposited in a special project fund now so that the fund would accrue to the required $15,000 exactly 5 years from now? If the special project fund pays interest at 9.2% per year, the required deposit would be

$$P = \$15,000(1 + 0.092)^{-5} = \$9660.03$$

Calculations with Uniform Series Present Worth Factor

The uniform series present worth factor is used to calculate the present worth equivalent, P, of a series of equal end-of-period amounts, A. Figure 11.3 shows

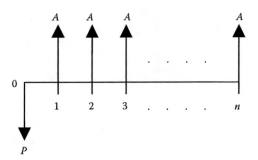

FIGURE 11.3
Uniform series cash flow.

the uniform series cash flow. The derivation of the formula uses the finite sum of the present worth values of the individual amounts in the uniform series cash flow as shown below:

$$P = \sum_{t=1}^{n} A(1+i)^{-t} = A\left[\frac{(1+i)^n - 1}{i(1+i)^n}\right]$$

Example

Suppose the sum of $12,000 must be withdrawn from an account to meet the annual operating expenses of a multiyear project. The project account pays interest at 7.5% per year compounded on an annual basis. If the project is expected to last 10 years, how much must be deposited in the project account now so that the operating expenses of $12,000 can be withdrawn at the end of every year for 10 years? The project fund is expected to be depleted to zero by the end of the last year of the project. The first withdrawal will be made 1 year after the project account is opened, and no additional deposits will be made in the account during the project life cycle. The required deposit is calculated in this way:

$$P = \$12,000\left[\frac{(1+0.075)^{10} - 1}{0.075(1+0.075)^{10}}\right] = \$82,368.92$$

Calculations with Uniform Series Capital Recovery Factor

The capital recovery formula is used to calculate the uniform series of equal end-of-period payments, A, that are equivalent to a given present amount, P.

This is the converse of the uniform series present amount factor. The equation for the uniform series capital recovery factor is obtained by solving for A in the uniform series present amount factor. That is

$$A = P\left[\frac{i(1+i)^n}{(1+i)^n - 1}\right]$$

Example

Suppose a piece of equipment needed to launch a project must be purchased at a cost of $50,000. The entire cost is to be financed at 13.5% per year and repaid on a monthly installment schedule over 4 years. It is desired to calculate what the monthly loan payments will be. It is assumed that the first loan payment will be made exactly 1 month after the equipment is financed. If the interest rate of 13.5% per year is compounded monthly, then the interest rate per month will be 13.5%/12 = 1.125% per month. The number of interest periods over which the loan will be repaid is 4(12) = 48 months. Consequently, the monthly loan payments are calculated to be

$$A = \$50,000\left[\frac{0.01125(1+0.01123)^{48}}{(1+0.01125)^{48} - 1}\right] = \$1353.82$$

Calculations with Uniform Series Compound Amount Factor

The series compound amount factor is used to calculate a single future amount that is equivalent to a uniform series of equal end-of-period payments. The cash flow is shown in Figure 11.4. Note that the future amount

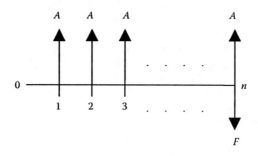

FIGURE 11.4
Uniform series compound amount cash flow.

occurs at the same point in time as the last amount in the uniform series of payments. The factor is derived as shown below:

$$F = \sum_{t=1}^{n} A(1+i)^{n-t} = A\left[\frac{(1+i)^n - 1}{i}\right]$$

Example

If equal end-of-year deposits of $5000 are made to a project fund paying 8% per year for 10 years, how much can be expected to be available for withdrawal from the account for capital expenditure immediately after the last deposit is made?

$$F = \$5000\left[\frac{(1+0.08)^{10} - 1}{0.08}\right] = \$72,432.50$$

Calculations with Uniform Series Sinking Fund Factor

The sinking fund factor is used to calculate the uniform series of equal end-of-period amounts, A, that are equivalent to a single future amount, F. This is the reverse of the uniform series compound amount factor. The formula for the sinking fund is obtained by solving for A in the formula for the uniform series compound amount factor. That is

$$A = F\left[\frac{i}{(1+i)^n - 1}\right]$$

Example

How large are the end-of-year equal amounts that must be deposited into a project account so that a balance of $75,000 will be available for withdrawal immediately after the twelfth annual deposit is made? The initial balance in the account is zero at the beginning of the first year. The account pays 10% interest per year. Using the formula for the sinking fund factor, the required annual deposits are

$$A = \$75,000\left[\frac{0.10}{(1+0.10)^{12} - 1}\right] = \$3507.25$$

Calculations with Capitalized Cost Formula

Capitalized cost refers to the present value of a single amount that is equivalent to a perpetual series of equal end-of-period payments. This is an extension of the series present worth factor with an infinitely large number of periods. This is shown graphically in Figure 11.5.

Using the limit theorem from calculus as n approaches infinity, the series present worth factor reduces to the following formula for the capitalized cost:

$$P = \frac{A}{i}$$

Example

How much should be deposited in a general fund to service a recurring public service project to the tune of $6500 per year forever if the fund yields an annual interest rate of 11%? Using the capitalized cost formula, the required one-time deposit to the general fund is

$$P = \frac{\$6500}{0.11} = \$59,090.91$$

Arithmetic Gradient Series

The gradient series cash flow involves an increase of a fixed amount in the cash flow at the end of each period. Thus, the amount at a given point in time is greater than the amount at the preceding period by a constant amount. This constant amount is denoted by G. Figure 11.6 shows the basic gradient

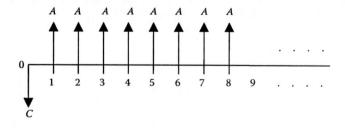

FIGURE 11.5
Capitalized cost cash flow.

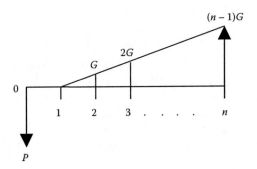

FIGURE 11.6
Arithmetic gradient cash flow with zero base amount.

series in which the base amount at the end of the first period is zero. The size of the cash flow in the gradient series at the end of period t is calculated as

$$A_t = (t-1)G, \quad t = 1, 2, ..., n$$

The total present value of the gradient series is calculated by using the present amount factor to convert each individual amount from time t to time 0 at an interest rate of $i\%$ per period and then summing up the resulting present values. The finite summation reduces to a closed form as shown below:

$$P = \sum_{t=1}^{n} A_t (1+i)^{-t} = G\left[\frac{(1+i)^n - (1+ni)}{i^2(1+i)^n}\right]$$

Example

The cost of supplies for a 10-year project increases by \$1500 every year starting at the end of year two. There is no cost for supplies at the end of the first year. If interest rate is 8% per year, determine the present amount that must be set aside at time zero to take care of all the future supplies expenditures. We have $G = 1500$, $i = 0.08$, and $n = 10$. Using the arithmetic gradient formula, we obtain

$$P = 1500\left[\frac{1 - (1 + 10(0.08))(1 + 0.08)^{-10}}{(0.08)^2}\right]$$

$$= \$1500(25.9768) = \$38,965.20$$

In many cases, an arithmetic gradient starts with some base amount at the end of the first period and then increases by a constant amount thereafter. The nonzero base amount is denoted as A_1. Figure 11.7 shows this type of cash flow.

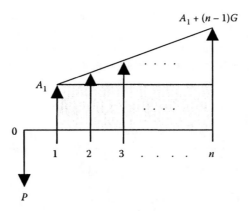

FIGURE 11.7
Arithmetic gradient cash flow with nonzero base amount.

The calculation of the present amount for such cash flows requires breaking the cash flow into a uniform series cash flow of amount A_1 and an arithmetic gradient cash flow with zero base amount. The uniform series present worth formula is used to calculate the present worth of the uniform series portion while the basic gradient series formula is used to calculate the gradient portion. The overall present worth is then calculated:

$$P = P_{\text{uniform series}} + P_{\text{gradient series}}$$

$$= A_1 \left[\frac{(1+i)^n - 1}{i(1+i)^n} \right] + G \left[\frac{(1+i)^n - (1+ni)}{i^2(1+i)^n} \right]$$

Internal Rate of Return

The IRR for a cash flow is defined as the interest rate that equates the future worth at time n or present worth at time 0 of the cash flow to zero. If we let i^* denote the IRR, then we have

$$\text{FW}_{t=n} = \sum_{t=0}^{n} (\pm A_t)(1+i^*)^{n-t} = 0$$

$$\text{PW}_{t=0} = \sum_{t=0}^{n} (\pm A_t)(1+i^*)^{-t} = 0$$

where "+" is used in the summation for positive cash-flow amounts or receipts and "−" is used for negative cash-flow amounts or disbursements.

A_t denotes the cash-flow amount at time t, which may be a receipt (+) or a disbursement (−). The value of i^* is referred to as *discounted cash-flow rate of return, internal rate of return,* or *true rate of return.* The above procedure essentially calculates the net future worth (NFW) or the net present worth (NPW) of the cash flow. That is

NFW = future worth of receipts − future worth of disbursements

NFW = FW (receipts) − FW (disbursements)

NPW = present worth of receipts − present worth of disbursements

NPW = PW (receipts) − PW (disbursements)

Setting the NPW or NFW equal to zero and solving for the unknown variable i determines the IRR of the cash flow.

Benefit–Cost Ratio Analysis

The benefit–cost ratio of a cash flow is the ratio of the present worth of benefits to the present worth of costs. This is defined below:

$$\frac{B}{C} = \frac{\sum_{t=0}^{n} B_t(1+i)^{-t}}{\sum_{t=0}^{n} C_t(1+i)^{-t}} = \frac{PW_{benefits}}{PW_{costs}}$$

where B_t is the benefit (receipt) at time t and C_t is the cost (disbursement) at time t. If the benefit–cost ratio is greater than one, then the investment is acceptable. If the ratio is less than one, the investment is not acceptable. A ratio of one indicates a break-even situation for the project.

Simple Payback Period

Payback period refers to the length of time it will take to recover an initial investment. The approach does not consider the impact of the time value of money. Consequently, it is not an accurate method of evaluating the worth of an investment. However, it is a simple technique that is used widely to perform a "quick-and-dirty" assessment of investment performance. Another limitation of the technique is that it considers only the initial cost. Other costs that may occur after time zero are not included in the calculation. The payback period is defined as the smallest value of n (n_{min}) that satisfies the following expression:

$$\sum_{t=1}^{n_{min}} R_t \geq C$$

where R_t is the revenue at time t and C_0 is the initial investment. The procedure calls for a simple addition of the revenues period by period until enough total has been accumulated to offset the initial investment.

Example

An organization is considering installing a new computer system that will generate significant savings in material and labor requirements for order processing. The system has an initial cost of $50,000. It is expected to save the organization $20,000 a year. The system has an anticipated useful life of 5 years with a salvage value of $5000. Determine how long it would take for the system to pay for itself from the savings it is expected to generate. Since the annual savings are uniform, we can calculate the payback period by simply dividing the initial cost by the annual savings. That is

$$n_{min} = \frac{\$50,000}{\$20,000} = 2.5 \text{ years}$$

Note that the salvage value of $5000 is not included in the above calculation since the amount is not realized until the end of the useful life of the asset (i.e., after 5 years). In some cases, it may be desired to consider the salvage value. In that case, the amount to be offset by the annual savings will be the net cost of the asset. In that case, we would have the following:

$$n_{min} = \frac{\$50,000 - \$5000}{\$20,000}$$

$$= 2.25 \text{ years}$$

If there are tax liabilities associated with the annual savings, those liabilities must be deducted from the savings before the payback period is calculated.

Discounted Payback Period

In this book, we introduce the *discounted payback period* approach, in which the revenues are reinvested at a certain interest rate. The payback period is determined when enough money has been accumulated at the given interest rate to offset the initial cost as well as other interim costs. In this case, the calculation is done by the following expression:

$$\sum_{t=1}^{n_{min}} R_t(1+i)^{n_{min}-1} \geq \sum_{t=0}^{n_{min}} C_t$$

Example

A new solar cell unit is to be installed in an office complex at an initial cost of $150,000. It is expected that the system will generate annual cost savings of $22,500 on the electricity bill. The solar cell unit will need to be overhauled every 5 years at a cost of $5000 per overhaul. If the annual interest rate is 10%, find the discounted payback period for the solar cell unit considering the time value of money. The costs of overhaul are to be considered in calculating the discounted payback period.

SOLUTION

Using the single payment compound amount factor for one period iteratively, the following set of solutions is obtained for cumulative savings for each time period:

Period 1: $22,500
Period 2: $22,500 + $22,500 (1.10)¹ = $47,250
Period 3: $22,500 + $47,250 (1.10)¹ = $74,475
Period 4: $22,500 + $74,475 (1.10)¹ = $104,422.50
Period 5: $22,500 + $104,422.50 (1.10)¹ − $5000 = $132,364.75
Period 6: $22,500 + $132,364.75 (1.10)¹ = $168,101.23

The initial investment is $150,000. By the end of period 6, we have accumulated $168,101.23, more than the initial cost. Interpolating between period 5 and period 6, results in n_{min} of 5.49 years. That is, it will take five-and-a-half years to recover the initial investment. The calculation is shown below:

$$n_{min} = 5 + \frac{150,000 - 132,364.75}{168,101.25 - 132,364.75}(6-5) = 5.49$$

Time Required to Double Investment

It is sometimes of interest to determine how long it will take a given investment to reach a certain multiple of its initial level. The "Rule of 72" is one simple approach to calculating the time required to for an investment to double in value, at a given interest rate per period. The Rule of 72 gives the following formula for estimating the time required:

$$n = \frac{72}{i}$$

where i is the interest rate expressed in percentage. Referring to the single payment compound amount factor, we can set the future amount equal to twice the present amount and then solve for n. That is, $F = 2P$. Thus

$$2P = P(1 + i)^n$$

Solving for n in the above equation yields an expression for calculating the exact number of periods required to double P:

$$n = \frac{\ln(2)}{\ln(1 + i)}$$

where i is the interest rate expressed in decimals. In general, the length of time it would take to accumulate m multiples of P is expressed as:

$$n = \frac{\ln(m)}{\ln(1 + i)}$$

where m is the desired multiple. For example, at an interest rate of 5% per year, the time it would take an amount, P, to double in value ($m = 2$) is 14.21 years. This, of course, assumes that the interest rate will remain constant throughout the planning horizon. Table 11.5 presents a tabulation of the values calculated from both approaches. Figure 11.8 shows a graphical comparison of the Rule of 72 to the exact calculation.

Effects of Inflation on Project Costing

Inflation can be defined as the decline in purchasing power of money, and as such, is a major player in the financial and economic analysis of projects. Multiyear projects are particularly subject to the effects of inflation. Some of the most common causes of inflation include the following:

- An increase in the amount of currency in circulation
- A shortage of consumer goods

TABLE 11.5

Evaluation of the Rule of 72

$i\%$	n (Rule of 72)	n (Exact Value)
0.25	288.00	277.61
0.50	144.00	138.98
1.00	72.00	69.66
2.00	36.00	35.00
5.00	14.20	17.67
8.00	9.00	9.01
10.00	7.20	7.27
12.00	6.00	6.12
15.00	4.80	4.96
18.00	4.00	4.19
20.00	3.60	3.80
25.00	2.88	3.12
30.00	2.40	2.64

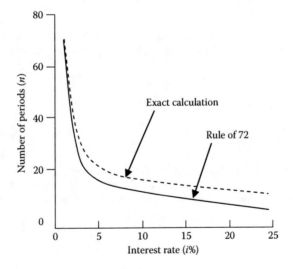

FIGURE 11.8
Evaluation of investment life for double return.

- An escalation of the cost of production
- An arbitrary increase in prices set by resellers

The general effects of inflation are felt in terms of an increase in the prices of goods and a decrease in the worth of currency. In cash-flow analysis, ROI for a project will be affected by time value of money as well as inflation. The real interest rate (d) is defined as the desired rate of return in the absence of inflation. When we talk of "today's dollars" or "constant dollars," we are referring to the use of the real interest rate. The combined interest rate (i) is the rate of return combining the real interest rate and the inflation rate. If we denote the inflation rate as j, then the relationship between the different rates can be expressed as shown below:

$$1 + i = (1 + d)(1 + j)$$

Thus, the combined interest rate can be expressed as follows:

$$i = d + j + dj$$

Note that if $j = 0$ (i.e., no inflation), then $i = d$. We can also define commodity escalation rate (g) as the rate at which individual commodity prices escalate. This may be greater than or less than the overall inflation rate. In practice, several measures are used to convey inflationary effects. Some of these are the consumer price index, the producer price index, and the wholesale price index. A "market basket" rate is defined as the estimate of inflation based on a weighted average of the annual rates of change in the costs of a wide range of representative commodities. A "then-current" cash flow is a cash flow that explicitly incorporates the impact of inflation. A "constant worth" cash flow is a cash flow that does not incorporate the effect of inflation. The real interest rate, d, is used for analyzing constant worth cash flows. Figure 11.9 shows constant worth and then-current cash flows.

The then-current cash flow in the figure is the equivalent cash flow considering the effect of inflation. C_k is what it would take to buy a certain "basket" of goods after k time periods if there was no inflation. T_k is what it would take to buy the same "basket" in k time period if inflation were taken into account. For the constant worth cash flow, we have

$$C_k = T_0, \quad k = 1, 2, \ldots, n$$

and for the then-current cash flow, we have

$$T_k = T_0(1 + j)^k, \quad k = 1, 2, \ldots, n$$

Constant-worth cash flow

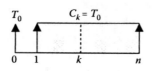

Then-current cash flow

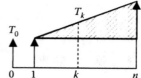

FIGURE 11.9
Cash flows for effects of inflation.

where j is the inflation rate. If $C_k = T_0 = \$100$ under the constant worth cash flow, then we have $\$100$ worth of buying power. If we are using the commodity escalation rate, g, then we will have

$$T_k = T_0(1+g)^k, \quad k = 1, 2, \ldots, n$$

Thus, a then-current cash flow may increase based on both a regular inflation rate (j) and a commodity escalation rate (g). We can convert a then-current cash flow to a constant worth cash flow by using the following relationship:

$$C_k = T_k(1+j)^{-k}, \quad k = 1, 2, \ldots, n$$

If we substitute T_k from the commodity escalation cash flow into the expression for C_k above, we get the following:

$$
\begin{aligned}
C_k &= T_k(1+j)^{-k} \\
&= T(1+g)^k(1+j)^{-k} \\
&= T_0\left[(1+g)/(1+j)\right]^k, \quad k = 1,2,\ldots,n
\end{aligned}
$$

Note that if $g = 0$ and $j = 0$, the $C_k = T_0$. That is, there is no inflationary effect. We can now define the effective commodity escalation rate (v):

$$v = [(1+g)/(1+j)] - 1$$

The commodity escalation rate (g) can be expressed as follows:

$$g = v + j + vj$$

Inflation can have a significant impact on the financial and economic aspects of an industrial project. Inflation may be defined, in economic terms, as the increase in the amount of currency in circulation. To a producer, inflation means a sudden increase in the cost of items that serve as inputs for the production process (equipment, labor, materials, etc.). To the retailer, inflation implies an imposed higher cost of finished products. To an ordinary citizen, inflation portends a noticeable escalation of prices of consumer goods. All these aspects are intertwined in a project management environment.

The amount of money supply, as a measure of a country's wealth, is controlled by the government. When circumstances dictate such action, governments often feel compelled to create more money or credit to take care of old debts and pay for social programs. When money is generated at a faster rate than the growth of goods and services, it becomes a surplus commodity, and its value (i.e., purchasing power) will fall. This means that there will be too much money available to buy only a few goods and services. When the purchasing power of a currency falls, each individual in a product's life cycle (that is, each person or entity that spends money on a product throughout its life cycle, from production through disposal) has to use more of the currency in order to obtain the product. Some of the classic concepts of inflation are discussed below:

1. In *cost-driven* or *cost-push inflation*, increases in producer's costs are passed on to consumers. At each stage of the product's journey from producer to consumer, prices are escalated disproportionately in order to make a good profit. The overall increase, in the product's price is directly proportional to the number of intermediaries it encounters on its way to the consumer.

2. In *demand-driven* or *demand-pull inflation, excessive* spending power of consumers forces an upward trend in prices. This high spending power is usually achieved at the expense of savings. The law of supply and demand dictates that the more the demand, the higher the price. This results in *demand-driven* or *demand-pull inflation*.

3. Impact of international economic forces can induce inflation on a local economy. Trade imbalances and fluctuations in currency values are notable examples of international inflationary factors.

4. In *wage-driven* or *wage-push inflation*, the increasing base wages of workers generate more disposable income and, hence, higher demands for goods and services. The high demand, consequently, creates a pull on prices. Coupled with this, employers pass the additional wage cost on to consumers through higher prices. This type of inflation is very difficult to contain because wages set by union contracts and prices set by producers almost never fall.

5. Easy availability of credit leads consumers to "buy now and pay later," thereby creating another opportunity for inflation. This is a dangerous type of inflation because the credit not only pushes prices up, but it also leaves consumers with less money later to pay for the credit. Eventually, many credits become uncollectible debts, which may then drive the economy toward recession.

6. Deficit spending results in an increase in money supply and, thereby, creates less room for each dollar to get around. The popular saying that indicates that "a dollar does not go far anymore," simply refers to inflation in laymen's terms. The different levels of inflation may be categorized as discussed below.

Mild Inflation

When inflation is mild (at 2–4%), the economy actually prospers. Producers strive to produce at full capacity in order to take advantage of the high prices to the consumer. Private investments tend to be brisk, and more jobs become available. However, the good fortune may only be temporary. Prompted by the prevailing success, employers are tempted to seek larger profits and workers begin to ask for higher wages. They cite their employer's prosperous business as a reason to bargain for bigger shares of the business profit. So, we end up with a vicious cycle where the producer asks for higher prices, the unions ask for higher wages, and inflation starts an upward trend.

Moderate Inflation

Moderate inflation occurs when prices increase at 5–9%. Consumers start purchasing more as a hedge against inflation. They would rather spend their money now than watch it decline further in purchasing power. The increased market activity serves to fuel further inflation.

Severe Inflation

Severe inflation is indicated by price escalations of 10% or more. Double-digit inflation implies that prices rise much faster than wages do. Debtors tend to be the ones who benefit from this level of inflation because they repay debts with money that is less valuable than when they borrowed.

Hyperinflation

When each price increase signals an increase in wages and costs, which again sends prices further up, the economy has reached a stage of malignant galloping inflation or hyperinflation. Rapid and uncontrollable inflation destroys the economy. The currency becomes economically useless as the government prints it excessively to pay for obligations.

Inflation can affect any industrial project in terms of raw materials procurement, salaries and wages, and/or cost tracking dilemmas. Some effects are immediate and easily observable while others are subtle and pervasive. Whatever form it takes, inflation must be taken into account in long-term project planning and control. Large projects, especially, may be adversely affected by the effects of inflation in terms of cost overruns and poor resource utilization. Managers should note that the level of inflation will determine the severity of the impact on projects.

Break-Even Analysis

Break-even analysis refers to the determination of the balanced performance level where project income is equal to project expenditure. The total cost of an operation is expressed as the sum of the fixed and variable costs with respect to output quantity. That is

$$TC(x) = FC + VC(x)$$

where x is the number of units produced, $TC(x)$ is the total cost of producing x units, FC is the total fixed cost, and $VC(x)$ is the total variable cost associated with producing x units. The total revenue resulting from the sale of x units is defined as

$$TR(x) = px$$

where p is the price per unit. The profit due to the production and sale of x units of the product is calculated as

$$P(x) = TR(x) - TC(x)$$

The break-even point of an operation is defined as the value of a given parameter that will result in neither profit nor loss. The parameter of interest may be the number of units produced, the number of hours of operation, the number of units of a resource type allocated, or any other measure of interest. At the break-even point, we have the following relationship:

$$TR(x) = TC(x) \text{ or } P(x) = 0$$

In some cases, there may be a known mathematical relationship between cost and the parameter of interest. For example, there may be a linear cost relationship between the total cost of a project and the number of units produced. The cost expressions facilitate a straightforward break-even analysis. Figure 11.10 shows an example of a break-even point for a single project. Figure 11.11 shows examples of multiple break-even points that exist when multiple projects are compared. When two project alternatives are compared, the break-even point refers to the point of indifference between the two alternatives. In Figure 11.11, $x1$ represents the point where projects A and B are equally desirable, $x2$ represents where A and C are equally desirable, and $x3$ represents where B and C are equally desirable. The figure shows that if we are operating below a production level of $x2$ units, then project C is the preferred project among the three. If we are operating at a level more than $x2$ units, then project A is the best choice.

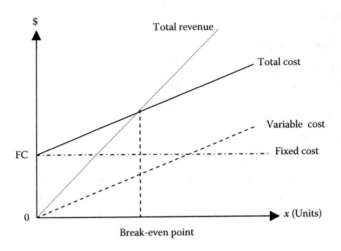

FIGURE 11.10
Break-even point for a single project.

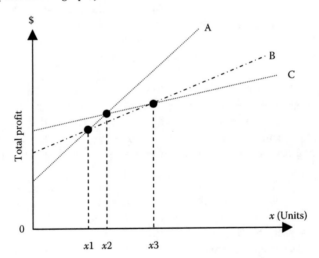

FIGURE 11.11
Break-even points for multiple projects.

Example

Three project alternatives are being considered for producing a new product. The required analysis involves determining which alternative should be selected on the basis of how many units of the product are produced per year. Based on past records, there is a known relationship between the number of units produced per year, x, and the net annual profit, $P(x)$, from each alternative. The level of production is expected to be between 0 and 250 units per year. The net annual profits (in thousands of dollars) are given below for each alternative:

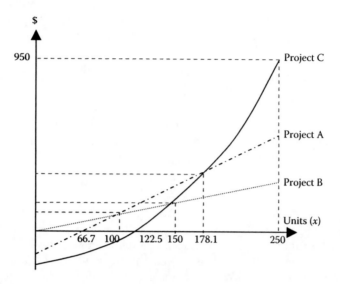

FIGURE 11.12
Plot of profit functions.

Project A: $P(x) = 3x - 200$
Project B: $P(x) = x$
Project C: $P(x) = (1/50)x^2 - 300$

This problem can be solved mathematically by finding the intersection points of the profit functions and evaluating the respective profits over the given range of product units. It can also be solved by a graphical approach. Figure 11.12 shows a plot of the profit functions. Such a plot is called a break-even chart. The plot shows that Project B should be selected if between 0 and 100 units are to be produced, Project A should be selected if between 100 and 178.1 units (178 physical units) are to be produced, and Project C should be selected if more than 178 units are to be produced. It should be noted that if less than 66.7 units (66 physical units) are produced, Project A will generate a net loss rather than a net profit. Similarly, Project C will generate losses if less than 122.5 units (122 physical units) are produced.

Profit Ratio Analysis

Break-even charts offer opportunities for several different types of analysis. In addition to the break-even points, other measures of worth or criterion measures may be derived from the charts. A measure called the *profit ratio* is presented here for the purpose of obtaining a further comparative basis for competing projects. A profit ratio is defined as the ratio of the profit area to the sum of the profit and loss areas in a break-even chart. That is

$$\text{Profit ratio} = \frac{\text{area of profit region}}{\text{area of profit region} + \text{area of loss region}}$$

For example, suppose that the expected revenue and the expected total cost associated with a project are given, respectively, by the following expressions:

$$R(x) = 100 + 10x$$

$$TC(x) = 2.5x + 250$$

where x is the number of units produced and sold from the project. Figure 11.13 shows the break-even chart for the project. The break-even point is shown to be 20 units. Net profits are realized from the project if more than 20 units are produced, and net losses are realized if less than 20 units are produced. It should be noted that the revenue function in Figure 11.13 represents an unusual case, in which a revenue of \$100 is realized when zero units are produced.

Suppose it is desired to calculate the profit ratio for this project if the number of units that can be produced is limited to between 0 and 100 units. From Figure 11.13, the surface area of the profit region and the area of the loss region can be calculated by using the standard formula for finding the

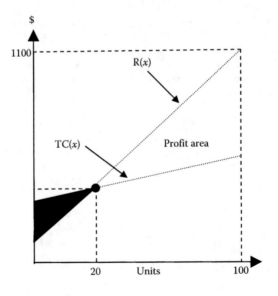

FIGURE 11.13
Area of profit versus area of loss.

area of a triangle: area = (1/2)(base)(height). Using this formula, we have the following:

$$\text{Area of profit region} = \frac{1}{2}(\text{base})(\text{height})$$

$$= \frac{1}{2}(1100 - 500)(100 - 20)$$

$$= 24,000 \text{ square units}$$

$$\text{Area of loss region} = \frac{1}{2}(\text{base})(\text{height})$$

$$= \frac{1}{2}(250 - 100)(20)$$

$$= 1500 \text{ square units}$$

Thus, the profit ratio is computed as follows:

$$\text{Profit ratio} = 24,000/(24,000 + 1500) = 0.9411 \equiv 94.11\%$$

The profit ratio may be used as a criterion for selecting among project alternatives. If this is done, the profit ratios for all the alternatives must be calculated over the same values of the independent variable. The project with the highest profit ratio will be selected as the desired project. For example, Figure 11.14 presents the break-even chart for an alternate project, say Project II. It can be seen that both the revenue and cost functions for the project are nonlinear. The revenue and cost are defined as follows:

$$R(x) = 160x - x^2$$
$$TC(x) = 500 + x^2$$

If the cost and/or revenue functions for a project are not linear, the areas bounded by the functions may not be easily determined. For those cases, it may be necessary to use techniques such as definite integrals to find the areas. Figure 11.14 indicates that the project generates a loss if less than 3.3 units (3 actual units) are produced or if more than 76.8 units (76 actual units) are produced. The respective profit and loss areas on the chart are calculated as shown below:

Area 1 (loss) = 802.80 unit-dollars
Area 2 (profit) = 132,272.08 unit-dollars
Area 3 (loss) = 48,135.98 unit-dollars

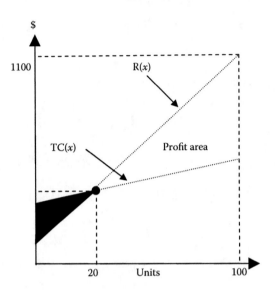

FIGURE 11.14
Break-even chart for revenue and cost functions.

Consequently, the profit ratio for Project II is computed as follows:

$$\text{Profit ratio} = \frac{\text{total area of profit region}}{\text{total area of profit region} + \text{total area of loss region}}$$

$$= \frac{132,272.08}{802.76 + 132,272.08 + 48,135.98}$$

$$= 72.99\%$$

The profit ratio approach evaluates the performance of each alternative over a specified range of operating levels. Most of the existing evaluation methods use single-point analysis with the assumption that the operating condition is fixed at a given production level. The profit ratio measure allows an analyst to evaluate the net yield of an alternative, given that the production level may shift from one level to another. An alternative, for example, may operate at a loss for most of its early life, but it may generate large incomes to offset those losses in its later stages. Conventional methods cannot easily capture this type of transition from one performance level to another. In addition to being used to compare alternate projects, the profit ratio may also be used for evaluating the economic feasibility of a single project. In such a case, a decision rule may be developed, such as the following:

If profit ratio is greater than 75%, accept the project.
If profit ratio is less than or equal to 75%, reject the project.

Project Cost Estimation

Cost estimation and budgeting help establish a strategy for allocating resources in project planning and control. Based on the desired level of accuracy, there are three major categories of cost estimation for budgeting: *order-of-magnitude estimates, preliminary cost estimates,* and *detailed cost estimates.* Order-of-magnitude cost estimates are usually gross estimates based on the experience and judgment of the estimator. They are sometimes called "ballpark" figures. These estimates are typically made without a formal evaluation of the details involved in the project. The level of accuracy associated with order-of-magnitude estimates can range from −50% to +50% of the actual cost. These estimates provide a quick way of getting cost information during the initial stages of a project. The estimation range is summarized as follows:

50% (actual cost) ≤ order-of-magnitude estimate ≤ 150% (actual cost)

Preliminary cost estimates are also gross estimates but with a higher level of accuracy. In developing preliminary cost estimates, more attention is paid to some selected details of the project. An example of a preliminary cost estimate is the estimation of expected labor cost. Preliminary estimates are useful for evaluating project alternatives before final commitments are made. The level of accuracy associated with preliminary estimates can range from −20% to +20% of the actual cost, as shown below:

80% (actual cost) ≤ preliminary estimate ≤ 120% (actual cost)

Detailed cost estimates are developed after careful consideration is given to all the major details of a project. Considerable time is typically needed to obtain detailed cost estimates. Because of the amount of time and effort needed to develop detailed cost estimates, the estimates are usually developed after a firm commitment has been made that the project will take off. Detailed cost estimates are important for evaluating actual cost performance during the project. The level of accuracy associated with detailed estimates normally ranges from −5% to +5% of the actual cost.

95% (actual cost) ≤ detailed cost ≤ 105% (actual cost)

There are two basic approaches to generating cost estimates. The first one is a variant approach, in which cost estimates are based on variations of previous cost records. The other approach is the generative cost estimation, in which cost estimates are developed from scratch without taking previous cost records into consideration.

Optimistic and Pessimistic Cost Estimates

Using an adaptation of the PERT formula, we can combine optimistic and pessimistic cost estimates. If O = optimistic cost estimate, M = most likely cost estimate, and P = pessimistic cost estimate, the estimated cost can be stated as follows:

$$E[C] = \frac{O + 4M + P}{6}$$

and the cost variance can be estimated as follows:

$$V[C] = \left[\frac{P - O}{6}\right]^2$$

Project Budget Allocation

Project budget allocation involves sharing limited resources among competing tasks in a project. The budget allocation process serves the following purposes:

- A plan for resource expenditure
- A project selection criterion
- A projection of project policy
- A basis for project control
- A performance measure
- A standardization of resource allocation
- An incentive for improvement

Top-Down Budgeting

Top-down budgeting involves collecting data from upper-level sources such as top and middle managers. The figures supplied by the managers may come from their personal judgment, past experience, or past data on similar project activities. The cost estimates are passed to lower-level managers, who then break the estimates down into specific work components within the project. These estimates may, in turn, be given to line managers, supervisors, and lead workers to continue the process until individual activity costs are obtained. Thus, top management provides the global budget, while the functional-level worker provides specific budget requirements for project items.

Bottom-Up Budgeting

In this method, elemental activities, and their schedules, descriptions, and labor skill requirements are used to construct detailed budget requests. Line workers familiar with specific activities are asked to provide cost estimates, and then make estimates for each activity in terms of labor time, materials, and machine time. The estimates are then converted to an appropriate cost basis. The dollar estimates are combined into composite budgets at each successive level up the budgeting hierarchy. If estimate discrepancies develop, they can be resolved through the intervention of senior management, middle management, functional managers, project manager, accountants, or standard cost consultants. Figure 11.15 shows the breaking down of a project into phases and parts in order to facilitate bottom-up budgeting and improve both schedule and cost control.

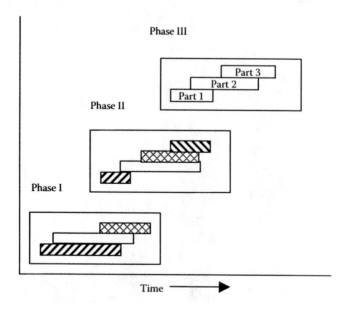

FIGURE 11.15
Budgeting by project phases.

Elemental budgets may be developed on the basis of the timed progress of each part of the project. When all the individual estimates are gathered, we can obtain a composite budget estimate. Figures 11.16 and 11.17 show an example of the various components that may be involved in an overall budget. The bar chart appended to a segment of the pie chart indicates the

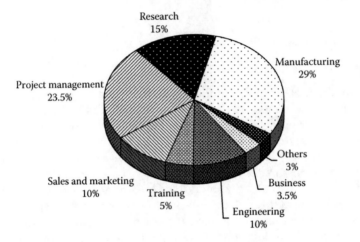

FIGURE 11.16
Pie chart of budget distribution.

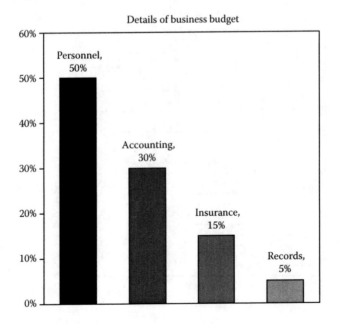

FIGURE 11.17
Bar chart of budget and distribution.

individual cost components making up that particular segment. To further aid in the process, analytical tools such as learning curve analysis, work sampling, and statistical estimation may be employed in the cost estimation and budgeting processes.

Budgeting and Risk Allocation for Types of Contract

Budgeting and allocation of risk are handled based on the type of contract involved. The list below carries progressively higher risk to the buyer (customer) while it carries progressively lower risk to the contractor (producer):

Type 1: Firm fixed price (FFP)

Type 2: FFP with economic adjustment

Type 3: Fixed price incentive fee (FPIF)

Type 4: Cost and cost sharing (CCS)

Type 5: Cost plus incentive fee (CPIF)

Type 6: Cost plus award fee (CPAF)

Type 7: Cost plus fixed fee (CPFF)

Type 8: Cost plus percentage fee (CPPF)

Type 9: Indefinite delivery

Type 10: Time and materials

Type 11: Basic agreements (blanket contract)

Type 1 contract carries the highest risk to the contractor (producer) whereas it carries the lowest risk to the buyer (customer). Type 11 contract carries the lowest risk to the contractor (producer) whereas it carries the highest risk to the buyer (customer). The risk level is progressive in each direction of the list.

Cost Monitoring

As a project progresses, costs can be monitored and evaluated to identify areas of unacceptable cost performance. Figure 11.18 shows a plot of cost versus time for projected cost and actual cost. The plot permits a quick identification of the points at which cost overruns occur in a project.

Plots similar to those presented above may be used to evaluate cost, schedule, and time performance of a project. An approach similar to the profit

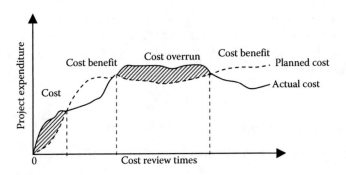

FIGURE 11.18
Evaluation of actual and projected cost.

ratio presented earlier may be used along with the plot to evaluate the overall cost performance of a project over a specified planning horizon. Presented below is a formula for cost performance index (CPI):

$$\text{CPI} = \frac{\text{area of cost benefit}}{\text{area of cost benefit} + \text{area of cost overrun}}$$

As in the case of the profit ratio, CPI may be used to evaluate the relative performances of several project alternatives or to evaluate the feasibility and acceptability of an individual alternative. In Figure 11.19, we present another cost-monitoring tool, referred to as a cost-control pie chart. The chart is used to track the percentage of the cost going into a specific

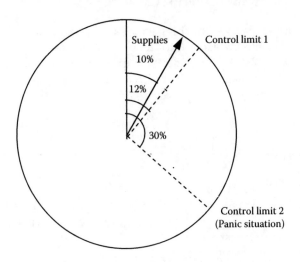

FIGURE 11.19
Cost-control pie chart.

component of a project. Control limits can be included in the pie chart to identify costs that have become out of control. The example in Figure 11.19 shows that 10% of total cost is tied up in supplies. The control limit is located at 12% of total cost. Hence, the supplies expenditure is within control (so far, at least).

Project Balance Technique

One other approach to monitoring cost performance is the project balance technique. The technique helps in assessing the economic state of a project at a desired point in time in the life cycle of the project. It calculates the net cash flow of a project up to a given point in time. The project balance is calculated as follows:

$$B(i)_t = S_t - P(1 + i)^t + \sum_{k=1}^{t} PW_{income}(i)_k$$

where
$B(i)_t$ = project balance at time t at an interest rate of i% per period
PW income $(i)_k$ = present worth of net income from the project up to time k
P = initial cost of the project
S_t = salvage value at time t

The project balance at time t gives the net loss or net profit associated with the project up to that time.

Cost and Schedule Control Systems Criteria

Contract management involves the process by which goods and services are acquired, utilized, monitored, and controlled in a project. Contract management addresses the contractual relationships from the initiation of a project to the completion of the project (i.e., completion of services and/or hand over of deliverables). Some of the important aspects of contract management that oil and gas practitioners should be familiar with include:

- Principles of contract law
- Bidding process and evaluation

- Contract and procurement strategies
- Selection of source and contractors
- Negotiation
- Worker-safety considerations
- Product liability
- Uncertainty and risk management
- Conflict resolution

In 1967, the U.S. Department of Defense (DOD) introduced a set of 35 standards or criteria with which contractors must comply under cost or incentive contracts. The system of criteria is referred to as the *cost and schedule control systems criteria* (C/SCSC). Although no longer in vogue, many government agencies still require compliance with modified and updated versions of C/SCSC, albeit under different "new" and trendy monikers. The primary goal of C/SCSC is to manage the risk of cost overrun to the government on major contracts. That goal is a desirable pursuit of any modern cost management and contract administration system; although actual implementation is often lamentable. The C/SCSC system presents an integrated approach to cost and schedule management. This "integrated approach" is in agreement with the premise of project management as presented in this book. C/SCSC has been widely used in major project undertakings. It is intended to facilitate greater uniformity and provide advance warning about impending schedule or cost overruns as well as performance risks. Some of the factors influencing schedule, performance, and cost problems are summarized below, with suggested lists of control actions:

Causes of schedule problems:
- Delay of critical activities
- Unreliable time estimates
- Technical problems
- Precedence structure
- Change of due dates
- Bad time estimates
- Changes in management direction

Schedule control actions:
- Use activity crashing
- Redesign tasks
- Revise milestones
- Update time estimates
- Change the scope of work

- Combine related activities
- Eliminate unnecessary activities (i.e., operate lean)

Causes of performance problems:

- Poor quality
- Poor functionality
- Maintenance problems
- Poor mobility (knowledge transfer)
- Lack of training
- Lack of clear objectives

Performance control actions:

- Use SMART job objectives (specific, measurable, aligned, realistic, timed)
- Use improved tools/technology
- Adjust project specifications
- Improve management oversight
- Review project priorities
- Modify project scope
- Allocate more resources
- Require higher level of accountability
- Improve work ethics (through training, mentoring, and education)

Causes of cost problems:

- Inadequate budget
- Effects of inflation
- Poor cost reporting
- Increase in scope of work
- High overhead cost
- High labor cost

Cost-control actions:

- Reduce labor costs
- Use competitive bidding
- Modify work process
- Adjust work breakdown structure
- Improve coordination of project functions
- Improve cost estimation procedures
- Use less expensive raw materials

- Mitigate effects of inflationary trends (e.g., use of price hedging in procurement)
- Cut overhead costs
- Outsource work

The topics covered by C/SCSC or any of its modern derivates include cost estimating and forecasting, budgeting, cost control, cost reporting, earned value analysis, resource allocation and management, and schedule adjustments. There is no doubt that the contemporary evolution of cost management as presented in PMI's PMBOK was influenced by the foundational contents of C/SCSC. The important link between all of these developments is the dynamism of the relationship between performance, time, and cost, as was alluded to earlier in this book. Figure 11.20 illustrates an example of the dynamism that exists in cost–schedule–performance relationships. The relationships represent a multiobjective problem. The resultant function, $f(p, c, t)$, in Figure 11.20 represents a vector of decision; taking into account the relative nuances of project cost, schedule, and performance. Because performance, time, and cost objectives cannot be satisfied equally well, concessions or compromises need to be worked out in implementing C/SCSC or other project control criteria.

To comply with the ideals of cost management, contractors must use standardized planning and control methods based on *earned value*. Earned

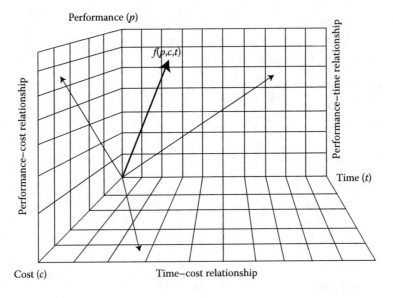

FIGURE 11.20
Cost–schedule–performance relationships.

value refers to the actual dollar value of work performed at a given point in time compared to planned cost for the work. This is different from the conventional approach of measuring actual versus planned, which is explicitly forbidden by C/SCSC. In the conventional approach, it is possible to misrepresent the actual content (or value) of the work accomplished. The work rate analysis technique can be useful in overcoming the deficiencies of the conventional approach. C/SCSC is developed on a work content basis using the following factors:

- The actual cost of work performed (ACWP), which is determined on the basis of the data from cost accounting and information systems
- The budgeted cost of work scheduled (BCWS) or baseline cost determined by the costs of scheduled accomplishments
- The budgeted cost of work performed (BCWP) or earned value, the actual work of effort completed as of a specific point in time

The following equations can be used to calculate cost and schedule variances for a work package at any point in time:

Cost variance = BCWP − ACWP

Percent cost variance = 100 × (cost variance/BCWP)

Schedule variance = BCWP − BCWS

Percent schedule variance = 100 × (schedule variance/BCWS)

ACWP and remaining funds = target cost (TC)

ACWP + cost to complete = estimated cost at completion (EAC)

The above characteristics of C/SCSC and R&M 2000 have undergone application modifications in recent years. Several new systems of cost control are now available in practice. The essential elements of cost control in any new approach are discussed in the section that follows.

Elements of Cost Control

Cost control, in the context of cost management, refers to the process of regulating or rectifying cost attributes to bring them within acceptable levels. Because of the volatility and dynamism often encountered in oil and gas projects, it is imperative to embrace the following project cost-control practices as presented in PMBOK:

- Influence the factors that create changes to the cost baseline
- Ensure requested changes are agreed upon
- Manage the actual changes when and as they occur
- Assure that potential cost overruns do not exceed authorized funding (by period and in total)
- Monitor cost performance to detect and understand variances from the cost baseline
- Record all appropriate changes accurately against the cost baseline
- Prevent incorrect, inappropriate, or unapproved changes from being included in cost reports
- Inform appropriate stakeholders or approved changes
- Act to bring expected cost overruns within acceptable limits
- Use earned value technique (EVT) to track and rectify cost performance

---

Contemporary Earned Value Technique

This section details the elements of a contemporary EVT. EVT is used primarily for cost-control purposes. The technique involves developing important diagnostic values for each schedule activity, work package, or control element. Although the definitions presented below are similar to those in the foregoing C/SCSC discussions, there are shades of differences that are important to highlight. The definitions according to PMI's PMBOK are summarized below:

Planned value (PV): This is the budgeted cost for the work scheduled to be completed on an activity or WBS element up to a given point in time.

Earned value (EV): This is the budgeted amount for the work actually completed on the schedule activity or WBS component during a given time period.

Actual cost (AC): This is the total cost incurred in accomplishing work on the schedule activity or WBS component during a given time period. AC must correspond in definition, scale, units, and coverage to whatever was budgeted for PV and EV. For example, direct hours only, direct costs only, or all costs including indirect costs.

The PV, EV, and AC values are used jointly to provide performance measures of whether or not work is being accomplished as planned at any given point in time. The common measures of project assessment are cost variance (CV) and schedule variance (SV).

Cost variance (CV): This equals earned value minus actual cost. The cost variance at the end of the project will be the difference between the budget at completion (BAC) and the actual amount expended.

$$CV = EV - AC$$

Schedule variance (SV): This equals earned value minus planned value. Schedule variance will eventually become zero when the project is completed because all of the planned values will have been earned.

$$SV = EV - PV$$

Cost performance index (CPI): This is an efficiency indicator relating earned value to actual cost. It is the most commonly used cost-efficiency indicator. CPI value less than 1.0 indicates a cost overrun of the estimates. CPI value greater than 1.0 indicates a cost advantage (under-run) of the estimates.

$$CPI = \frac{EV}{AC}$$

Cumulative CPI (CPI^C): This is a measure that is widely used to forecast project costs at completion. It equals the sum of the periodic earned values (Cum. EV) divided by the sum of the individual actual costs (Cum. AC).

$$CPI^C = \frac{EV^C}{AC^C}$$

Schedule performance index (SPI): This is a measure that is used to predict the completion date of a project. It is used in conjunction with CPI to forecast project completion estimates.

$$SPI = \frac{EV}{PV}$$

Estimate to complete (ETC) based on new estimate: Estimate to complete equals the revised estimate for the work remaining as determined by the performing organization. This is an independent noncalculated estimate to complete for all the work remaining. It considers the performance or production of the resources to date. The calculation of ETC uses two alternate formulas based on earned value data.

ETC based on atypical variances: This calculation approach is used when current variances are seen as *atypical* and the expectations of the project team are that similar variances will *not* occur in the future.

$$ETC = BAC - EV^C$$

where BAC = budget at completion.

ETC based on typical variances: This calculation approach is used when current variances are seen as *typical* of what to expect in the future.

$$ETC = \frac{BAC - EV^C}{CPI^C}$$

Estimate at completion (EAC): This is a forecast of the most likely total value based on project performance. EAC is the projected or anticipated total final value for a schedule activity, WBS component, or project when the defined work of the project is completed. One EAC forecasting technique is based upon the performance organization providing an estimate at completion. Two other techniques are based on earned value data. The three calculation techniques are presented below. Each of the three approaches can be effective for any given project because it can provide valuable information and signal if the EAC forecasts are not within acceptable limits.

EAC using a new estimate: The approach calculates the actual costs to date plus a new ETC that is provided by the performing organization. This is most often used when past performance shows that the original estimating assumptions were fundamentally flawed or that they are no longer relevant due to a change in project operating conditions.

$$EAC = AC^C + ETC$$

EAC using remaining budget: In this approach, EAC is calculated as cumulative actual cost plus the budget that is required to complete the remaining work, where the remaining work is the budget at completion minus the earned value. This approach is most often used when current variances are seen as *atypical* and the project management team expectations are that similar variances will not occur in the future.

$$EAC = AC^C + (BAC - EV)$$

where (BAC – EV) = remaining project work = remaining PV.

EAC using cumulative CPI: In this approach, EAC is calculated as actual costs to date plus the budget that is required to complete the remaining project work, modified by a performance factor. The performance factor of choice is usually the cumulative CPI. This approach is most often used when current variances are seen as *typical* of what to expect in the future.

$$EAC = AC^C + \frac{(BAC - EV)}{CPI^C}$$

Other important definitions and computational relationships among the earned value variables are

Earned $\rightarrow$ Budgeted cost of work actually performed

Planned $\rightarrow$ Budgeted cost of work scheduled

Actual $\rightarrow$ Cost of actual work performed

Ending CV = budget at completion − actual amount spent at the end
= BAC − EAC
= VAC (variance at completion)
EAC = ETC + AC
= (BAC − EV) + AC
= AC + (BAC − EV)
ETC = EAC − AC
= BAC − EV

Figure 11.21 illustrates the relationships among the earned value variables discussed above.

Activity-Based Costing

Activity-based costing (ABC) has emerged as an effective costing technique for industrial projects. The major motivation for ABC is that it offers an improved method to achieve enhancements in operational and strategic decisions. ABC offers a mechanism to allocate costs in direct proportion to the activities that are actually performed. This is an improvement over the traditional way of generically allocating costs to departments. It also improves the conventional approaches to allocating overhead costs. In general, ABC is a method for estimating the resources required to operate an organization's business activities, produce its products, and provide services to its clients.

The ABC methodology assigns resource costs through activities to the products and services provided to its customers. It is generally used as a tool for understanding product and customer costs with respect to project profitability. ABC is also frequently used to formulate strategic decisions such as product pricing, outsourcing, and process improvement efforts.

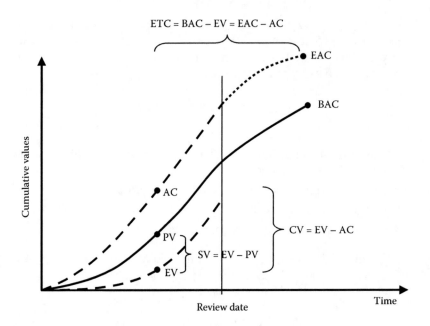

FIGURE 11.21
Graphical plot of earned value performance analysis.

The use of PERT/CPM, precedence diagramming, the critical resource diagramming method, and WBS can facilitate the decomposition or breakdown of a task to provide information for ABC. Some of the potential impacts of ABC on a production line include the following:

- Identification and removal of unnecessary costs
- Identification of the cost impact of adding specific attributes to a product
- Indication of the incremental cost of improved quality
- Identification of the value-added points in a production process
- Inclusion of specific inventory carrying costs
- Provision of a basis for comparing production alternatives
- Ability to assess "what-if" scenarios for specific tasks

ABC is just one component of the overall activity-based management (ABM) in an organization, and thus has its limitations, as well. ABM involves a more global management approach to the planning and control of organizational endeavors. This requires consideration for product planning, resource allocation, productivity management, quality control, training, line balancing, value analysis, and a host of other organizational responsibilities. In the implementation of ABC, several issues must be considered:

- Level and availability of resources committed to developing activity-based information and cost
- Duration and level of effort needed to achieve ABC objectives
- Level of cost accuracy that can be achieved by ABC
- Ability to track activities based on ABC requirements
- Challenge of handling the volume of detailed information provided by ABC
- Sensitivity of the ABC system to changes in activity configuration

From ABM to ABC, there are both qualitative and quantitative aspects of tracking, managing, and controlling costs. Unfortunately, many attempts to use ABC often degenerate into conceptual arm-waving rather than real quantitative accountability. To be successful, the SMART principle can be applied for developing ABC strategies. Under ABM and ABC, cost tracking must satisfy the following SMART requirements:

Specific: Cost tracking must be specific so as to facilitate accountability.

Measurable: Cost tracking must be measurable.

Aligned: Cost tracking must be aligned with the organization's goals.

Realistic: Cost tracking must be realistic and within the organization's capability.

Timed: Cost tracking must be timed in order to avoid ambiguities.

Also, to increase the effectiveness of ABC, an organization should use parametric cost techniques, which utilize project characteristics (parameters) to develop mathematical models for cost management. In summary, oil and gas project cost management requires more prudent approaches compared to conventional cost management practices. Frequent changes in science, technology, and engineering undertakings lead to dynamism of cost scenarios. Consequently, step-by-step tractable approaches must be used.

Strategic Capital Rationing

Strategic budget allocation involves sharing limited resources between several operational objectives in a strategic cost management challenge. Considering each budget allocation as an investment, the valuation of such investment represents a measure of outcome relative to the input of budgetary resources in the presence of risks. Even in a nonprofit environment, investment analysis and management are essential for ensuring operational effectiveness. Budget and investment analysis can serve any of the following purposes:

- A plan for resources expenditure
- A project selection criterion
- A projection of project policy
- A basis for project control
- A performance measure
- A standardization of resource allocation
- An incentive for process improvement

General Formulation of Budget Allocation Problem

A general formulation for capital rationing involves selecting a combination of projects that will optimize the ROI or maximize system effectiveness. A general formulation of the capital budgeting (Badiru and Omitaomu, 2007) problem is presented below:

$$\text{Maximize } z = \sum_{i=1}^{n} v_i x_i$$

$$\text{Subject to } \sum_{i=1}^{n} c_i x_i \leq B$$

$$x_i = 0,1; \quad i = 1,...,n$$

where
n = number of projects
v_i = measure of performance for project i (e.g., present value)
c_i = cost of project i
x_i = indicator variable for project i
B = budget availability level

A solution of the above model will indicate what projects should be selected in combination with other projects. The example that follows illustrates a capital rationing problem. Planning a portfolio of projects is essential in resource-limited projects. The capital-rationing formulation that follows demonstrates how to determine the optimal combination of project investments (or budget allocations) so as to maximize total ROI or total system effectiveness. Suppose a project analyst is given N projects, $X_1, X_2, X_3,..., X_N$, with the requirement to determine the level of investment in each project so that total investment return is maximized subject to a specified limit on available budget. We assume that the projects are not mutually exclusive. The investment in each project starts at a base level b_i ($i = 1, 2, ..., N$) and increases by variable increments

k_{ij} ($j = 1, 2, 3, \ldots, K_i$), where K_i is the number of increments used for project i. Consequently, the level of investment in project X_i is defined as follows:

$$x_i = b_i + \sum_{j=1}^{K_i} k_{ij}$$

where

$$x_i \geq 0, \quad \forall i$$

For most cases, the base investment will be zero. In those cases, we will have $b_i = 0$. In the modeling procedure used for this problem, we have

$$X_i = \begin{cases} 1 & \text{if the investment in project } i \text{ is greater than zero} \\ 0 & \text{otherwise} \end{cases}$$

and

$$Y_{ij} = \begin{cases} 1 & \text{if } j\text{th increment of alternative } i \text{ is used} \\ 0 & \text{otherwise} \end{cases}$$

The variable x_i is the actual level of investment in project i, while X_i is an indicator variable indicating whether or not project i is one of the projects selected for investment. Similarly, k_{ij} is the actual magnitude of the jth increment while Y_{ij} is an indicator variable that indicates whether or not the jth increment is used for project i. The maximum possible investment in each project is defined as M_i, such that

$$b_i \leq x_i \leq M_i$$

There is a specified limit, B, on the total budget available to invest, such that

$$\sum_i x_i \leq B$$

There is a known relationship between the level of investment, x_i, in each project and the expected return, $R(x_i)$. This relationship will be referred to as the *utility function, f(.)*, for the project. The utility function may be developed through historical data, regression analysis, and forecasting models. For a given project, the utility function is used to determine the expected return, $R(x_i)$, for a specified level of investment in that project. That is

$$R(x_i) = f(x_i) = \sum_{j=1}^{K_i} r_{ij} Y_{ij}$$

where r_{ij} is the incremental return obtained when the investment in project i is increased by k_{ij}. If the incremental return decreases as the level of investment increases, the utility function will be concave. In that case, we will have the following relationship:

$$r_{ij} \geq r_{ij+1}$$

$$r_{ij} - r_{ij+1} \geq 0$$

Thus

$$Y_{ij} \geq Y_{ij+1}$$

$$Y_{ij} - Y_{ij+1} \geq 0$$

so that only the first n increments ($j = 1, 2, \ldots, n$) that produce the highest returns are used for project i. Figure 11.22 shows an example of a concave investment utility function.

If the incremental returns do not define a concave function, $f(x_i)$, then one has to introduce the inequality constraints presented above into the optimization model. Otherwise, the inequality constraints may be left out of the model, since the first inequality, $Y_{ij} \geq Y_{ij+1}$, is always implicitly satisfied for concave functions. The objective is to maximize the total return. That is

$$\text{Maximize } Z = \sum_i \sum_j r_{ij} Y_{ij}$$

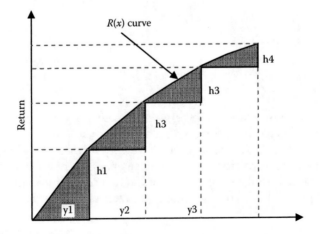

FIGURE 11.22
Utility curve for investment yield.

Subject to the following constraints:

$$x_i = b_i + \sum_j k_{ij} Y_{ij} \quad \forall i$$

$$b_i \leq x_i \leq M_i \quad \forall i$$

$$Y_{ij} \geq Y_{ij+1} \quad \forall i, j$$

$$\sum_i x_i \leq B$$

$$x_i \geq 0 \quad \forall i$$

$$Y_{ij} = 0 \text{ or } 1 \quad \forall i, j$$

The modeling approach presented above is an illustration of how operations research can be applied to complex budget allocation problems. The model can be adapted and modified for budget decision and capital rationing applications in oil and gas projects. Oil and gas analysts in unique operating environments can customize the basic modeling approach to fit their respective budget allocation practices and operational requirements.

References

Badiru, A. B. 1993. *Managing Industrial Development Projects: A Project Management Approach*, Van Nostrand/John Wiley, New York, NY.

Badiru, A. B. 1996. *Project Management in Manufacturing and High Technology Operations*, 2nd edition, John Wiley & Sons, New York, NY.

Badiru, A. B. 2009. *STEP Project Management: Guide for Science, Technology, and Engineering Projects*, Taylor & Francis/CRC Press, Boca Raton, FL.

Badiru, A. B., A. Badiru, and A. Badiru. 2008. *Industrial Project Management: Concepts, Tools, and Techniques*, Taylor & Francis/CRC Press, Boca Raton, FL.

Badiru, A. B. and O. A. Omitaomu. 2007. *Computational Economic Analysis for Engineering and Industry*, Taylor & Francis/CRC Press, Boca Raton, FL.

Badiru, A. B. and P. S. Pulat. 1995. *Comprehensive Project Management: Integrating Optimization Models, Management Principles, and Computers*, Prentice-Hall, Englewood Cliffs, NJ.

12

Project Risk Analysis

Risk management is an essential and integral part of project management in the oil and gas industry. For an oil and gas infrastructure project, risk management can be carried out effectively by investigating and identifying the sources of risks associated with each activity of the project. These risks can be assessed or measured in terms of likelihood and impact. Because of the exploration basis of the oil and gas industry, a different and diverse set of risk concerns will be involved. So, as risks are assessed for managerial processes, technical and exploration risks must also be assessed. Risk and estimation of reserves constitute a major portion of project risk analysis in the oil and gas industry. The major activities in oil and gas risk analysis consist of feasibility studies, design, transportation, utility, survey works, construction, permanent structure works, mechanical and electrical installations, maintenance, and so on. This chapter addresses only selected topics from the list.

Definition of Risk

Risk is often ambiguously defined as a measure of the probability, level of severity, and exposure to all hazards for a project activity. Practitioners and researchers often debate the exact definition, meaning, and implications of risk. Two alternate definitions of risk are presented below:

Risk is an uncertain event or condition that, if it occurs, has a positive or negative effect on a project objective.

Risk is an uncertain event or set of circumstances that, should it occur, will have an effect on the achievement of the project's objectives.

In this book, we present the following definition of risk management:

Risk management is the state of having a contingency ready to respond to the impact (good or bad) of occurrence of risk, such that risk mitigation or risk exploitation becomes an intrinsic part of the project plan.

For any oil and gas project, there is always a chance that things will not turn out exactly as planned. Thus, project risk pertains to the probability of uncertainties of the technical, schedule, and cost outcomes of the project. All oil and gas projects are complex and they involve risks in all the phases of the project starting from the feasibility phase to the operational phase. These risks have a direct impact on the project schedule, cost, and performance. These projects are inherently complex and volatile with many variables. A proper risk mitigation plan, if developed for identified risks, would ensure better and smoother achievement of project goals within the specified time, cost, and technical requirements. Conventional project management techniques, without a risk management component, are not sufficient to ensure time, cost, and quality achievement of a large-scale project, which may be mainly due to changes in scope and design, changes in government policies and regulations, changes in industry agreement, unforeseen inflation, underestimation and improper estimation. Projects, which are exposed to such risks and uncertainty, can be effectively managed with the incorporation of risk management throughout the projects' life cycle.

Sources of Project Uncertainty

Project risks originate from the uncertainty that is present in all projects to one extent or another. A common area of uncertainty is the size of project parameters, such as time, cost, and quality with respect to the expectations of the project. For example, we may not know precisely how much time and effort will be required to complete a particular task. Possible sources of uncertainty include the following:

- Poor estimates of time and cost
- Lack of a clear specification of project requirements
- Ambiguous guidelines about managerial processes
- Lack of knowledge of the number and types of factors influencing the project
- Lack of knowledge about the interdependencies among activities in the project
- Unknown events within the project environment
- Variability in project design and logistics
- Project scope changes
- Varying direction of objectives and priorities

Using a fishbone diagram, Figure 12.1 illustrates examples of various pathways to project risks leading to project failure. Notice how uncertainty in one factor at one level can influence the outlook of another factor at a different level.

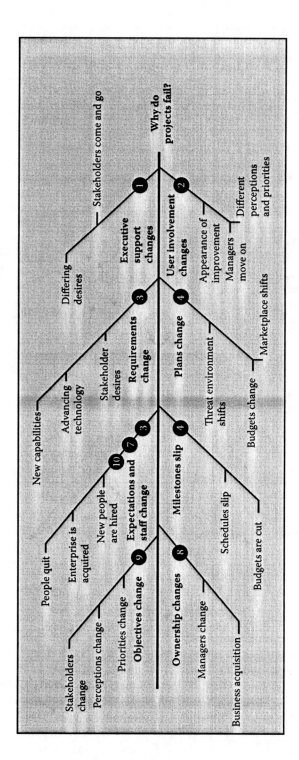

FIGURE 12.1
Fishbone diagram of risk events leading to project failure.

Impact of Government Regulations

Risks can be mitigated, not eliminated. In fact, risk is the essence of any enterprise. In spite of government regulations designed to reduce accident risks in the energy industry, accidents will occasionally happen. Government regulators can work with oil and gas producers to monitor data and operations. This will only preempt a fraction of potential risks of incidents. For this reason, regulators must work with operators to ensure that adequate precautions are taken in all operating scenarios. Government and industry must work together in a risk mitigation partnership, rather than in an adversarial "lording" relationship. There is no risk-free activity in the oil and gas business. For example, many of the recent petroleum industry accidents involved human elements—errors, incompetence, negligence, and so on. How do you prevent negligence? You can encourage non-negligent operation or incentivize perfect record, but human will still be human when bad things happen. Operators and regulators must build on experiences to map out the path to risk reduction in operations. Effective risk management requires a reliable risk analysis technique. Below is how to deal with risk management:

- Avoid
- Assign
- Assume
- Mitigate
- Manage

Below is a four-step process of managing risk

- Step 1—Identify the risks
- Step 2—Assess the risks
- Step 3—Plan risk mitigation
- Step 4—Communicate risk

We must venture out on the risk limb in order to benefit from what the project offers. Consider the quote below:

> Behold the lowly Turtle—he only makes
> progress when he sticks his neck out
>
> **James Conan Bryan, 1954***

---

* http://www.quotationspage.com/quote/2830.htm

Let us take another look at the basic definition:

Risk—"Potential realization of an unwanted negative consequence"
Reward—"Potential realization of a desired positive consequence"

A master list of risk management involves the following:

- New technology
- Functional complexity
- New versus replacement
- Leverage on company
- Intensity of business need
- Interface existing applications
- Staff availability
- Commitment of team
- Team morale
- Applications knowledge
- Client information systems (IS) knowledge
- Technical skills availability
- Staff conflicts
- Quality of information available
- Dependability on other projects
- Conversion difficulty
- End-date dictate
- Conflict resolution mechanism
- Continued budget availability
- Project standards used
- Large/small project
- Size of team
- Geographic dispersion
- Reliability of personnel
- Availability of support organization
- Availability of champion
- Vulnerability to change
- Stability of business area
- Organizational impact
- Tight time frame

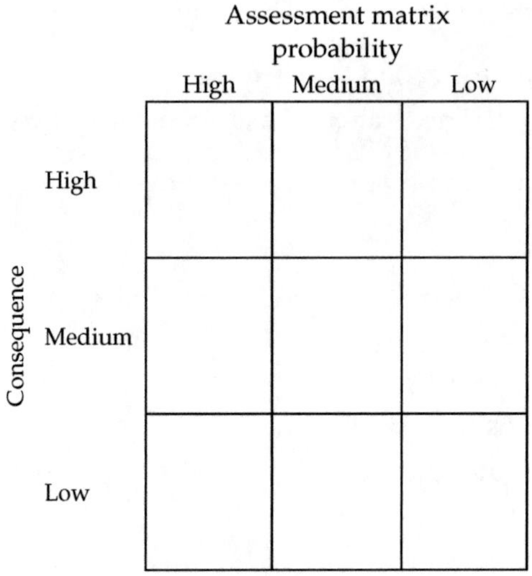

Assessment matrix
probability

FIGURE 12.2
Risk assessment matrix.

- Turnover of key people
- Change budget accepted
- Change process accepted
- Level of client commitment
- Client attitude toward IS
- Readiness for takeover
- Client design participation
- Client participation in acceptance test
- Client proximity to IS
- Acceptance process

A potential layout for risk assessment matrix is presented in Figure 12.2. Possible risk response planning can follow the following options:

- Accept—Do nothing because the cost to fix is more expensive than the expected loss
- Avoid—Elect not to do part of the project associated with the risk
- Contingency planning—Frame plans to deal with risk consequence and monitor risk regularly (identify trigger points)
- Mitigate—Reduce the probability of occurrence, the loss, or both
- Transfer—Outsource

Case Example of Risk Analysis

Following on the heels of the chapter on construction management, this section presents a case example of project risk management for an underground construction of metro rail in the capital city of a developing nation in South Asia (Sarkar and Dutta [1]). Although this pertains to the construction of a city transportation system, the problem scenario is not unlike what an oil and gas organization might face. The project considered for analysis is the construction of an underground corridor for metro rail operations in the capital city of an emerging economic nation in South Asia. Phase I of the project is about 65 km with 59 stations. The estimated capital cost of Phase I is about INR 105 billion. The project under study for this research work is a part of Phase I. The scope of work is the design and construction of a 6.6 km underground metro corridor with six underground stations and a twin tunnel system. The underground stations are referred to as S1, S2, ..., S6. Here, S6 is the terminal station equipped with an overrun tunnel (where an up train can be converted to a down train). The client is a public sector company floated jointly by the state and central government. The principal contractor is a joint venture (JV) of three foreign contractors and two domestic contractors. The type of contract is a design build turnkey (DBT) where the principal contractor is required to design the underground corridor and execute the project. The project cost for the execution of 6.6 km is about INR 18 billion. The contract period is about 5 years (exclusively for execution). The feasibility phase of the project is an additional 5 years. The activity chart of the sample stretch under analysis consisting of the tunnel connecting two stations S5 and S6; S6 station box and the overrun tunnel succeeding S6 station box is provided in Table 12.1. The corresponding network diagram is given in Figure 12.3.

Risk Analysis by Expected Value Method

Reviewing the available literature, we observed that no well-defined technique is available for quantitative risk analysis for a complex infrastructure transportation project like construction of underground corridor for metro rail operations. Also, we observed that the expected value method (EVM) has the potentiality of quantifying the risks in terms of likelihood, impact, and severity. This would enable the project authorities to classify the risks according to the severity, adopt mitigation measures, and allocate contingency funds accordingly.

Thus, this method appears to be quite suitable for risk analysis for the underground corridor metro rail construction, which has risks and uncertainties

TABLE 12.1

Major Activities and Their Time Estimates in the Underground Corridor
Construction Project (Terminal Station S_6)

Activity	Description	Immediate Predecessors	Duration (Days)	ES	EF	LS	LF
A	Feasibility studies	—	1875	0	1875	0	1875
B	Design	A	295	1875	2170	1985	2280
C	Technology selection	A	90	1875	1965	1875	1965
D	Traffic diversion	B,E	475	2280	2755	2280	2755
E	Utility diversion	C	315	1965	2280	1965	2280
F	Survey works	B,E	290	2280	2570	2821	3111
G	Shoulder/king piles	D	356	2755	3111	2755	3111
H	Timber lagging	C	240	1965	2205	2871	3111
I	Soil excavation	G,F,H	330	3111	3411	3111	3441
J	Rock excavation	L,R	165	2655	2820	3276	3441
K	Fabrication and erection of construction decks	C	170	1965	2135	2941	3111
L	Fabrication and erection of steel struts	C	690	1965	2655	2421	3111
M	Rock anchor installation	N,O	285	2280	2565	3156	3441
N	Shotcreting and rock bolting	L,R	120	2655	2775	2871	2991
O	Subfloor drainage	Q	170	2110	2280	2821	2991
P	Water proofing	I,K,J,M	120	3441	3561	3441	3561
Q	Diaphragm wall construction	C	145	1965	2110	2604	2749
R	Top down construction	Q	122	2110	2232	2749	2871
S	Permanent structure	N,O	570	2280	2850	2991	3561
T	Mechanical/electrical installations and services	P,S	225	3561	3786	3561	3786
U	Backfilling and restoration works	N,O	225	2280	2505	3561	3786

Note: ES: early start; EF: early finish; LS: late start; LF: late finish.

involved in all phases of the project. We assume a network of deterministic
time and cost. We also assume that the critical path model network has "*N*"
activities that are indicated by $j = (1,\ldots,N)$ and there are "*M*" risk sources
indicated by $i = (1,\ldots,M)$. Define the variables as follows:

L_{ij}: Likelihood of ith risk source for jth activity

W_{ij}: Weightage of ith risk source for jth activity

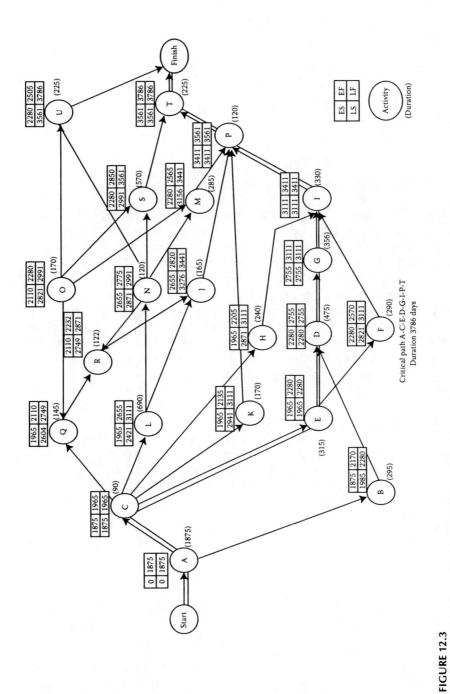

FIGURE 12.3
Network diagram for underground corridor construction project.

I_{ij}: Impact of ith risk source for jth activity

CLF_j: Composite likelihood factor for jth activity

CIF_j: Composite impact factor for jth activity

BTE_j: Base time estimate for jth activity

BCE_j: Base cost estimate for jth activity

CC_j: Corrective cost for jth activity

CT_j: Corrective time for jth activity

RC_j: Risk cost for jth activity

RT_j: Risk time for jth activity

EC_j: Expected cost for jth activity

ET_j: Expected time for jth activity

Base time estimate (BTE) of the project is the estimated basic project duration determined by critical path method of the project network. Similarly, the estimated basic cost of project determined by the cost for each activity is termed as the base cost estimate (BCE). The BTE and BCE data of all the major activities of the project have been obtained as per the detailed construction drawings, method statement and specifications for the works collected from the project. The corresponding corrective time (CT) or the time required to correct an activity in case of a failure due to one or more risk sources for each activity and their corresponding corrective cost (CC) have been estimated based on the personal experiences of the first author and have been tabulated. An activity may have several risk sources each having its own likelihood of occurrence. The value of likelihood should range between 0 and 1. The likelihood of failure (L_{ij}) defined above, of the identified risk sources of each activity were obtained through a questionnaire survey. The target respondents were experts and professionals involved in and associated with the project under analysis and also other similar projects. The corresponding weightage (W_{ij}) of each activity has also been obtained from the feedback of the questionnaire survey circulated among experts. The summation of the weightages should be equal to 1.

$$\sum_{i=1}^{M} W_{ij} = 1, \quad \text{for all } j\,(j = 1, \ldots, N)$$

The weightages can be based on local priority (LP) where the weightages of all the subactivities of a particular activity equal 1. Also, weightages can be based on global priority (GP) where the weightages of all the activities of the project equal 1. The mean of all the responses should desirably be considered for analysis. Inconsistent responses can be modified using a second-round questionnaire survey using the Delphi technique. The next step is to compute the risk cost (RC) and risk time (RT) of the activities of the project. RC and RT for an activity can be obtained from the following relationship:

$$\text{Risk cost for activity } j \ (RC)_j = (CC)_j \times L_j \quad \text{for all } j$$

$$\text{Risk time for activity } j \ (RT)_j = (CT)_j \times L_j \quad \text{for all } j$$

The total risk time for an activity is the summation of the risk time of all the subactivities along the critical path. The likelihood (L_{ij}) of all risk sources for each activity j can be combined and expressed as a single composite likelihood factor $(CLF)_j$. The weightages (W_{ij}) of the risk sources of the activities are multiplied with their respective likelihoods to obtain the CLF for the activity. The relationship of computing the CLF as a weighted average is given below:

$$\text{Composite likelihood factor } (CLF)_j = \sum_{i=1}^{M} L_{ij} W_{ij} \quad \text{for all } j$$

$$0 \leq L_{ij} \leq 1 \quad \text{and} \quad \sum_{i=1}^{M} W_{ij} = 1 \quad \text{for all } j$$

The impact of a risk can be expressed in terms of the effect caused by the risk to the time and cost of an activity. This time impact and cost impact can be considered as the risk time and risk cost of the activity. A similar computation as that of likelihood can be done for obtaining a single combined composite impact factor (CIF) by considering the weighted average as per the relationship given below:

$$\text{Composite impact factor } (CIF)_j = \sum_{i=1}^{M} I_{ij} W_{ij}$$

$$0 \leq I_{ij} \leq 1 \quad \text{and} \quad \sum_{i=1}^{M} W_{ij} = 1 \quad \text{for all } j$$

Risk consequence or severity can be expressed as a function of risk likelihood and risk impact. Thus the numerical value will range from 0 to 1. This severity can also be expressed in terms of qualitative rating as "no severity" for value 0 and "extremely high severity" for value 1. The numerical value of the risk severity (RS) is obtained from the below-mentioned relationship:

$$\text{Risk consequence/severity } (RS)_j = L_j \times I_j \quad \text{for all } j$$

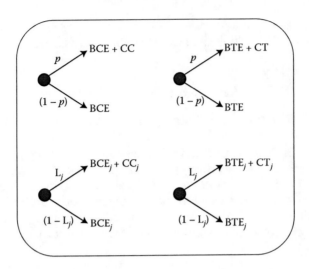

FIGURE 12.4
Decision tree structure.

The risk consequence derived from this equation measures how serious the risk is to project performance. Small values represent unimportant risks that might be ignored and large values represent important risks that need to be treated. The expected cost $(EC)_j$ and expected time $(ET)_j$ for each project activity and subsequently the computation of the expected project cost and time was carried out from the concept of the expected value (EV) of a decision tree analysis. The expected value is calculated as follows and is shown in Figure 12.4:

$$EV = \text{probability of occurrence } (p) \text{ [higher payoff]} + (1 - p) \text{ [lower payoff]}$$

$$\text{Expected cost } (EC)_j = L_j (BCE_j + CC_j) + (1 - L_j) BCE_j$$

$$= BCE_j + CC_j (L_j)$$

$$= BCE_j + RC_j \quad \text{for all } j$$

$$\text{Expected time } (ET)_j = L_j (BTE_j + CT_j) + (1 - L_j) BTE_j$$

$$= BTE_j + CT_j (L_j)$$

$$= BTE_j + RT_j \quad \text{for all } j$$

Risk Analysis

The sample stretch under analysis consists of a 530 m cut-and-cover tunnel connecting stations S_5 and S_6, a 290 m S_6 station box, and a 180 m cut-and-cover overrun tunnel adjoining the S_6 station box. S_6 station being the terminal station, the down trains toward this station after leaving station S_5 will travel through the 530 m cut-and-cover tunnel and enter the platforms of the terminal station S_6. After the commuters vacate the train at this terminal station, this down train will travel through the 180 m overrun tunnel and will be converted into an up line train that will travel from station S_6 to S_1. The activities of the sample stretch under analysis consist of the installation and erection of temporary supporting and retaining structures to enable construction by cut-and-cover technology and for the construction of permanent structures like tunnels and station boxes, which are RCC single boxes/twin boxes for tunnels and RCC boxes with intermediate concourse slab for station boxes.

We have considered some basic assumptions during the analysis. These assumptions are (i) the maximum cost overrun permissible is 25% of the basic cost estimate beyond which the project becomes less feasible and (ii) the maximum permissible time overrun for infrastructure projects is about 30% of the base time estimate, beyond which the feasibility of the project reduces. The common risk sources that are identified for all the activities in Table 12.1 and Figure 12.3 are provided in Table 12.2. This questionnaire was circulated among 67 experts having adequate experience in underground construction projects or similar infrastructure projects. These experts were required to respond with respect to the likelihood of occurrence and the weightages associated with each risk based on their experience. The methodology for receiving the filled-up questionnaires from the respondents was through personal approach, telephonic conversation, e-mails, and post.

The experts were designers, consultants, deputy project leaders, project managers, deputy project managers, CEOs, managing directors, area managers, people in charge of quality assurance/quality control, and safety, senior engineers, and project engineers of the principal contractor of the above project, the client organization, the consulting organization, major subcontractors of the above project, and other ongoing metro rail projects within the country. Of around 67 experts, 45 had responded to this study and the mean of all the responses of respective risk likelihoods and their associated weightages in the related activities have been considered. The inconsistent responses were revised by conducting a second-round questionnaire survey using the Delphi technique.

A sample of a part of a filled-up questionnaire consisting of the likelihood of risks and the weightage associated with the identified risks for the feasibility project risk (FPR) is presented later on. The value of likelihood (L_{ij}) varies from 0 to 1 and the sum of the weightages (W_{ij}) on LP basis is equal to 1. That is

$$(0.121 + 0.185 + 0.155 + 0.295 + 0.075 + 0.169)$$

TABLE 12.2

Identification and Classification of Risks Involved in the Project

S. No.	Risk Classification Nomenclature	Risk Description
1	FPR	Feasibility project risk
2	PEPR 1	Preexecution project risk—design risks
3	PEPR 2	Preexecution project risk—technology risks
4	EPR 1	Execution project risk—risks in traffic diversion works
5	EPR 2	Risks in utility diversion works
6	EPR 3	Risks in survey works
7	EPR 4	Risks in soldier piling and king piling works
8	EPR 5	Risks in timber lagging works
9	EPR 6	Risks in soil excavation works
10	EPR 7	Risks in rock excavation works
11	EPR 8	Risks in installation of construction decks
12	EPR 9	Risks in installation of steel struts
13	EPR 10	Risks in installation of rock anchors
14	EPR 11	Risks in shotcreting and rock bolting works
15	EPR 12	Risks in subfloor drainage works
16	EPR 13	Risks in waterproofing works
17	EPR 14	Risks in diaphragm wall construction
18	EPR 15	Risks in top down construction
19	EPR 16	Risks in permanent structure works
20	EPR 17	Risks in mechanical and electrical installation works
21	EPR 18	Risks in backfilling and restoration works

The corresponding composite likelihood factor is calculated as follows:

$$(\text{CLF})_j = \sum_{i=1}^{M} L_{ij} W_{ij} \quad \text{for all } j \quad (j = 1, \ldots, N) = 0.348$$

Similar tables were formulated for preexecution project risk (PEPR 1 and PEPR 2) and execution project risk (EPR 1 to EPR 18). The common risk sources of the project activities are shown in Table 12.3.

Expected Value Method for Project Risk Assessment

The network diagrams consisting of the major activities of the project have been drawn and their activity times (early start, early finish, late start, and

TABLE 12.3

Common Risk Sources of the Project Activities

S. No.	Risk Source Description
1	Risks due to delay in approval of detailed project report (DPR)
2	Land acquisition risks
3	Design risks
4	Technology selection risks
5	Approval and permit risks
6	Joint venture risks
7	Financial and investment risks
8	Political risks
9	Environment related risks
10	Geotechnical risks
11	Major or minor accidents during execution
12	Unforeseen heavy rain
13	Force majeure risks like flood, fire earthquake, and so on
14	Labor agitation and strikes
15	Inflation risks
16	Risks due to delayed payment from client
17	Risks due to delayed payment to subcontractor

late finish) have been calculated by forward and backward pass and then their critical path has been tracked out. The duration along the critical path is the longest duration path and is considered as the duration of the project. The BCE and BTE of each activity and subactivity of the project have been calculated as per the actual site data. The corrective cost and time for each activity have been assumed as a certain percentage (25–75%) of BCE and BTE, respectively, depending upon the severity and casualty caused by that risk. Each activity of the project as presented in Figure 12.3 has been analyzed at the subactivity level for computation of RC, RT, EC, ET, and risk severity. The detailed analysis for computation of risk cost and time for all the activities of the project is presented in Table 12.4. As per Figure 12.1, which represents the critical path diagram of the entire project of the underground corridor construction, and Table 12.4, for activity A (feasibility studies), the CLF is 0.348 as obtained from the feedback of the questionnaire survey.

The base cost estimate $(BCE)_j$ for the activity feasibility studies (A) is INR 240 million, the corrective cost $(CC)_j$ is INR 60 million (assumed in consultation with experts), the base time estimate $(BTE)_j$ is 1875 days, and the corrective time $(CT)_j$ is 1130 days (assumed in consultation with experts).

$$\text{Risk cost } (RC)_j = 0.348 \times 60 \times 10^6 = \text{INR } 20.88 \times 10^6$$
$$\text{Risk time } (RT)_j = 0.348 \times 1130 \text{ days} = 393.24 \text{ days}$$

TABLE 12.4

Expected Cost and Time Analysis for the Project

Activity	CLF_j	Base Cost Estimate (BCE)_j INR Million	Corrective Cost (CC)_j INR Million	Risk Cost (RC)_j INR Million	Base Time Estimate (BTE)_j Days	Corrective Time (CT)_j Days	Risk Time (RT)_j Days	Expected Cost (EC)_j INR Million	Expected Time (ET) Days	EC % Higher than BCE	ET % Higher than BTE
A	0.348	240	60	20.88	1875	1130	393.24	260.88	2268.24	8.7	20.97
B	0.356	110	32	11.39	295	245	87.22	121.39	382.22	10.36	29.57
C	0.27	40	10	2.7	90	85	22.95	42.7	112.95	6.75	25.5
D	0.319	50	11.9	3.80	475	355	113.25	53.80	588.25	7.59	23.84
E	0.262	100	82.4	21.59	315	267	69.95	121.59	384.95	21.59	22.21
F	0.186	10	8.66	1.61	290	247	45.94	11.61	335.94	16.11	15.84
G	0.28	220	176.46	49.41	356	356	99.68	269.41	455.68	22.46	28
H	0.252	20	15.97	4.03	240	180	45.36	24.03	285.36	20.13	18.9
I	0.377	150	122	45.99	330	205	77.29	195.99	407.29	30.66	23.42
J	0.419	80	56	23.46	165	140	58.66	103.46	223.66	29.33	35.55
K	0.398	120	108	42.98	170	113	44.97	162.98	214.97	35.82	26.46
L	0.367	300	245	89.92	690	485	178	389.92	868	29.97	25.8
M	0.345	50	49.2	16.97	285	250	86.25	66.97	371.25	33.95	30.26
N	0.343	80	70.3	24.11	260	185	63.46	104.11	323.46	30.14	24.41
O	0.306	60	58	17.75	170	130	39.78	77.78	209.78	29.58	23.4
P	0.384	120	83.2	31.98	120	95	36.48	151.95	156.48	26.62	30.4
Q	0.278	60	59.2	16.46	145	115	31.97	76.46	176.97	27.43	22.05
R	0.227	80	77.2	17.52	122	88	19.98	97.52	141.98	21.91	16.37
S	0.223	800	596.5	133.02	570	415	92.55	933.02	662.55	16.63	16.24
T	0.398	300	217.7	86.64	225	180	71.64	386.64	296.64	28.88	31.84
U	0.354	250	189.3	67.01	225	163	57.7	317.01	282.7	26.8	25.65
		3240	2329	729.20	3786		884.47	3969.20	4670.47	22.51	23.36

Note: Base time estimate and risk time is considered as the time estimate along the critical path (refer to Figure 12.3).

TABLE 12.5

Project Expected Cost and Time Analysis (Based on Questionnaire Survey Response)

Base Cost Estimate (INR Million)	Risk Cost (INR Million)	Base Time Estimate (Days)	Risk Time (Days)	Expected Cost (INR Million)	Expected Time (Days)
3240	729.2	3786	884.47	3969.2	4670.47

Then, we have the following:

$$\text{Expected cost (EC)}_j = \text{BCE}_j + \text{RC}_j = \text{INR } 260.88 \text{ million}$$

$$\text{Expected time (ET)}_j = \text{BTE}_j + \text{RT}_j = 2268.24 \text{ days}$$

A similar computation has been carried out for activities B, C, D, ..., and U (refer to Table 12.4). Table 12.5 shows the expected project cost and time analysis. The expected cost $(\text{EC})_{\text{Project}}$ of the entire project of underground corridor construction has been calculated as follows:

$$\text{Expected cost (EC)}_{\text{Project}} = \sum_{j=A}^{u} \text{EC}_j = \text{INR } 3969.20 \text{ million}$$

$$\text{Expected time (ET)}_{\text{Project}} = (\text{BTE})_{\text{Project}} + (\text{RT})_{\text{Project}}$$

$$= 3786 + 884.47 \text{ days} = 4670.47 \text{ days}$$

Thus, as per the analysis, the EC of the project is 22.51% higher than the BCE of the project. The ET of the project is 23.36% higher than the BTE. As per the basic assumptions considered for risk management analysis, the cost overrun should not exceed 25% of the estimated base cost and the time overrun should not be more than 30% of the estimated base time. Exceeding these limits would increase the chances of the project becoming less feasible. The risk management analysis predicts that the expected cost of the project is 22.51% higher than the estimated base cost. This situation is highly alarming as it is the upper limit of the permissible cost overrun. It requires meticulous planning and proper risk mitigation measures to enhance the probability of success of the project. The expected time predicted from the analysis is 23.36% higher than the estimated base time, which is close to the upper limit of the permissible time overrun. Thus, it is essential to judiciously follow the risk mitigation measures to ensure that the project is completed within the scheduled time frame.

Risk Severity Analysis Using the Concept of CLF and CIF

The product of the likelihood and impact of a risk can be considered as the severity of that risk. This concept can be extended for multiple risk sources in a work package, the likelihood and impact of which can be expressed in terms of CLF_j and CIF_j, respectively. Thus, for the underground corridor construction project, the scale for the classification of the risks is presented in Table 12.6 and computation of risk severity based on the equations presented earlier. Table 12.7 shows risk severity analysis.

TABLE 12.6

Risk Severity Classification

Severity	Classification
0.00–0.02	Very low
0.03–0.05	Low
0.06–0.15	Medium
0.16–0.20	High
0.21–1.00	Very high

TABLE 12.7

Risk Severity Analysis of Total Project Using the Concept of Composite Likelihood Factor (CLF) and Composite Impact Factor (CIF)

Description of Project Risk (Activity)	Composite Likelihood Factor (CLF)$_j$	Composite Impact Factor (CIF)$_j$	Severity Quantitative $CLF_j \times CIF_j$	Qualitative
FPR (A)	0.348	0.875	0.305	Very high
PEPR 1 (B)	0.393	0.868	0.341	Very high
PEPR 2 (C)	0.27	0.829	0.224	Very high
EPR 1 (D)	0.319	0.784	0.25	Very high
EPR 2 (E)	0.262	0.809	0.212	Very high
EPR 3 (F)	0.186	0.832	0.155	Medium
EPR 4 (G)	0.28	0.827	0.232	Very high
PER 5 (H)	0.252	0.818	0.206	High
PER 6 (I)	0.377	0.863	0.325	Very high
EPR 7 (J)	0.419	0.816	0.342	Very high
EPR 8 (K)	0.398	0.842	0.335	Very high
EPR 9 (L)	0.367	0.828	0.303	Very high
EPR 10 (M)	0.345	0.86	0.298	Very high
EPR 11 (N)	0.343	0.827	0.284	Very high
EPR 12 (O)	0.306	0.806	0.247	Very high
EPR 13 (P)	0.384	0.858	0.329	Very high
EPR 14 (Q)	0.278	0.872	0.242	Very high
EPR 15 (R)	0.227	0.837	0.19	High
EPR 16 (S)	0.223	0.811	0.181	High
EPR 17 (T)	0.513	0.845	0.433	Very high
EPR 18 (U)	0.254	0.544	0.138	Medium

TABLE 12.8

Outcome of Risk Severity Analysis by Expected Value and PERT

Very High	High	Medium	Low
Design	Traffic diversion	Survey	Nil
Technology selection	Top down construction	Backfilling and restoration	
Utility diversion	Timber lagging		
Soldier piles	Mechanical and		
King piles	electrical works		
Soil/rock excavation	Permanent structure		
Diaphragm wall			
Steel struts			
Rock anchors			
Shotcreting and rock bolting			

The risk severity analysis was also carried out by PERT analysis and the outcome of both the EVM and PERT analysis in terms of the severity of the major activities of the project is presented in Table 12.8.

Application of Monte Carlo Simulation

We apply the Monte Carlo simulation to predict the outcome of the ET and EC of all the possible paths of activities as represented in the network diagram of the project (Figure 12.3). The Monte Carlo simulation also takes into account the effects of the near critical paths becoming critical. By carrying out a detailed path analysis of the project network diagram, we observed that the path A–C–E–D–G–I–P–T has the longest duration of 3786 days. Hence, this path is considered as the critical path of the project network (refer Figure 12.3). The corresponding cost for the completion of activities along this path is INR 1220 million. It is also observed that the probability of the successful completion of the project within the stipulated time and cost frame is only 4%: (0.625 × 0.730 × 0.738 × 0.681 × 0.720 × 0.623 × 0.616 × 0.602 = 0.040).

Path A–B–D–G–I–P–T is a near critical path with a probability of about 4.8% for successful completion within the stipulated time and cost frame. There are chances of this path becoming critical. The application of the Monte Carlo simulation to the above path analysis resulted in Table 12.9.

From the above analysis, we observed that path 2 (A–C–E–D–G–I–P–T) has the longest duration of 3785.98 days and remains critical. The corresponding cost for the completion of all the activities along the critical path is INR 1222.8 million. The probability of the completion of path 2 or the critical path within the scheduled time is 50%. The probability of the successful completion of the near critical path or path 1 within the scheduled time is 84.13% ($Z = 1.009$, $P = 0.8413$). Also, the probability of the successful completion of all the paths within the scheduled time is 42.05%

TABLE 12.9

Outcome of Path Analysis of the Project Network Diagram Applying Monte Carlo Simulation

Path	Activity/Node	Path Duration (Days)	Cost (INR in Crores)
1	A–B–D–G–I–P–T	3676.17	119.28
2	A–C–E–D–G–I–P–T	3785.98	122.28
3	A–C–E–F–I–P–T	3244.88	96.17
4	A–C–H–I–P–T	2879.88	87.11
5	A–C–K–P–T	2479.67	82.09
6	A–C–L–J–P–T	3164.79	108.19
7	A–C–Q–R–J–P–T	2741.60	92.20
8	A–C–Q–O–S–T	3074.89	150.10
9	A–C–Q–O–U	2504.95	65.07

($P = 0.8413 \times 0.5 \times 1 \times 1 \times 1 \times 1 \times 1 \times 1 \times 1 = 0.4205$). Carrying out about 10,000 runs of the Monte Carlo simulation, the EC was found to have a value of INR 3532.9 million and the ET of the project was found to be 4351.12 days. The generalized risk management model for the underground corridor construction for the metro rail is proposed on the basis of the detailed analysis carried out. This model can be effectively implemented in the ongoing and upcoming metro rail projects across the nation. As a part of the formulation of risk mitigation strategies, the following risk response planning can be adapted by the project authority:

- Risk transfer
- Risk sharing
- Risk reduction
- Risk contingency planning
- Risk mitigation through insurance

Project risk management, which primarily comprises schedule and cost uncertainties and risks, should be essentially carried out for complex urban infrastructure projects such as the construction of an underground corridor for metro rail operations. In the current research work, we found that the number of major and minor risks involved during the construction of the project, from the feasibility to the completion of the execution, are large, and if not treated or mitigated properly, the probability of successful completion of the project within the stipulated time and cost frame will reduce. This will have a direct impact on the efficiency and profitability of the organization. As per the analysis carried out by EVM, based on the expert questionnaire survey, the expected project cost for the sample stretch under analysis (530 m tunnel from station S_5 to S_6, S_6 station box and 180 m overrun tunnel) is about 22.51% higher than the base cost estimate of the project. According to the basic assumptions made for the analytical procedure adopted, the maximum permissible cost overrun for

TABLE 12.10

Additional Project Details

Project Description	Details
Length of route	6569 m
Tunnel (by tunnel boring machine)	3811 m
Tunnel (by cut and cover method)	937 m
Station boxes	1821 m
Average depth of stations	15–20 m below ground level
Typical width of stations	Average 20 m
Typical length of stations	275–300 m
Design life	120 years for underground structures and 50 years for superstructures
Major Scope of Civil Engineering Works	
Excavation (soil)	1,090,000 cubic meters
Excavation (rock)	215,000 cubic meters
Concreting	300,000 cubic meters
Reinforcement	47,500 metric tons
Strutting	24,500 metric tons

the project is 25%. Thus, if proper project risk management is not carried out by the authority, the project may result in a cost and time overrun, which will ultimately reduce the feasibility of the successful completion of the project. The expected project time as obtained by the analysis is about 23.36% higher than the base time estimate of the project, the maximum permissible time overrun as per the basic assumptions being 30% of the base time estimate. This value is also quite alarming making the concerned authority feel the need for carrying out proper risk management for such complex infrastructure projects.

Hence, considering the results of all the analyses carried out in this case example, it can be concluded that for complex infrastructure projects like that of an underground corridor construction, based on EVM, about INR 0.82 million extra per day per station would be incurred if proper risk management is not followed to mitigate the anticipated risks. Thus, for six underground stations for this 6.6 km underground metro corridor package, approximately INR 4.92 million extra per day will have to be incurred by the project authorities. A major limitation of the model adopted for analysis is that the entire model being probabilistic, the outcome of the analysis is largely dependent on the opinion of the likelihood and weightages of the identified risks obtained from the expert questionnaire survey. Also, any sort of misinformation provided will result in erroneous results. Although at present, a very nominal percentage of identified risks can be insured under the existing "Contractors All Risk Policy," the potentiality of insurance and the means of making insurance a strong risk mitigation tool for the construction industry provide scope for future exploitation of this risk management approach. Table 12.10 presents additional project details for this illustrative example. Table 12.11 presents the sample questionnaire for the feasibility project risk.

TABLE 12.11

Sample Questionnaire for Feasibility Project Risk (FPR)

Risk Description	Likelihood (L_{ij})	Weightage $(LP)(W_{ij})$	Impact (I_{ij})
FPR 1: Feasibility Project Risk 1—Risks in Preparation of Feasibility Report			
Delay in submission of preliminary feasibility report	0.15	0.029	0.65
Delay in approval for carrying out detailed feasibility study	0.20	0.030	0.75
Delay in preparation and submission of detailed project report (DPR)	0.20	0.018	0.85
Delay in approval of DPR	0.30	0.044	0.90
CLF = 0.027			
CIF = 0.096		Total: 0.121	
FPR 2: Resettlement and Rehabilitation Risks			
Resettlement site not accepted by affected parties	0.35	0.085	0.95
Resettlement site very costly	0.15	0.055	0.80
Litigation by affected parties	0.45	0.035	0.95
Resistance and agitation by political parties	0.5	0.01	0.90
CLF = 0.059			
CIF = 0.167		Total: 0.185	
FPR 3: Preinvestment Risks			
Cancellation of project after bidding	0.1	0.023	0.90
Delay in setting of consortium (JV)	0.35	0.052	0.95
Prolonged delay in project finalization	0.3	0.08	0.80
CLF = 0.045			
CIF = 0.134		Total: 0.155	
FPR 4: Land Acquisition Risks			
Political interference	0.55	0.013	0.9
Delay in finalizing temporary rehabilation schemes	0.4	0.055	0.85
Public interference for changing the alignment	0.25	0.055	0.9
Interference of environmental activists	0.4	0.012	0.9
Delay due to interdepartmental issues	0.35	0.03	0.9
Delay in construction of diversion roads for existing traffic	0.2	0.014	0.85
Problems with the physical possession of land	0.65	0.116	0.95
CLF = 0.136			
CIF = 0.264		Total: 0.295	
FPR 5: Financial Closure Risks			
Project not bankable	0.2	0.035	0.85
Lenders not comfortable with project viability	0.15	0.005	0.75
Adverse investment climate	0.1	0.035	0.80
CLF = 0.011			
CIF = 0.061		Total: 0.075	

TABLE 12.11 (continued)

Sample Questionnaire for Feasibility Project Risk (FPR)

Risk Description	Likelihood (L_{ij})	Weightage $(LP)(W_{ij})$	Impact (I_{ij})
FPR 6: Permit and Approval Risks			
Delay in contractual clearances	0.2	0.023	0.80
Delay in project specific orders and approvals	0.25	0.019	0.85
Delay in the approval of major utilities (telecom cables, electrical cables, storm water drains, sewer lines, filtered and unfiltered water lines)	0.45	0.049	0.90
Delay in clearance from environmental and forest departments	0.5	0.078	0.95
CLF = 0.070			
CIF = 0.153		Total: 0.169	
$CLF_{Feasibility} = 0.348\ (0.027 + 0.059 + 0.045 + 0.136 + 0.011 + 0.070)$			
$CIF_{Feasibility} = 0.875(0.096 + 0.167 + 0.134 + 0.264 + 0.061 + 0.153)$		Grand total: 1	

Case Research of Spatial Risk Analysis of Oil Refineries within the United States

Zachary L. Schiff and William E. Sitzabee, Energy Risk Research Report, Department of Systems and Engineering Management, Air Force Institute of Technology, Dayton, OH, 2011.

Abstract

A risk analysis methodology is necessary to manage potential effects of oil refinery outages to the increasingly connected, interdependent critical infrastructure of the United States. This paper outlines an approach to develop a risk analysis methodology that incorporates spatial and coupling elements in order to develop a better understanding of risk. The methodology proposed in this paper utilizes a three-phase approach to look at both natural disaster and terrorist risk. Understanding the uncertainty involved with events that could shut down the petroleum energy sector can help make better decisions to manage risk to the government, people, and economy.

Index terms: geographic information systems (GIS), risk analysis, oil refineries, petroleum industry, critical infrastructure, interdependency, autocorrelation, coupling, spatial relationships.

Introduction

In the last decade, the United States has experienced first hand the devastating impacts of disasters, both natural and terrorist, to critical infrastructure. The events of the September 11, 2001 attacks (9/11); Hurricanes Ike, Katrina, and Rita; and British Petroleum's Deep Horizon oil accident illustrate the effects of a major disaster to the United States. The monetary costs of 9/11, Hurricane Katrina, and Deep Horizon oil accident are estimated at $110 billion, $81 billion, and $40 billion, respectively [2–4]. The nation's security, economy, and health are dependent on critical infrastructure to provide key services in order for the government, people, and businesses to function properly.

During Hurricanes Katrina and Rita, refinery capability was reduced by 13% and 14%, respectively. Due to reduced capacity, the hurricanes influenced gas prices to rise from $1.10 to $2.55 after the disasters [5]. The cost is an increase that has not been recovered from and has contributed to the economic recession. In addition, increased petroleum demand in the last 20 years has increased at a faster rate than refining capability to provide gas, diesel, and other petroleum products. According to GAO-09-87, refineries are producing at a level very near their maximum capacity across the United States [6]. As a result, a disaster, either natural or terrorist, could potentially result in large shortages for a given time period.

The Department of Defense (DoD) fuel costs represented nearly 1.2% of the total DoD spending during FY2000 and increased to nearly 3.0% by FY2008 [7]. Andrews [7] stated that over the same period, total defense spending doubled and fuel costs increased 500% from $3.6 billion to $17.9 billion. Nearly 97.7 billion barrels of jet fuel were consumed in FY2008 and represents nearly 71% of all fuel purchased in the DoD. According to the Air Force Infrastructure Energy Plan, the fuel bill for the Air Force exceeds $10 million dollars per day and every $10 per barrel fuel price increase drives costs up $600 million dollars per year [8]. In 2007, the Air Force spent $67.7 million on ground fuel energy and consumed 31.2 million gallons of petroleum. The ground fuel energy only accounts for 4% of all fuel costs [8]. The military is a large customer of oil refinery products and is dependent on petroleum to complete military operations.

In the last decade, the petroleum industry has experienced several examples of cascading failures, including Hurricanes Katrina and Rita. These experiences can provide useful data with regard to outages and consequences of the events. Integrating spatial analysis into the research provides two opportunities to advance risk management: (1) utilize spatial tools to analyze relationships that provide insight into how the system functions and (2) visually identify trends that are not obvious within data analysis. This paper outlines an approach to develop a modified risk equation incorporating interdependency and spatial relationships utilizing critical infrastructure analysis and geographical information systems and sciences.

Background

Critical Infrastructure

The USA Patriot Act of 2001 (P.L. 107-56 Section 1016e) contains the federal government's definition of critical infrastructure. It stated that critical infrastructure is the

> set of systems and assets, whether physical or virtual, so vital to the United States that the incapacity or destruction of such systems and assets would have a debilitating impact on security, national economic security, national public health or safety, or the combination of those matters.

The National Strategy for Homeland Security categorized critical infrastructure into 13 different sectors and they are as follows: agriculture, food, water, public health, emergency services, government, defense industrial base, information and telecommunications, energy, transportation, banking and finance, chemical industry and hazardous materials, and postal and shipping [9].

Approximately 85% of the national infrastructure is owned by private industry [10]. The relationship between government and private industry is complicated by government acting as both regulator and consumer. This is especially true within the energy sector, which is composed of electrical power, oil, and gas infrastructure [11]. The energy sector is connected physically and virtually to all other sectors and has been shown to cause cascading failures to other sectors.

The petroleum industry was split into five petroleum administration for defense districts (PADDs) based on geographic location during WWII [12]. Parformak [13] discussed geographic concentration of critical infrastructure across numerous sectors and policy methods for encouraging dispersion. Specifically, Texas and Louisiana (PADD 2) refineries account for over 43% of the total U.S. refining capacity [13]. Rinaldi, Peerenboom, and Kelly [14] discussed interdependencies, coupling and response behavior, and types of failures with respect to critical infrastructure across the United States.

Risk Analysis Methods

The Department of Homeland Security (DHS) introduced the risk function as a combination of threat, vulnerability, and consequence, displayed below as Equation 12.1 [15]. Lowrance [16] introduced risk as a measure of the probability and severity of adverse effects. Chertoff [15] defined consequence as the effect and loss resulting from event, vulnerability as the physical feature that renders an entity open to exploitation, and threat as a natural or man-made occurrence that has the potential to harm life, operations, or property.

$$\text{Risk} = f(\text{threat, vulnerability, consequence}) \qquad (12.1)$$

Solano [17] investigated vulnerability assessment methods for determining the risk of critical infrastructure, and spatial distribution appeared to be an area where research can be expanded. Rinaldi, Peerenboom, and Kelly [14] discussed the challenges of modeling multiple interdependent infrastructures due to volume of data required and that isolation of infrastructure does not adequately analyze the behavior of the system. Ahearne [18] discussed the appropriateness of the multiplicative use of the risk function and found that it is generally accepted for natural disasters. Chai et al. [19] utilized a social network analysis to evaluate the relationship between infrastructure risk and interdependencies. The study utilized a node and arc approach to determine the number of in and out degrees to show dependencies and coupling. Expanding this approach could potentially result in better quantification of coupling effects on critical infrastructure.

Mohtadi [20] presented extreme value analysis as a method to predict large-scale terrorism events. In the study, methods for measuring terrorism as a probabilistic risk were developed for terrorism risk, which is extreme and occurs infrequently. Paté-Cornell and Guikema [21] presented a model that utilized risk analysis, decision analysis, and elements of game theory to account for both the probabilities of scenario and objectives between the terrorists and the United States. In their research, they highlighted the importance of utilizing a multisource method for collecting data on terrorism risk, which includes expert opinion, output of other system analysis, and statistics from past events. Leung, Lambert, and Mosenthal [22] utilized the risk filtering, ranking, and management (RFRM) and hierarchal holographic modeling (HHM) to conduct a multilevel analysis of protecting bridges against terrorist attacks.

Geographic Information Systems Spatial Tools

Nearly 40 years ago, Tobler [23] stated that "nearly everything is related to everything else, but near things are more related than distant things." This became Tobler's First Law of Geography and is acknowledged as the foundation of geographic information systems and science. Longley, Goodchild, Maguire, and Rind [24] discussed spatial autocorrelation as a tool that allows us to describe the interrelatedness of events and relationships that exist across space. Griffith [25] discussed spatial autocorrelation as "a dependency exists between values of a variable ... or a systematic pattern in values of a variable across the locations on a map due to underlying common factors."

Methodology

The goal of this study is to establish a process and develop techniques that can be expanded to look at the risk to both the critical infrastructure system and critical components of the system. This is a three-phase study and is organized in the following manner: (1) assess and compile inventory of assets, risk components, and characteristics, (2) validate the natural disaster quantitative risk model with spatial and coupling effects, and (3) qualitatively assess terrorism risk utilizing coefficients from the quantitative model. Figure 12.5 shows the research process and provides an outline of the phase progression.

The first phase analyzed the factors that contribute to risk, the data available to characterize infrastructure, and the methodologies that are currently used to quantify risk. This chapter presents the first phase of the study, which resulted in the identification of two additional variables: (1) spatial relationship and (2) coupling effect. Equation 12.2 shows the modified risk equation, which is the focus of Phase II and Phase III.

$$\text{Risk} = f(\text{threat, vulnerability, consequence, spatial relationship,} \\ \text{coupling effect}) \tag{12.2}$$

The goal of the second phase is to better quantify the cumulative risk of cascading failures by including the spatial relationship and coupling effects.

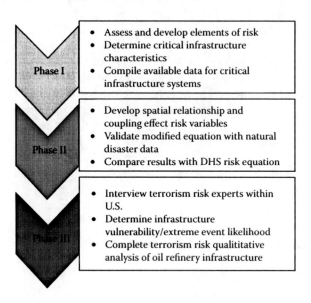

FIGURE 12.5
Overview of research phases and risk analysis methodologies.

To determine the spatial relationship, spatial autocorrelation will be utilized to develop a quantitative relationship of distance between critical infrastructures. The coupling effect will utilize a node-arc analysis to determine the number of connections to other infrastructures and expand on the research effort by Chai et al. [18]. Natural disaster data will be utilized to develop a case study to compare the results of the second phase to already established and validated methods of quantifying risk.

The third and final phase of this research is to utilize the spatial and coupling effect information to qualitatively assess terrorism risk. This phase will include phenomenological methods that will be utilized to interview experts in the terrorism field in order to develop threat and vulnerability data for petroleum infrastructure. The combination of results from the second and third phases will provide the foundation to complete a qualitative terrorism risk assessment. The goal of the third phase is to determine the highest terrorism risk to oil refinery infrastructure that could potentially result in cascading failures and large impacts to the United States.

Conclusion

The relationships between critical infrastructures are complicated and interdependencies that exist between infrastructures are not well defined. Incorporating spatial relationships and coupling effects into the risk equation proposes a better way to predict the effect of interdependencies, which have been shown to cause cascading failures during disaster events. Understanding and analyzing risk provides the decision and policy-making process better information in order to protect critical infrastructure across the United States. This chapter presents a new risk equation and a methodology to analyze and validate risk based on the modified risk equation. While spatial relationships and coupling have been identified as key factors to quantifying infrastructure risk, it appears that this is an area of study that requires further investigation. This research intends to further define the interdependencies of the infrastructure system in order to better quantify the overall risk to both the infrastructure system and individual parts of the system.

References

1. Sarkar, D. and Dutta, G. 2012, A framework for project risk management for the underground corridor construction of metro rail, *International Journal of Construction Project Management (IJCPM)*, 4(1), 1–19.

2. Berg, R. 2009, *Tropical Cyclone Report Hurricane Ike* (TCR- AL092008).
3. Knabb, R., Rhome, J., and Brown, D. 2005, *Tropical Cyclone Report Hurricane Katrina* (AL122005).
4. Thompson, W. C. 2002, *One Year Later: Fiscal Impact of 9/11 on New York City*, Report of the Comptroller of The City of New York, New York, NY, 2002.
5. Seesel, J. 2006, *Investigation of Gas Price Manipulation and Post-Katrina Gas Price Increases*, Report of Federal Trade Commission (FTC), Washington, DC, Spring 2006.
6. Rusco, F. 2008, *Refinery Outages Can Impact Petroleum Product Prices, but No Federal Regulations to Report Outages Exist*, US Department of Energy, Energy Information Administration, Washington, DC, March 2007.
7. Andrews, A. 2009, *Department of Defense Fuel Spending, Supply, Acquisition, and Policy*, Congressional Research Service (CRS) Report for Congress, Washington, DC, September 2009.
8. Air Force Energy Infrastructure Plan 2010. Available at: http://www.safie.hq.af.mil, accessed.
9. Bush, G. W. 2003, *The National Strategy for the Physical Protection of Critical Infrastructure*, White House Report, Washington, DC, February 2003.
10. Robinson, C., Woodard, J., and Varnado, S. 1999, Critical infrastructure: Interlinked and vulnerable, *Issues in Science and Echnology*, 10(4), 61–67.
11. Simonoff, J., Restrepo, C., Zimmerman, R., and Naphtali, Z. 2008, Analysis of electrical power and oil and gas pipeline failures, *International Federation for Information Processing*, E. Goetz and S. Shenoi, eds., Springer, Boston, vol. 253, pp. 381–294.
12. Cheryl Trench. 2001 *Oil Market Basics* [Online]. http://www.eia.doe.gov/pub/oil_gas/petroleum/analysis_publications/oil_market_basics/full_contents.htm.
13. Parformak, P. 2007, *Vulnerability of Concentrated Critical Infrastructure: Background and Policy Options*, Congressional Report, Washington, D.C.
14. Rinaldi, S., Peerenboom, J., and Kelly, T. 2001, Identifying, understanding, and analyzing critical infrastructure interdependencies, *IEEE Control Systems Magazine*, 21(4), 11–25, 2001.
15. Chertoff, M. 2009, *National Infrastructure Protection Plan*, Washington, D.C.
16. Lowrance, W. 1976, *Of Acceptable Risk*, William Kaufmann, Los Altos, CA.
17. Solano, E. 2010, *Methods for Assessing Vulnerability of Critical Infrastructure*, Institute for Homeland Security Solutions, RTI International, Research Triangle Park, NC.
18. Ahearne, J. F. 2010, *Review of the Department of Homeland Security's Approach to Risk Analysis*, National Research Council (NRC), Washington, D.C.
19. Chai, L. et al. 2008, Social network analysis of the vulnerabilities of interdependent critical infrastructures, *International Journal of Critical Infrastructures*, 3(4), 256–273.
20. Mohtadi, H. and Murshid, A. P. 2009, Risk of catastrophic terrorism: An extreme value approach, *Journal of Applied Econometrics*, 24(4), 537–559.
21. Pate-Cornell, E. and Guikema, S. 2002, Probabilistic modeling of terrorist threats: A systems analysis approach to setting priorities among countermeasures, *Military Operations Research*, 7(4), 5–20.
22. Leung, M., Lambert, J., and Mosenthal, A. 2004, A risk-based approach to setting priorities in protecting bridges against terrorist attacks, *Risk Analysis*, 24(4), 963–984.

23. Tobler, W. R. 1970, A computer movie simulating urban growth in the Detroit region, *Economic Geography*, 46(2), 234–240.
24. Longley, P., Goodchild, M., Maguire, D., and Rhind, D. 2011, *Geographic Information Systems and Science*. John Wiley & Sons, Inc., Hoboken, NJ.
25. Griffith, D. A. 2009, Spatial autocorrelation, *Journal of Geographical Systems*, 11(2), 117–140.

Appendix A

Reservoir Monitoring and Management: A Project Management Approach

Prof. Adedeji B. Badiru
Prof. Samuel O. Osisanya

<u>To accompany</u>:
Project Management for the Oil and Gas Industry: A World Systems Approach

---

Fundamentals of Project Management
==========O==========

"With project management, everything is possible." – Adedeji Badiru, 1987

What is a Project?

A unique one-of-kind endeavor with a specific goal that has a definite beginning and a definite end. – Traditional definition

"A temporary endeavor undertaken to create a unique product, service, or result." – PMBOK®

What is Project Management?

"The process of managing, allocating, and timing resources to achieve a given goal in an efficient and expeditious manner." – Adedeji Badiru, 1988

"Project management as the application of knowledge, skills, tools, and techniques to project activities to achieve project objectives." – PMBOK®

2

Project Life Cycle

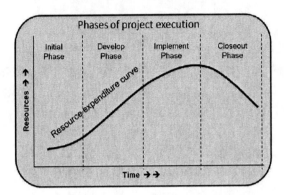

3

Systems View of Project Management

Project Management System

A project management system (PMS) is the set of interrelated project elements whose collective output, through synergy, exceeds the sum of the individual outputs of the elements.

Project Systems Logistics

Project systems logistics is the planning, implementation, movement, scheduling, and control of people, equipment, goods, materials, and supplies across the interfacing boundaries of several related projects

What is a System?

A system is a collection of interrelated elements (subsystems) working together towards a common goal. The synergistic output of a system is larger than the sum of the individual outputs of its components.

4

Application of Systems Engineering

"Systems engineering is the application of engineering principles to solutions of a multi-faceted problem through a systematic collection and integration of parts of the problem with respect to the lifecycle of the problem. It involves the development, implementation, and use of large or complex systems. It focuses on specific goals of a system considering the specifications, prevailing constraints, expected services, possible behaviors, and structure of the system. It also involves a consideration of the activities required to assure that the system's performance matches the stated goals. Systems engineering addresses the integration of tools, people, and processes required to achieve a cost-effective and timely operation of the system."

This approach is very much needed in the Oil and Gas industry.

5

Systems Constraints

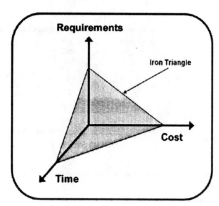

6

Systems Model for Project Management

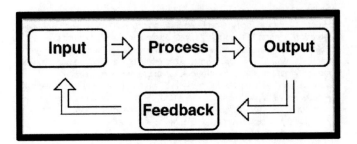

- This model framework is applicable to Reservoir Monitoring and Management in Oil and Gas Operations

7

Integrated Systems Flow-through Model*

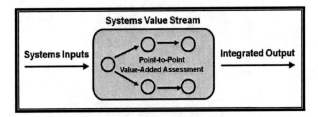

*Badiru, A. B. (2012), **Project Management: Systems, Principles, and Applications**, Taylor & Francis CRC Press, Boca Raton, FL.

8

Applying the **DEJI** Model

Systems **D**esign, **E**valuation, **J**ustification, and **I**ntegration*

*Badiru, A. B. (2010), "Half-life of Learning Curves for Information Technology Project Management," *International Journal of IT Project Management,* 1(3), 28-45, July-September 2010.

9

Tools and Techniques for **DEJI** Model

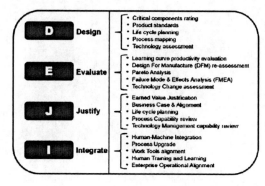

10

A Familiar Example: The Gas Pump

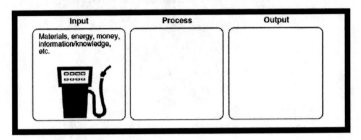

Input	Process	Output
Materials, energy, money, information/knowledge, etc.		

An Exercise in Systems View:

a) View the project management Youtube posting at the following link:
   http://www.youtube.com/watch?v=vU16cgVWD34
b) Pay particular attention to the gas pump example in the video.
c) Fill in the appropriate entries in the Process and Output blocks above.

Note: Copy and paste the link into your Internet Browser.   11

Summary of Systems View

Feedback	A response to a given process or output. In most technological systems, feedback is using information or a signal.	Process	The conversion of resources (inputs) into products or services (outputs) in an organized manner.
Input	The resources required for a system to function as designed. Typical inputs are energy, materials, equipment, people, information, money, and time.	System	A collection of parts working together for a common purpose or goal.
Output	The result of a formal, designated system. Outputs are often categorized as expected or unexpected and desirable or undesirable.	Systems model	A graphic that illustrates the inputs, processes, and outputs for a given system. Systems models may be two-dimensional graphics (such as a diagram or computer model) or a physical display.

12

Introduction to Reservoir Management

- The fundamentals of petroleum reservoir management are based on sound reservoir engineering.

- Reservoir engineering deals with the movement of fluids out of, into, and through the porous geological formations of the earth by means of wells and well systems.

- But reservoir engineering and reservoir management are not identical.

- Reservoir engineering is one of the elements that are used to achieve reservoir management.

13

Reservoir Life Process

A reservoir's life starts with exploration that leads to discovery which is followed by delineation (drilling and completion) of the reservoir, development of the field, production by primary, secondary, and tertiary means and finally to abandonment. A sound integrated reservoir management is the key to a successful operation throughout the reservoir life.

14

Integrated Petroleum Reservoir Analysis

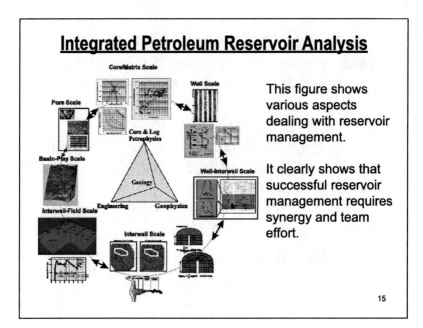

This figure shows various aspects dealing with reservoir management.

It clearly shows that successful reservoir management requires synergy and team effort.

15

Introduction to Reservoir Management

- Reservoir management infers the existence of goals towards which reservoir engineering is directed. Thus, the idea of reservoir management is that goals and implementation of the technology to achieve the goals that are specific to individual reservoirs.

- Reservoir management is defined as the allocation of resources to optimize HC recovery from a reservoir while minimizing capital investments and operating expenses.

- The primary objective of reservoir management is to determine the optimum conditions needed to maximize the economic recovery of HC from a prudently operated field. Reservoir modeling is the tool used for this objective.

16

Why Reservoir Management?

- A vast amount of HC remains unrecovered in producing areas of the world. Fortunately now, there are many new leading-edge technologies that have been made in geophysics (3-D and 4-D seismic), geology, petrophysics, drilling and well completion, production, and reservoir engineering.

- The world reserve is declining everyday, and a small percentage increase in recovery efficiency due to sound reservoir management would result in significant additional reserves.

17

Why Need Sound Reservoir Management?

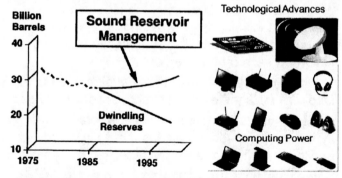

The world reserve is declining everyday, and a small percentage increase in recovery efficiency due to sound reservoir management would result in significant additional reserves. These incentives and challenges provide the motivation to practice better reservoir management.

18

Why Need Sound Reservoir Management?

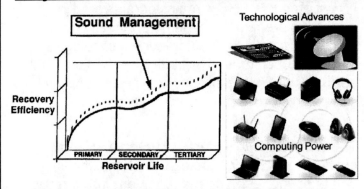

The incentives for significant additional reserves and challenges provide the motivation to practice better reservoir management.

19

Reservoir Management Concept

- Reservoir management concepts are concerned with the definition of reservoir management, computer tools used; synergy and team work, integration of geo-science, and engineering, and time for reservoir management.
- The goal (prime objective) of reservoir management is to maximize the economic recovery of oil and gas
- Reservoir management can be interpreted as the judicious use of various means available in order to maximize profit from a reservoir.
- Some definitions are appropriate in order to understand reservoir management concepts

20

What is Reservoir Management?

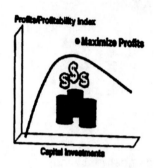
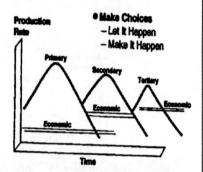

Reservoir management is the utilization of available resources (i.e.. human, technological & financial) in order to maximize profits from a reservoir by optimizing recovery while minimizing capital investments and operating expenses

21

Reservoir Management Approach

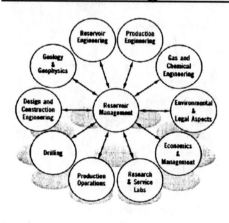

- The team member must work as a well-coordinated team player; like "basket-ball team" rather than "relay team".
- Reservoir engineers should not wait on the geologists to complete their work. Rather a constant interaction between the functional groups should take place.

22

Reservoir Management Team Members

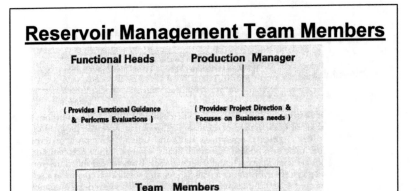

Figure shows team members working under functional heads and a production manager, where the functional heads provide functional guidance and perform evaluations, and the production manager provides project direction and focuses on business needs.

23

Old Reservoir Management System- Conventional Organization

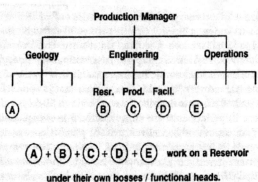

Here various members of the team (geologists, reservoir, production, facilities engineers, operational staff and others) work on a reservoir under their own bosses/functional heads.

24

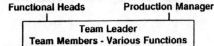

New Reservoir Management System- Multidisciplinary Team Approach

Functional Heads   Production Manager

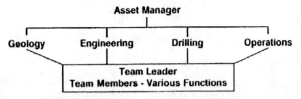

Team Leader
Team Members - Various Functions

Self-Managed Team on a Particular Reservoir
(A Team Leader manages members from various functions
working on a given Reservoir.)

Asset Manager

Geology   Engineering   Drilling   Operations

Team Leader
Team Members - Various Functions

Here the team members from various functions work on a given reservoir under
a team leader and sometimes operate as a self-managed team. The team leader
generally provide the day-to-day guidance and occasionally by so-called "guru"

25

Reservoir Management

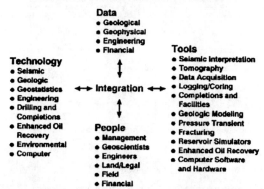

Data
- Geological
- Geophysical
- Engineering
- Financial

Technology
- Seismic
- Geologic
- Geostatistics
- Engineering
- Drilling and Completions
- Enhanced Oil Recovery
- Environmental
- Computer

← → Integration ← →

Tools
- Seismic Interpretation
- Tomography
- Data Acquisition
- Logging/Coring
- Completions and Facilities
- Geologic Modeling
- Pressure Transient
- Fracturing
- Reservoir Simulators
- Enhanced Oil Recovery
- Computer Software and Hardware

People
- Management
- Geoscientists
- Engineers
- Land/Legal
- Field
- Financial

Synergy and team concepts are the essential elements for
integration of geoscience and engineering. It involves people,
technology, tools, and data

26

Reservoir Management Concepts

The primary objective of reservoir surveillance is the economic optimization of oil and gas recovery. The following steps are used to achieve this objective:

1. Identification and definition of individual reservoirs in a particular field (i.e. reservoir characterization)

2. Deduction of past and prediction of future reservoir performance

3. Optimization of the number of wells to be drilled

4. Modification of wellbore and surface systems based on sound engineering

5. Initiation of operating controls at the right time

6. Consideration of all pertinent economic (price of oil, taxation etc.) and legal (environment, safety, etc.) factors.

27

Reservoir Management Concepts

The above steps involve acquisition of facts, information, and knowledge. From the above steps it can be said that the engineering system essential to the petroleum engineer can be divided into three subsystems:

1. Drilling, completion, and operation of wells.

2. Surface processing of produced fluids.

3. Characterization of fluids and their behavior within the reservoir.

Questions and answers that provide the philosophies for reservoir management are as follows:

- When should reservoir management start?

- What, how, and when to collect data?

- What are the questions to be asked to ensure the right answers for reservoir management?

28

Reservoir Management Concepts

Some example questions that need answers are:

- What does the answer mean?

- Does the answer fit all the facts; why or why not?

- Are there other possible interpretations of the data?

- Were the assumptions reasonable?

- Are the data reliable?

- Are additional data needed and necessary?

- Has there been an adequate geological study?

- Has the reservoir been adequately defined?

29

Reservoir Management Concepts

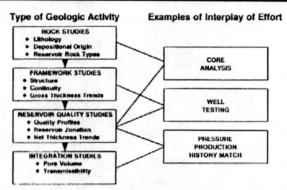

Figure shows the general geological activities in reservoir description and input from engineering studies. The geologist requires input and feedback from the engineer. Well test studies aid in recognizing the barriers, fractures, and variations in permeability.

30

Integration of Exploration and Development Technology

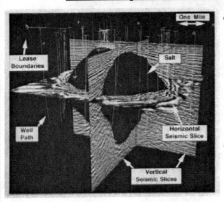

- 3-D computer visualization via a video monitor of a reservoir at a micro- or macro-scale is the latest major breakthrough in computer technology.
- Figure on the left shows an example visualization of a Gulf Coast salt dome., blending many types of information.

31

Integration of Exploration and Development Technology

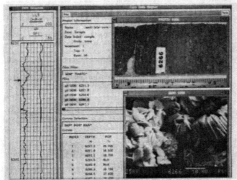

Figure shows computer visualization of electron microscope pictures of rock samples alongside classic rock displays. 3-D visualization technique will enhance our understanding of the reservoir, providing better reservoir description and simulation of reservoir performance.   **32**

Two Types of Organization

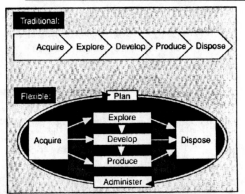

Traditionally, finding and producing HCs were considered the essence of success in the upstream end of the petroleum industry. Now, companies are viewing their options as far more flexible, and a diversified portfolio of skills within an integrated and flexible business framework is emerging.   **33**

Reservoir Management Process

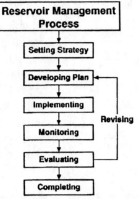

Figure shows modern reservoir management process which involves establishing a purpose or strategy and developing a plan, implementing and monitoring the plan, and evaluating the results.   **34**

Reservoir Knowledge

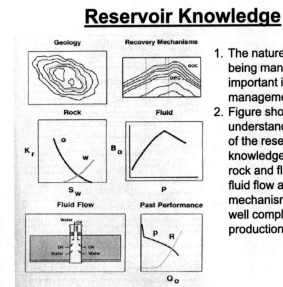

Geology

Recovery Mechanisms

Rock

Fluid

Fluid Flow

Past Performance

1. The nature of the reservoir being managed is vitally important in setting its management strategy.
2. Figure shows the understanding the nature of the reservoir requires a knowledge of the geology, rock and fluid properties, fluid flow and recovery mechanisms, drilling and well completions, and past production performance.

35

Reservoir Management Development Plan

Developing Plan

Development & Depletion Strategies

Environmental Considerations

Data Acquisition & Analyses

Geological & Numerical Model Studies

Production & Reserves Forecasts

Facilities Reqirements

Economic Optimization

Management Approval

- Figure shows that formulating a comprehensive reservoir management plan is essential for the success of a project.
- It needs to be carefully worked out involving many time-consuming development steps,

36

Steps for Data Acquisition and Analysis

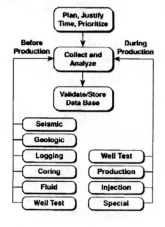

- Figure shows a list of data needed before and during production.
- Data analysis require a great deal of effort, scrutiny, and innovation. The key steps are:
1. Plan, justify, time, and prioritize
2. Collect and analyze,
3. Validate and store (data base)

37

Components of Reservoir Management:

An integrated reservoir and production analysis approach

1

Useful Definitions

- **Integration** - a combination and coordination of separate and diverse elements or units into a more complete or harmonious whole.

- **Process of Integration** - can involve a number of disciplines; different professionals; different professional cultures; different software and hardware platforms teamwork

- **Vertical Integration** – this is integration within a discipline

- **Horizontal Integration** – this is integration across disciplines

- **Loose integration** – this is a low level of operability between two distinct elements

- **Tight Integration** - this is a high level of operability between two distinct elements

2

Definitions contd. . .

- **Accuracy** - conformity to truth or some standard or model

- **Precision** - the degree of agreement of repeated measurements of a quantity

  - Typical measurements available in a reservoir study span all possible combinations of accuracy and prediction.

  - In many instances reservoir parameters can be estimated by means of different techniques.

  - Integrating different sources of independently determined data each of which with certain degree of accuracy and precision will go a long way in providing unbiased estimates of the reservoir parameters.

3

Systems Analysis

- **Systems analysis** - is a way to analyze a process which is based on the study of all the factors internal or external that have an influence on the development and outcomes of a process. The overall principles of systems analysis involves the following:

1. Understanding the process of change instead of focusing on the individual constituent parts of the process itself;

2. Understanding the existing interrelationships amongst all the constituent parts rather than the linear cause-effect linkages

3. Concentrating on the dynamic complexity of the process rather than on the static complexity of its details.

4

Systems Analysis

- An integrated reservoir study is by definition a complex systems analysis, which results from the integration of different disciplines and which has a definite objective.

- Systems analysis may help us to see through the static (intrinsic) complexity of the geological or engineering work and to identify those parameters that have an impact over the global objectives of the reservoir study.

5

Reservoir Management Team

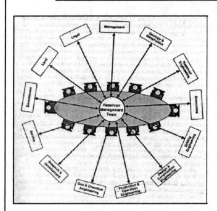

- Successful reservoir management requires synergy and team efforts.
- Success requires multidisciplinary, integrated team efforts.
- The players are everybody who has anything to do with the reservoir.

6

Reservoir Management – Synergy and Team

- All reservoir and operating decision should be made by the reservoir management team, which recognizes the dependence of the entire system upon the nature and behavior of the reservoir.

- It will help tremendously if each team member has background knowledge of reservoir engineering, geology, production and drilling, well completion and performance, and surface facilities.

7

Reservoir Management – Synergy and Team

- The team effort in reservoir management cannot be overemphasized. It is even more necessary now than it has ever been because of the current trend of the oil industry not one of expansion. Most companies are carrying on the production activities with a staff much smaller than had existed about 20 years ago.

- Also, the advent of technology and the complex nature of our reservoirs make it difficult for anyone to become expert in all areas. Hence, it is the reduction of talent and the increasingly complex technologies that must be offset by an increase in quality, productivity, and emphasis on the team effort.

8

Reservoir Management – Synergy and Team

- A team effort approach in reservoir management involves the following steps or procedures
  - Facilitate communication among various engineering discipline , geology, and operations staff by (a) meeting periodically, (b) interdisciplinary cooperation in teaching each others functional objectives, and © building trust and mutual respect. Also, each member of the team should learn to be a good teacher.
  - The engineer must develop the geologist's knowledge or rock characteristics and depositional environment and a geological must cultivate knowledge in well completion and other engineering tasks, as they relate to the project at hand.
  - Each member should subordinate the ambitions and egos to the goals of the reservoir management team. 9

Reservoir Management – Synergy and Team

- Each member must maintain a high level of technical competence
- The team member must work as a well-coordinated team player. Reservoir engineers should not wait on the geologists to complete their work. Rather a constant interaction between the functional groups should take place.

- Organization and management of the reservoir management team requires special attention. Formation of the team, selection of team members, appropriate motivational tools, and composition of the team (as the needs of the reservoir change) should be carefully considered.
- Other aspects such as team leadership, establishment of team goals and objectives and performance appraisals of the team members are some matters that play a vital role in effective reservoir management.

10

Issues of Integrated Reservoir Management

1. Identify the critical features of the field, with respect to the overall objectives of the study. This in turn allows for the identification of the main requirements of the project in terms of both human and technical resources.

2. Rank the critical parameters. This process aims at identifying those parts of the study that do not have a considerable impact in the final results.

3. Define the degree of complexity of the critical phases of the study, compatible with the project constraints.

11

Construction & Management of Integrated Database

This involves the following :

- **Integrated Database -** a data repository system to interactively store, retrieve and share exploration and production (E&P) data.

- **Data Warehouse (Data Storage) -** an integrated non-volatile time variant collection of data to support management needs. It has a reduced degree of interaction with the user.

- **Data Management** - process of storing, organizing and delivering information from a database or data warehouse.

12

Exploration & Production Data Base

- E&P data base can be divided into 3 levels as follows:
- Corporate Data Base
    - Stores the official data of the company.
    - Data quality is high and the
    - Rate of change (volatility) is low.
    - No new data is created within corporate database.
- Project Data Base
    - Contains the data relevant to a particular project or asset.
    - All the team members can access and modify the database. Hence, the volatility is high.
- Application Data Base
    - Contains data relevant to a single application.
    - Information is highly volatile.

13

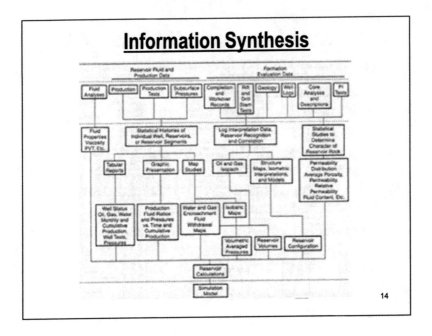

Reservoir Characterization

- In the area of reservoir description following disciplines are used to define characterize the reservoir:
- Geological Studies: rock facies, depositional environment, reservoir continuity
- Geophysics: reservoir boundaries, structure and faults
- Petrophysics: various reservoir parameters (from well logs)
- Well Testing: effective permeability, fracture characterization, well connectivity reservoir pressure
- Reservoir Engineering: material balance, volumetric oil and gas in place reservoir continuity
- Production Engineering: gas/oil, water/oil contacts, well productivity indices

15

Reservoir Characterization

- It can be described as the integration of three independent components:
  - Fluid characterization
  - Rock characterization
  - Geologic modeling
- Reservoir characterization has four principal objectives:
  - To identify key reservoir futures
  - To identify main drive mechanisms
  - To determine the reservoir volumetrics
  - To monitor performance

16

Reservoir Characterization

- Reservoir characterization has both a static and dynamic description.
  - Static description addresses the question of how much hydrocarbon there is in place.
  - Dynamic description addresses the question how much of it and at what rate can be recovered.
- The answers to these questions require a detailed description of geological and petrophysical features that affect reservoir flow.
  - For example, a detailed description of capillary pressure characteristics and the variation of porosity may be required for an accurate estimate of original oil in place but would have very little effect on how much oil can be recovered.
  - Questions of recovery depend upon a different set of parameters such as vertical permeability and the mount of pressure support.

17

Reservoir Characterization

- Engineering data plays a major role in reservoir characterization, thus engineering control and judgement are necessary.
- Any error can be costly in terms of engineering results since reservoir characterization forms the foundation for the other analysis steps.

18

Reservoir Characterization As a Process

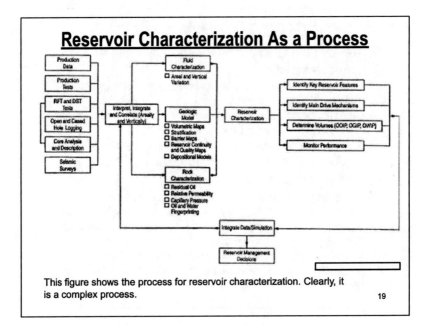

This figure shows the process for reservoir characterization. Clearly, it is a complex process.

19

Geological Aspects of Reservoir Characterization

- Reservoir characterization (RC) is the development of an understanding of the reservoir geology and the control which the geology places on fluid flow and reservoir behavior

- RC requires also the quantification of rock and fluid properties and their distribution to permit modeling and prediction of reservoir behavior

- RC gradually evolves a reservoir description from an initial idealized model towards a more detailed realistic picture

- The effort integrates all available data, including reservoir behavior as well as geological data

20

Geological Aspects of Reservoir Characterization

- Key questions faced by geoscientists and reservoir engineers in RC are those which directly control reservoir behavior.

- Fundamental questions to be answered are:
  - What is the depositional environment?
  - What rock types are present and what are their properties?
  - What is the effective reservoir to be managed?

21

Geological Aspects of Reservoir Characterization
Depositional Environment

- Sandstone Environment –
  Braided Streams
  - Results from high energy river systems carrying coarse-grained sand and gravels
  - Deposits are ideally characterized by thick extensive complexes of coarse-grained sand and gravels with good reservoir quality and continuity interrupted only by thin discontinuous shale lenses
  - Channel shifting and erosion cause most real deposits to consist of many partial depositional cycles stack on top of each other.

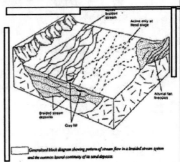

22

Geological Aspects of Reservoir Characterization
Depositional Environment

- Deltaic Environment
  - Results from high energy river systems carrying coarse-grained sand and gravels
  - Deposits are ideally characterized by thick extensive complexes of coarse-grained sand and gravels with good reservoir quality and continuity interrupted only by thin discontinuous shale lenses
  - Channel shifting and erosion cause most real deposits to consist of many partial depositional cycles stack on top of each other.

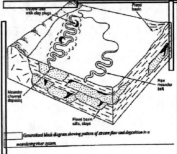

23

<u>Geological Aspects of Reservoir Characterization</u>
<u>Depositional Environment</u>

- **Deltaic Environment**
  - Results from high energy river systems carrying coarse-grained sand and gravels
  - Deposits are ideally characterized by thick extensive complexes of coarse-grained sand and gravels with good reservoir quality and continuity interrupted only by thin discontinuous shale lenses
  - Channel shifting and erosion cause most real deposits to consist of many partial depositional cycles stack on top of each other.

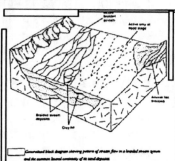

24

Geological Aspects of Reservoir Characterization
Depositional Environment

Carbonate Reservoir Environment
  - Results from high energy river systems carrying coarse-grained sand and gravels
  - Deposits are ideally characterized by thick extensive complexes of coarse-grained sand and gravels with good reservoir quality and continuity interrupted only by thin discontinuous shale lenses
  - Channel shifting and erosion cause most real deposits to consist of many partial depositional cycles stack on top of each other.

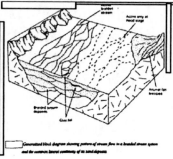

25

Geophysical Aspects of Reservoir Characterization

- Developing a picture of the large scale structure of the reservoir is the primary objective of geophysics within the context of reservoir management.
- An image of the reservoir structure is obtained by initiating a disturbance that propagates through the earth's crust and is reflected at the subsurface boundary interfaces. The reflected signal is acquired, processed, and interpreted

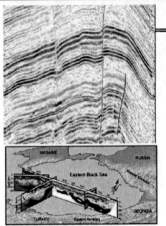

26

Geological Model

- One thing to bear in mind with regard to subsurface maps is that they are never finished. They may be thought of as progress maps or contemporary maps, only as complete as the data that are available when they are made."
- The geologic model contains the "raw" data obtained from several sources such as: Cores; Well logs, well tests; seismic surveys

27

Geological Model

- In a good geologic model, all relevant and reliable information is tallied either on a map, cross section, or in
- the computer as matrices of spatially related properties. Because there are so many types of information available on the average reservoir, the geologic model integrates the raw data with geological assumptions and geological concepts such as depositional models

28

Geological Model

- The geologic model is an *interpretation*. There are few "rights" and "wrongs" in geology and the interpretation is the result of geologic judgement. The geologic model is not an ending point but a milepost on the way to obtaining something else. For example, based up[on the geologic model, new well locations may be selected or workovers of existing wells may be indicated.

29

Geological Model

- The geologic model is often used to determine the amount and distribution of hydrocarbons-in-place because of its areal and vertical depiction of reservoir rock distribution.

- An improved reservoir depletion strategy is frequently developed given the impetus of a new geologic model.

- The geologic model is a tool for predicting certain events in the life of a well or reservoir. Also it provides information for predicting equipment sizes needed to prudently manage future reservoir depletion.

30

Why Geological Model?

- Provides a rationale for the areal and vertical variations in rock properties and production history observed in the field.

- Predicts the rock properties and production characteristics in areas outside or between control points.

- Defines vertical and horizontal barriers and restrictions to fluid flow.

- Gives the simulation engineer a starting point for the simulation model.

31

Why Geological Model?

- The geologic models are usually too complex for simulation models but through scale-up techniques they can provide a basis for block properties.

- The geologic model enables the engineer to make geologically reasonable changes to the simulation models. This helps with reservoir characterization through history matching and with development of prediction scenarios

- A geologic model is instrumental for visualization purposes.

32

Construction of a Geological Model

- A geologic model must be able to capture features that directly affect in-situ flow such as:
  - Identify reservoir stratification
  - Identify degree of vertical communication
  - Define what constitutes pay and reservoir rock
  - Establish areal connectivity
  - Establish reservoir quality
  - Identify contrasting lithological zones
  - Identify reservoir boundary conditions
  - Distinguish between localized and regional geologic features

33

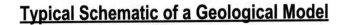

Typical Schematic of a Geological Model

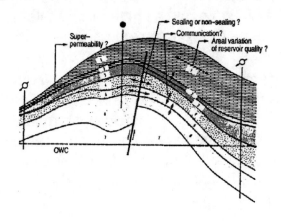

34

Process of Constructing a Geologic Model

- In the integration of the depositional model and reservoir performance data one needs to go through the following steps:
  - Correlation/zonation
  - Layering
  - Mapping
- The conceptual reconstruction of a sedimentary depositional environment through the development of a depositional model can be achieved by geologic analysis of
  - Lithological sequences and sedimentary structures observed in cores, cuttings, and well logs
  - Other similar reservoirs in the region
  - Regional geology and geologic history
  - Distant outcrop exposures
  - Modern analogs

35

Construction of a Geological Model

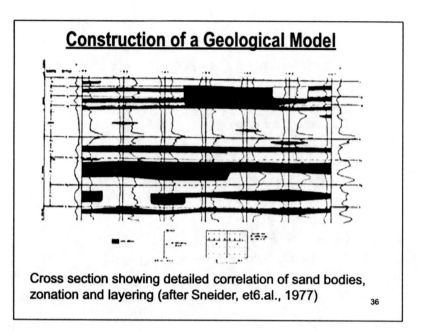

Cross section showing detailed correlation of sand bodies, zonation and layering (after Sneider, et6.al., 1977)

36

Construction of a Geological Model

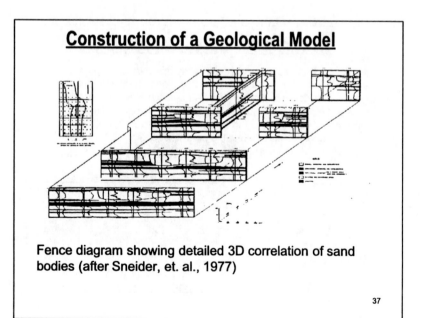

Fence diagram showing detailed 3D correlation of sand bodies (after Sneider, et. al., 1977)

37

Construction of a Geological Model

- The following maps are needed for constructing a geological model:
  - Structure top maps
  - Structure base maps
  - Isopach maps
  - Gross reservoir isopach maps
  - Net sand isopach maps
  - Net oil and/or gas sand isopach maps
  - Iso-porosity maps
  - Iso-permeability maps
  - Average saturation maps
  - Net-to-gross-ratio maps
  - Shale maps

38

Construction of a Geological Model

- Merak field is located approximately sixty miles west of Jakarta. External boundaries of this reservoir is the contour of zero thickness as shown in the first map below. The isoporosity and isopermeability maps are also shown below. Isopermeability map represents the permeability values along the southwest-northeast (SW-NE) direction. The permeability values along the northwest-southeast (NW-SE) direction can be taken as 130% of the SW-NE values

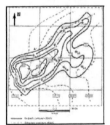 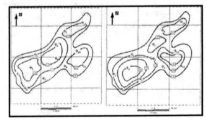

39

Petrophysics (Rock Physics)

- The study of the mechanical and acoustical properties of reservoir rocks and fluids is the focus of petrophysics.

- Petrophysical information is valuable in well logging and of growing importance in time-lapse seismology and reservoir modeling.

- Reservoirs can be classified according to their petrophysical characteristics, which control their performances.

  - Reservoirs classified by petrophysical characteristics are: sandstone reservoirs carbonate reservoirs tight-sand reservoirs shale reservoirs fractured reservoirs

40

Engineering Studies

- Engineering studies that purport to be an integrated study must incorporate all of the following engineering studies and analyses – well testing, reservoir engineering, reservoir simulation, production analysis

- **_WELL TESTING:_** Well tests provide information about reservoir structure, reservoir characteristics and expected flow performance of the reservoir. Well tests help refine the operator's understanding of the field and often motivate changes in the way the well or the field is managed. Well testing is an important resource in reservoir characterization studies.

- **_RESERVOIR ENGINEERING:_** Reservoir engineering studies focus on understanding the reservoir structure and reservoir characteristics as well as the existing reservoir drive mechanisms. The ultimate goal of a reservoir engineering study is to understand the feasibility of developing a field and optimization of the development process. Material balance techniques, improved oil recovery schemes, reserve calculations are examples of reservoir engineering studies.

41

Engineering Studies (cont'd)

- ***RESERVOIR SIMULATION:*** The most sophisticated technology available today for making reservoir performance predictions is reservoir simulation. The process of applying a reservoir flow simulator to the study of a reservoir development project requires a detailed reservoir description.

- ***PRODUCTION ANALYSIS:*** Field production data can be used as an independent means of verifying the reservoir description and volumes in place that were obtained using static data.

  - Static reservoir descriptions provide an initial representation of the reservoir that must be evaluated using dynamic flow information.

  - Decline curves, produced fluid ratios and tracer production are examples of dynamic data.

  - Material balance techniques, improved oil recovery schemes, reserve calculations are examples of reservoir engineering studies.

                                                                          42

Sources of Geological and Engineering Data

- ## Cores:

  - Provide opportunity to directly observe and measure reservoir materials

  - Help improve correlations

  - Help identify rock types

  - Help identify the depositional nature

  - Help calibrate well logs through core-to-log-transforms

- ## Well Logs:

  - Provides the most extensive data source

  - Electric Logs: Spontaneous Potential (SP); Resistivity (Conductivity) Porosity Logs; Sonic logs; Neutron logs; Density logs; Gamma-ray Logs; Dipmeter Logs; Caliper Logs; Thermal Decay Logs; Mud Logs

                                                                          43

<u>Reservoir Monitoring & Production Engineering</u>

- Reservoir Loop (months-years)
  - Reservoir Characterization
  - Development Planning & Execution (includes economic)
- Production Loop (days-weeks)
  - Identify Key Performance Index (KPI)
  - Update Operating Plan
- Production Monitoring
  - Review Production Loop Performance and Reservoir Loop Performance

44

<u>Reservoir Management</u>

- From Reservoir Characterization
- Development Plan
- Meets Economic & Tech
- Objectives?
- Forecast Performance
- Well Plan
- To Production Loop

45

Summary

- Integrated reservoir and production analysis is a powerful managing tool for our most important asset.
- Reservoir characterization as a process:
  - Begins prior to acquisition
  - Continues throughout all phases of the life cycle
  - Describes the critical set of activities and decisions for effective reservoir asset management
  - Uses multi-discipline teams
  - Helps solving the big picture puzzle by:
  - Planning strategies
  - Defining uncertainties
  - Applying cost/benefit analysis
  - Making decisions and taking risks
  - Valuing team concept

46

Basic Reservoir Geology and World's Hydrocarbon Resources

47

Basic Reservoir Geology and World's Hydrocarbon Resources

- **Oil and Gas Resource Reserve Terminology**
- Petroleum - *Rock-oil*
- HC occurs widely-*gas, liquid, semisolid or solid*
- Chemically *complex mixture of hydrogen and carbon*
- *Minor amounts of nitrogen, oxygen, and sulfur*
- Petroleum gas;   *Natural gas*
- Petroleum liquid;   *Crude oil Semi-solid and*
- Solid petroleum   *Asphalt, tar, pitch, bitumen*

48

A Glossary of Terms

- **Asphalt** - *A brown-to-black, solid or semisolid bituminous substance. Melting point is usually between 150 and 200 F.*
- **Bitumen** - *Native mineral pitch, tar or asphalt.*
- **Crude oil** - *The liquid petroleum as it comes out of wells.*
- **Kerogen** - *Has an indefinite composition, consisting of complex organic debris and forming the hydrocarbon content of kerogen shales.*
- **Natural gas** - *A petroleum that is a gaseous mixture under surface conditions of temperature and pressure but some of which becomes liquid under ground with high pressure and temperature.*
- **Oil shale** - *Properly called kerogen shale. Oil shales are organic shales that yield petroleum hydrocarbons upon destructive distillation.*
- **Petroleum** - *A gaseous, liquid, or solid mixture of many hydrocarbons.*
- **Pitch** - *Asphalt or mineral tar*
- **Tar** - *A thick, black or dark-brown, viscous liquid obtained by distilling coal, wood, peat or other organic matter.*

49

Reserves and Resources

- The petroleum *reserves* of any region should be distinguished from its petroleum *resources*.
  - The reserves consist of oil and gas that are now available for use.
  - The resources are always far in excess of reserves, include the reserves, the prospective undiscovered reserves, and any substances from which petroleum could be derived, either by one or both of (1) present or improved technology and (2) present or more favorable economic conditions.
- Every geoscientist and petroleum engineer working in the petroleum industry should bear in mind that the ultimate objective is *to find oil and gas that are profitable to produce.*

50

Reserves and Resources

RESOURCES

1. Known and recoverable oil and gas deposits.

2. Oil and gas known to have been left behind in pools but not recoverable at present.

3. Undiscovered and undeveloped petroleum pools

4. Tar and asphalt deposits, outcropping oil pools.

5. Oil or kerogen, shales, and coals

HOW ARE THE RESOURCES ABOVE TRANSFORMED INTO RESERVES

1. Now available.

2. In part by improved oil recovery methods, but chiefly by new technology and more favorable economic conditions.

3. By discovery and development through present and improved technology, with present or more favorable economic conditions.

4. By present and improved technology, together with more favorable economic conditions.

5. By present and improved technology, together with more favorable economic conditions.

51

Classification of Reserves

- All reserves involve some degree of uncertainty, depending chiefly on the amount and reliability of geologic and engineering data available at the time of the estimate and interpretation of these data.
- The relative degree of uncertainty may be conveyed by placing reserves in one of two broad classifications, either *proved* or *unproved*.
- Unproved reserves are less certain to be recovered than proved reserves and may be sub-classified as *probable* or *possible* to denote progressively increasing uncertainty.

52

Reserve Determination Methods

- There are basically three methods: analogy; volumetric; & performance analysis
1. Analogy – similarities in well spacing, geologic age, rock and fluid properties, depth, pressure, temperature, reservoir size, pay thickness and reservoir drive mechanism
2. Volumetric method – requires the estimation of bulk volume of the reservoir; estimation of rock and fluid properties; and determination of the portion of the hydrocarbon in place which is recoverable.
3. Performance Analysis: uses material balance calculations; decline curve analysis; and reservoir simulation models

53

Types of Resources

- Conventional and Unconventional
- Conventional
- Unconventional Resources
  - Little primary production
  - Technology less well developed or experimental
  - Infrastructure often limited or absent
  - Long time (years) between delineation and development
  - Low permeability necessitates stimulation
  - Low rates, long life
  - Major ($billion) projects

54

Nature of Reserve Estimates

- Estimated value = actual value ± measurement error ± model error ± bias
- Reserves – production (history)
- Reserves – resources (future)
- Estimate:
  - In-place volumes
  - Technically recoverable volume
  - Economically recoverable volumes
  - Commercially recoverable volumes
  - AT A POINT IN TIME UNDER SPECIFIED TECHNOLOGY AND ECONOMIC CONDITIONS

55

Definition of Commerciality

- Evidence to support a reasonable development timetable
- Reasonable assessment of future economics and operating criteria
- Reasonable expectation of a market for the expected sales quantities required to justify development
- Evidence of adequate production and transportation facilities are or will be made available
- Evidence that legal, contractual, environmental and other social and economic concerns will permit development

56

Occurrence of Petroleum

- The occurrence of petroleum is widespread but very uneven. In some rocks it occurs only in infinitesimal amounts and rocks of other areas contain enormous accumulations measured in billions of barrels.
- Petroleum occurs on all the continents of the world, although some continents are much richer in petroleum than others.
- The unevenness of the occurrence of petroleum is due in part to the unevenness of the exploration effort. This in turn depends on such variables as current geologic thought about the occurrence of petroleum, and on economic technical and political factors that either aid or hinder exploration activities.

57

Mode of Occurrence of Petroleum

1. Surface occurrences, such as seepages, springs, mud volcanoes, vug and vein fillings, various kinds of oil kerogen and bituminous shales.
2. Subsurface occurrences such as minor showings of oil and gas and oil and gas pools and fields and basins.
   - Pool: The simplest unit of commercial
   - Field: When several pools are related to a single geologic feature, either structural or stratigraphic, the group of pools is termed a field
- For oil and gas to accumulate in a pool, four essential requirements must be met:
   1. There must be a reservoir rock to hold the hydrocarbons and this rock must be permeable.
   2. The reservoir rock must be overlain by an impermeable roof rock to prevent upward escape of oil/gas.
   3. The reservoir rock and roof rock forma trap that holds the hydrocarbons and prevents it from moving any farther under the pressure of the water beneath it.
   4. There must be a source rock to provide oil

58

The Origin of Oil

- Oil is believed to be a product of the decomposition of organic matter, both plant and animal, not because it consists of hydrocarbons but because of two other characteristics:
- (1) It possesses optical properties known only in organic substances.
- (2) It contains nitrogen and certain compounds that chemists believe could come only from organic sources.
- Furthermore, oil is nearly always found in marine sedimentary rocks. Indeed, in places on the sea floors of the continental shelves, sampling has shown that fine-grained sediments now accumulating contain up to 7% organic matter, chemically good potential oil substance.

59

<u>Simplified Theory of Origin of Oil</u>

1. The raw material consists of simple marine organisms, mostly plants, living in multitudes at and near the sea surface.
2. The organic matter accumulates on the bottom, mostly in basins where the water is stagnant and deficient oxygen and where the substance is neither devoured by scavengers, nor destroyed by oxidation.
3. Deep burial beneath further fine sediment destroys the bacteria and provides pressure, heat and time for further chemical changes that convert the substance into oil and gas.
4. Gradual compaction of the inclosing sediments under the pressure of their increasing weight reduces the space between the rock particles and squeezes out oil and gas into nearby layers of sandstone where open spaces are larger.
5. Aided by their buoyancy and by artesian water circulation oil and gas migrate generally upward through the sand until are reached the surface and lost, or until they are caught in a trap and from a pool.

60

<u>Organic Origin of Hydrocarbons</u>

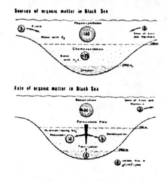

- Photosynthesis is the Origin basis for the mass production of organic matter. About two billion years ago in Precambrian time, photosynthesis appeared a worldwide phenomenon. The enrichment of molecular oxygen in the atmosphere is a result of photosynthesis and the mass production of organic matter.

- The average preservation rate of the primary organic production, expressed as organic carbon is estimated to be less than 0.1%. The upper limit of the preservation rate of organic carbon to be found in certain oxygen- deficient environments favorable for deposition of source rock sediments is about 4%.

61

Organic Origin of Hydrocarbons

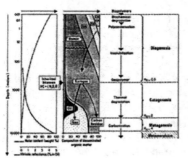

The three main stages of the evolution of organic matter in sediments are:

1. **Diagenesis** begins in recently deposited sediments where microbial activity occurs. At the end of diagenesis, the organic matter consists mainly of a fossilized, insoluble organic residue called kerogen.
2. **Catagenesis** results from an increase in temperature during burial in sedimentary basins. Thermal breakdown of kerogen is responsible for the generation of most hydrocarbons.
3. **Metagenesis** is reached only at great depth, where pressure and temperature are high. At this stage, organic matter is composed of only of methane and carbon residue. The constituents of residual kerogen are converted top graphitic carbon.

62

Sources of Hydrocarbons in Geological Situations

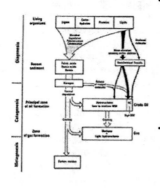

- The sources of hydrocarbons in in geological situations is related to the evolution of organic matter. Geochemical fossils represent a first source of hydrocarbons in the subsurface, while degradation of kerogen represents a second source of hydrocarbons. The bulk of petroleum is generated at a depth where thermal degradation of kerogen becomes important. Hydrocarbons having their source in relatively recent sediments are derived from living organisms and can be regarded as geochemical fossils

- **Kerogen** is the organic constituent of sedimentary rocks that is not soluble in aqueous alkaline solvents or in common organic solvents. The part which is extractable with organic solvents is bitumen. Kerogen is the most important form of organic carbon on earth, and is 100 times more abundant than bitumen.

63

Reservoir Traps

- One of the most essential elements of a petroleum reservoir is the trap – the place where oil and gas are barred from further movement.

- The impervious stratum that overlies the reservoir is called the *roof rock.* A roof rock that is concave as viewed from below prevents the oil and gas from escaping either vertically or laterally. Such an external barrier is a *structural trap.*

- A lateral lessening of permeability due to facies changes, truncations, and other stratigraphic changes will, together with the roof of the rock form an interior barrier, or *stratigraphic trap.* In other words, the stratigraphic traps are chiefly the result of a lateral variation in the lithology of the reservoir rock.

64

Petroleum Reservoir Traps (or Shapes)

Numerous classifications of reservoir traps have been proposed. But the most common classification is: structural or stratigraphic or combination. Geologists classify reservoir shapes or traps, into three types

1. Structural traps - form because of a rock deformation - examples of structural traps are fault traps and anticlinal traps

2. Stratigraphic traps - formed when other beds seal a reservoir bed or when the permeability changes within the reservoir bed itself

3. Other types of traps are the combination traps, where more than one kind of trap forms a reservoir – lenses, salt dome

65

Reservoir Traps

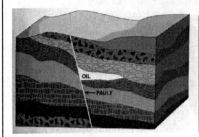 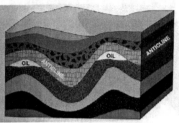

A structural trap - fault trap

A structural trap - anticlinal trap

66

Reservoir Traps

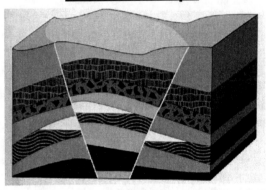

A faulted-anticline combination trap

67

Salt Domes

- The intrusion of deep-seated rocks into the overlying sediments may form a great variety of traps, structural, stratigraphic, and combination. Some of these traps are associated with igneous rocks.
- The great majority, however, of the commercially important traps of this class are in sediments associated with the rock-salt intrusions of the Gulf Coastal region of the United States, the North Sea, so that the pools of this group are commonly called *salt-plug pools* or *salt-dome pools.* The source of the salt in the intrusive salt dome is deep-seated.

68

Reservoir Traps

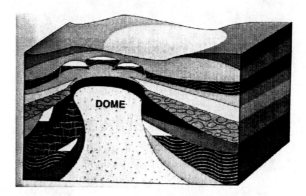

The piercement dome combination trap

69

Salt Domes

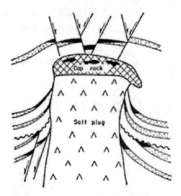

Idealized section through a Gulf Coast salt-dome Field, showing some of the common types of pools (black) found in combination traps associated with the salt intrusion. Many of the flanking traps are wedge-shaped.

70

Structural and Stratigraphic Traps

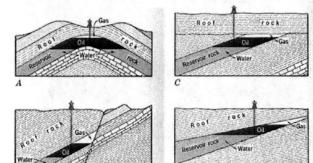

Four of the many kinds of oil traps. A, B, C are structural traps; D is a stratigraphic trap. Gas (white) overlies oil (black), which floats on ground water (blue), saturates reservoir rock, and is held down by roof of claystone. Oil fills only the pore spaces in rock

71

Lens-Type Traps

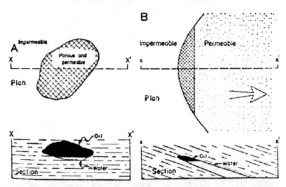

Sketches showing (A) typical lens-type traps completely surrounded by impermeable rocks, and (B) an irregular up-dip edge of permeability on a homoclinal dip. Arrow shows direction of dip.

72

Combination Traps

- These are the traps that combine structural, stratigraphic, and fluid barriers in varying proportions. A combination trap generally has a two- or three-stage history:

1. A stratigraphic element caused the edge of permeability of the reservoir rock,

2. A structural element caused the deformation that combines with the stratigraphic element to complete the rock portion of the trap, and

3. A down-dip flow of formation water increased the trapping effect

73

Combination Traps

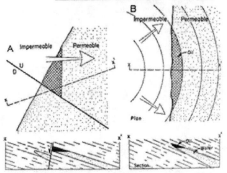

sketches showing characteristic combination traps: A,
intersection of a fault with the up-dip edge of permeability;
B, arching across an up-dip edge of permeability. Arrows
show direction of fluid flow.

74

Migration and Accumulation of HCs

- Most hydrocarbons are found in relatively coarse-grained
  porous, permeable rocks that contain little or no
  insoluble organic matter. It is unlikely that the petroleum
  found in these rocks could have originated in them as no
  trace of the solid organic matter remains. Therefore,
  most oil and gas reservoirs are traps for migrating
  hydrocarbons.

- The release of petroleum compounds from kerogen in
  source beds and their transport within the narrow. Pores
  of a fine-grained source bed are called *primary
  migration*.

- The oil and/or gas expelled from a source rock passes
  through the wider pores of more permeable rock units.
  This is called *secondary migration*.

75

Migration and Accumulation of HCs

- Since most permeable rocks in the subsurface are water saturated, movement of hydrocarbons has to be due to active water flow or occur independently of the aqueous phase either by displacement or by diffusion.

- Since the gas and oil densities are usually lower than that of water, accumulation in traps is usually at a structural high. Finally, relatively impermeable cap rock limits further migration.

- The processes of primary and secondary migration are still poorly understood. data on pore geometry, porosity and permeability relationships and distribution of water in buried source rocks are rare. Likewise, information on hydrocarbon migration is largely theoretical and needs further quantification.

76

Migration and Accumulation of HCs

- One can speculate on distance of secondary migration. One would expect short range secondary migration in isolated sand lenses. Long distance secondary migration would be expected for the very large hydrocarbon deposits since an approximate material balance suggests that there should be a relationship between the size of the accumulation and the drainage area.

- Secondary migration over vertical distances should be only possible through faults, fracture systems and other preferred paths such as dikes, thrust planes and mud volcanoes.

77

Migration and Accumulation of HCs

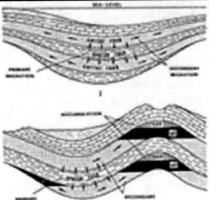

Primary and secondary migration of hydrocarbons and their ultimate trapping

78

Sandstone Formation

- About 60% of the oil and gas reserves in the world's giant fields occur in sandstone reservoirs.

- It should be remembered that large reservoir size, high porosity and permeability and low minimum water saturation yield high producing rates and long producing life. Poor permeability can be offset by a sufficiently thick productive interval. Low producing rate can be tolerated if the pricing structure is adequate. Of course, the continuity and regularity of the pore system needs to be considered.

- Most typical sands are mixtures of sand, silt and clay in varying proportions.

79

Sandstone Formations

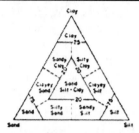

- In some sandstones the grains are virtually all of quartz, with little or no cement. From this extreme, there are all gradations, into sandstones that contain grains of other minerals in greater or lesser abundance, and Into those that contain various kinds of matrix or cement, or both, in various quantities.
- There are more pools producing from sandstone reservoirs than from any other single rock type. The total production and ultimate reserves, however, are probably less than from carbonate reservoirs.

80

Carbonate Formations

- About one-third of the oil and gas reserves in the world's giant fields occur in carbonate reservoirs. Although averaging only about one-half of the effective porosity of sandstones, carbonate reservoirs often offset this large disadvantage by thicker productive intervals and, on occasion, higher permeabilities.
- Some shallow continental shelves were protected from the influx of sediments that provided an opportunity for the development of abundant marine life. many of the organisms living in that environment have hard parts of calcium carbonate, most frequently of the unstable variety, aragonite. When the organisms die, the accumulation of their skeletons forms the original limestone deposits. Later on, diagenetic processes during burial change the aragonite into the more stable calcite.

81

Carbonate Formations

- There are many types of depositional porosity, but the main ones are intergranular and growth framework. Secondary chemical leaching producing vugs and channels coupled with vertical fractures caused by bending stresses from tectonic movements, often enhance the original rock properties. On the other hand circulating solutions may cause the infill of the pores by calcite.

- The general shape and depositional settings will lead to the following types of carbonate buddies are:

  - Reefs and other organic buildups

  - Bedded platform grainstones

  - Limestones and secondary dolomites

82

Carbonate Formations

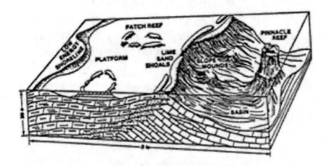

Block diagram illustrating the configuration of carbonate bodies in various depositional settings.

83

Unconventional Oil and Gas

- Unconventional Oil: Extra-heavy oil; Conventional heavy oil; Bitumen; Shale oil Syncrude
- Physical Properties of Unconventional Oil

	Heavy Crude	Extra Heavy Crude	Bitumen(Tar)
– Gravity (API)	10-20	<10	<10
– Viscosity (cp.)	100-10,000	100-10,000	>10,000
– Sulfur (% wt.)	<0.5	0.5-3.0	>3.0

- World Oil Endowment - 9 - 13 trillion barrels
- Conventional Oil - 3 - 4 trillion barrels
- Heavy Oil - 1.5 - 2 trillion barrels
- Extra Heavy and Bitumen - 4.5 - 7 trillion barrels
- Unconventional Gas: Low quality gas; coalbed gas; tight sand gas; Devonian shale gas; high pressure aquifers gas hydrates

84

Resources Triangle

Conventional Reservoirs
Small volumes that are
easy to develop

Unconventional
Large volumes
difficult to
develop

Increased pricing

Improved technology

High-Medium Quality

Low Perm Oil

Tight Gas Sands

Gas Shales

Heavy Oil

Coalbed Methane

Gas Hydrates

Oil Shale

85

Unconventional Oil Resources

- Unconventional oil is an umbrella term for oil resources that are typically more challenging to extract than conventional oil. While many unconventional oil resources cannot be economically produced at the present time, two exceptions are extra-heavy oil from Venezuela's Orinoco oil belt region and bitumen – a tar-like hydrocarbon that is abundant in Canada's tar sands.

- These resources are already being economically produced and are likely, in coming years, to become increasingly more important to global oil supplies.

86

Unconventional Oil Resources

- Both of these Canadian and Venezuelan unconventional oils are characterized by the fact that they are nearly as dense as, or denser than, water. Venezuelan extra-heavy crude is significantly more viscous than conventional crude, while Canadian bitumen is even more so. The high densities and viscosities of these hydrocarbons pose significant challenges for extraction and transport while their high levels of sulfur and other characteristics can make them challenging and energy intensive to refine.

- Today, Canada and Venezuela both produce approximately 3 million barrels per day of oil. As of 2000, they produced approximately 600,000 STB/D and 300,000 STB/D, respectively of this total from unconventional oil. By 2015, government forecasts project that Canada and Venezuela combined will produce nearly 3.5 million barrels per day of extra-heavy oil or diluted bitumen and syncrude.

87

Unconventional Oil Resources

Are Unconventional Oil Resources - Economically Recoverable?

- Most unconventional oil resources, like oil hales, are not currently economically recoverable. The most probable exceptions to this statement are Canadian tar sands and Venezuelan extra-heavy oil.

- If these unconventional oil resources are recognized as established reserves at some point in the future, overall estimates of globally recoverable oil would increase substantially. If Canada's classification of 174 billion barrels of oil sands is considered as establishes reserves and Canada will become the second largest oil-reserve holding nation in the world after Saudi Arabia. Similarly if the 225 billion barrels of extra-heavy oil of Venezuela is considered as recoverable then Venezuela would be credited with the largest oil reserves in the world.

88

Unconventional Oil Resources

Are Unconventional Oil Resources - Economically Recoverable?

- These changes in the classification would tilt the balance of global oil reserves away from the Middle East, though neither of these unconventional oil producing regions is likely to displace the Middle-East as the lowest-cost or highest volume supplier of oil to world markets anytime soon.

- The energy required and environmental impacts incurred in extracting and utilizing unconventional, extra-heavy and bitumen based oils are generally greater than in the case of conventional oil. Tar sands require substantial amount of energy mining and separating of for heating underground reservoirs. Similarly, extra heavy oil requires significant effort to bring to the surface and transport for processing.

- Besides their greater upstream energy requirements, producing extra-heavy oil and bitumen may entail greater environmental impacts than conventional oil production. large amounts o water are necessary to separate bitumen from the tar sand and solids. Similarly the production of extra-heavy oil in Venezuela will yield high sulfur at the surface

89

Role of Technology on Unconventional HC Recovery

- In the 1990's extraordinary technology improvements in exploration, drilling and production contributed to the near record low energy prices. There is every reason to believe that technology advances will continue unabated. The following are a few examples of technologies on the verge of breakthrough (if not yet commercial), that will affect the development of traditional as well as unconventional oil and gas resources.

- Cross-borehole seismic tomography and the associated miniature (0.5 inch) geophone tools permitting improved reservoir characterization on the interwell scale

- Through-casing resistivity logging that locates bypassed pays old, steel-cased wells

- Recent advances in the design of downhole-or subsea-installed water separators which will allow removal of produced water from oil in the wellbore with subsequent reinjection of the same in a different formation within the same well.

90

Role of Technology on Unconventional HC Recovery

- Gas-to-liquids (GTL) technology that converts natural gas into synthetic liquid fuels.

- The improvements in the horizontal well technology.

- The development of laser drilling provides an example of a really high-tech approach which will result in increased rate of penetration, reduced or eliminated rig day rates, enhanced well control, and perforating and side tracking. The rate of penetration through a sand-shale sequence is expected to be around 450 ft./hr.

  - A wave of business-to-business electronic commerce, smart well technology.

  - 3D and 4D seismic technology.

  - Fuel cell technology.

91

Unconventional HC Resources

- **Which Unconventional Oil and Gas Options? When? How Much? What Price?**

- Answering these questions, even in a qualitative sense, is no less difficult than predicting today's desktop computing power 20 years ago. In order to approach to a correct set of answers one must consider: the rate of technology development; the world's economic activity; climate change scenarios; population growth; & politics

- However, in general there is agreement that the world's supply of conventional crude oil will peak between 2020 and 2030. Most experts do not consider he contribution of unconventional oil and gas of the kind discussed in this lecture, because it seems too speculative. However, it is reasonable to state that any assumption related to decline in oil production will result in increase in oil price is necessarily not correct as oil supply from conventional sources will be replaced by oil production from unconventional resources. This observation implies that global peak oil production will probably be not achieved probably for another 50 years.

92

Reservoir Rock Properties (Petrophysics)

93

Reservoir Rock Properties

- Information on reservoir rock properties is required by reservoir engineers for defining reservoir rock and fluid volumes distributions, and understanding and predicting reservoir behavior.
- It is vital for reservoir engineers not only to have representative average values for rock properties but also, often more importantly, to understand the nature of their distribution and variation throughout a reservoir.
- The most fundamental reservoir rock properties that appear in reservoir analysis and characterization are:
  - Porosity; Fluid Saturations
  - Compressibility; Permeability; Wettability
  - Relative Permeability; Capillary Pressure

94

Porosity

- Porosity is a measure of the void space in rock, hence measures how much HC in rock

- Porosity $\phi = V_p/V_b = (V_b-V_m)/V_b$; $V_b = V_p + V_m$

  - theoretically, ϕ varies from 0% - 47.6%

  - in practice, ϕ varies between 3% and 37%

- Porosity is a function of particle size distribution:

  - framework materials (sandstone) - high ϕ

  - interstitial materials (shay-sand) - low ϕ

95

Various Packing of Spheres: (a) Cubic, (b) Rhombohedra

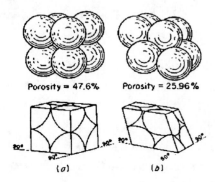

Porosity = 47.6%   Porosity = 25.96%

(a)   (b)

96

Classification of Porosity

- Porosity is classified based on the following: pore connection and geology

- Pore connection:
    - Absolute or total porosity
    - Effective porosity

- Geology
    - Primary or intergranular porosity
    - Secondary porosity - due to fractures or dolotomization (chemical reaction b/w $CaCO_3$ and $MgCl_2$)

97

Factors Affecting Porosity & Measuring Techniques

- Porosity values are affected by particle shape, vugs, arrangement of particles, particle size distribution, cementing material, & fractures

- Techniques for measuring porosity are:
  - Logging: from density and sonic logs. Must know the matrix type: sandstone or shale or limestone. Logs must be calibrated
  - Surface method: determine V_p & V_b or V_m & V_b
    - Core analysis (displacement method) - Porosimeter determines V_b by using Boyle's law

                                                                    98

Use of Porosity

- Basic use is to calculate volumetrically the quantity of hydrocarbon (HC) in the rock
- $N = 7758 \times A_s \times H \times \phi \times S_{oi}$
  - N = HC volume in the reservoir, res. bbl
  - A_s = surface area, acres
  - H = thickness of formation, ft.
  - ϕ = porosity, fraction
  - S_{oi} = initial oil saturation $(1.0 - S_{wi})$, fraction
- If N is divided by Bo, we will get the volume on surface. Since oil shrinks as it comes to the surface due to gas coming out, $N_{surface} < N_{reservoir}$

                                                                    99

Determination of Porosity

- Several methods: involves only the determination of two out of 3 (V_p, V_m, & V_B)
- Bulk volume by the following methods
  - Coated sample immersed in water, or
  - Water-saturated immersed in water, or
  - Dry sample immersed in Hg method (no more Hg in the labs)

- Grain volume: by Melcher-Nutting method in which the sample is crushed and its volume measured with a pycnometer

100

Example Porosity Calculation - Determination of V_B - Coating Method

- A = mass dry sample in air = 20.0 gm
- B = mass dry sample coated with paraffin=20.9 gm ($SG_{paraffin}$ = 0.9).
- C= mass coated sample immersed in H_2O at 40ºF =10 gm (SG_{water}=1.0).
- Mass of paraffin = B - A = 20.9 - 20.0 = 0.9 gm
- Volume of paraffin = 0.9/0.9 = 1 cc
- Mass of water displaced = B - C = 20.9 - 10.0 = 10.9 gm
- Vol. of water displaced = mass of water/ρ of water=10.9/1.0 = 10.9 cc
- Bulk volume = volume of water displaced – volume of paraffin
- = 10.9 – 1.0 = 9.9 cc
- Bulk volume of rock = 9.9 cc

101

__Example - Porosity Calculation__

- Determine porosity from previous example
- Mass of dry sample in air   = 20 gm
- Bulk volume of sample   = 9.9 cc
- Grain volume of sample = (mass of dry sample in air)/(sand-grain density)
- = 20/2.67 = 7.5 cc
- Total porosity = ϕ_t = [(bulk volume – grain volume)/bulk volume] x 100
- = [(9.9 – 7.5)/9.9] x 100 = 24.2 per cent

102

__Rock Compressibility__

- Porosity is a function of compaction. It is generally reduced by increase in compaction

- Compaction is a function of depth of burial. Sediments such as shales which have been deeply buried exhibit lower porosity

- Three types of compressibility
    - Rock-matrix (grain) compressibility, C_s
    - Rock-bulk compressibility, C_b
    - Pore-volume compressibility, C_p

103

<u>Definition of Compressibility</u>

- Rock-matrix (grain) compressibility $\boxed{C_{s} = \frac{1}{V_{s}}\frac{dV_{s}}{dP}}$

- Rock-bulk compressibility $\boxed{C_{b} = \frac{1}{V_{b}}\frac{dV_{b}}{dP}}$

- Pore-volume compressibility $\boxed{C_{p} = \frac{1}{V_{p}}\frac{dV_{p}}{dP}}$
  - Most important to the reservoir engineer
  - There is a correlation between porosity and compressibility; C_p decreases as ϕ increases
- C_p varies between 5.0 to 40.0 x 10^{-6} psi^{-1}
- For sandstone, between 5.0 to 10.0 x 10^{-6} psi^{-1}

104

<u>A generalized sketch of the effect of natural compaction on porosity - do not use for numerical calculations</u>

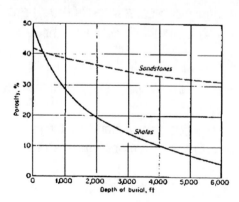

105

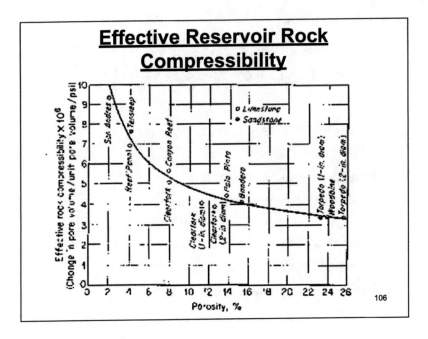

Fluid Saturations

- Saturation defines the fluid content of the reservoir

- Initially majority of reservoir rocks (90%) are completely saturated with saline water

- Two methods for determining saturation
  - Direct - involves core analysis on the surface. Subject of this section
  - Indirect - other physical properties of the rock are measured. Examples are well logs, and capillary pressure P_c data

107

Permeability

- Permeability is the fluid conductance capacity of a rock.

- A French hydrologist named Darcy did the first work on permeability. He was concerned about flow of water through filters. He found that flow rate Q, is proportional to area of flow A, h, and 1/L. This is expressed mathematically as follows:

$$Q = \frac{KA\left(h_1 - h_2\right)}{L}$$

- K is a constant of proportionality called permeability. K – f(size of the sand pack) both of the arrangement and size. K is a constant for a particular sand pack.

108

Permeability

- **Two types of permeability - absolute permeability & effective permeability**
- Absolute permeability is the measure of the ability of a porous media to transmit a fluid through its interconnected pores when the media is 100% saturated with that fluid.
- The dimension of permeability is square of length, which is the dimension of area.
- The practical field unit of permeability is Darcy or milli-Darcy. One Darcy is approximately 10-8 cm2.
- *Heterogeneous Versus Homogeneous Property Distribution*
- *Homogeneous* system refers to a system that has uniform property distribution.
- *Heterogeneous* system refers to a system with properties vary in spatial domain from one point or region to another.
- When the reservoir is described as having homogeneous or heterogeneous property distribution, it is important to specify the property that is referred to.

109

Reservoir Rock Properties

- **Reservoir Rock Properties**
- *Isotropic vs. Anisotropic Systems*
- If the property value is the same regardless of the direction, the reservoir is said to be *isotropic* with respect to that property.
- On the contrary, if the property changes its value with the direction, then that reservoir property is described to have *anisotropic* distribution.

110

Reservoir Characterization

- *Sealed Reservoirs*
  - A sealed reservoir refers to a reservoir surrounded by an impermeable formation.
  - The drive mechanism for sealed reservoirs can be dissolved-gas drive, or gas-cap drive.
- *Reservoirs with constant pressure boundaries*
  - Some reservoirs are connected to a large aquifer, which provides the much needed energy to the hydrocarbon reservoir. This helps sustain production and maintain pressure in the reservoir.
  - A reservoir with the behavior described above is known to have a "water-drive" mechanism which results in significantly higher recovery than other reservoir drive mechanisms.

111

Reservoir Characterization

- *Single-phase reservoirs*
  - Single-phase reservoirs are reservoirs that have only one fluid phase flowing in the reservoir.
  - This type of reservoir rarely exists in reality; however, some reservoirs can be considered as single- phase reservoirs if only one phase is mobile.
- *Multi-phase reservoirs*
  - In a naturally occurring reservoir, there are always more than one fluid phase flowing together in the reservoir, with the exception of some special cases as described in the previous section.
  - Reservoirs with more than one fluid phase are known as *multi-phase* reservoirs. The fluids flowing in such reservoirs can be two-phase (oil-water, oil-gas, or gas-water), or three-phase (oil-water-gas).

112

An Overview of Geostatistics

- Geostatistics is the statistics of spatially or temporally correlated data. Most geostatisticians are concerned with estimation of expected values such as initial pressure at a well site or average thickness of a reservoir rock under a property.
- Geostatistics is not limited to simple calculation of expected values. Resource evaluation and mapping form the core of geostatistics.
- The tools used by geostatisticans are: univariate statistics, means, variances and histograms; scattergrams; semivariogram calculation; interactive curve fitting; plotting scattergrams; grid searching; equation solving; contouring and map drawing
- A straightforward mapping problem includes five steps: data gathering and cleanup; univariate analysis; semivariogram calculations; modeling; & estimation mapping

113

Control Volume Concept

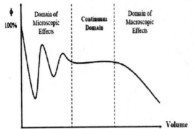

The Porous Media as a Continuum

- The microscopic scrutiny of a porous medium reveals that its local properties may vary widely depending on the volume over which the scrutiny is performed.

- Instead of a microscopic description, the usual way of approaching a description of a porous media and the fluids within it is to use the *continuum approach*.

Fluid properties and porous medium properties are treated as varying "continually" in space.

114

Fundamental Rock Properties

- The three fundamental properties of reservoir rock are:
  - Porosity, ϕ
  - Permeability, k
  - Fluid saturations within the reservoir pore space.
- These properties are important with respect to two basic questions:
  - How much hydrocarbon do we have in a given reservoir?
  - At what rate can we produce this reservoir?

115

Characterization of Reservoir Rocks to Form a Reservoir

- It must have storage capacity - this property is characterized by porosity.
- The fluid must be able to flow through the rock: this property is characterized by permeability.
- It must contain a sufficient volume of hydrocarbons, with a sufficient concentration.
- The porosity of interest to the reservoir specialist is the one which allows fluids in the pores to circulate. That is the effective porosity ϕ_{eff}
- The effective porosity of rocks varies between less than 1% to over 40%.

116

Porosity

- Classification of effective porosity (-)
  - Low: $< 5\%$
  - Mediocre: $5\% < \ < 10\%$
  - Average: $10\% < \ < 20\%$
  - Good: $20\% < \ < 30\%$
  - Excellent: $>30\%$
- Measurement of Porosity
  1. Direct Method
  2. Indirect Methods
  - Sonic (or Acoustic) Log
  - Density Log

117

Darcy's Law

- **Basic Assumptions of Darcy's Law**
  - Steady-state flow conditions exist
  - The porous rock is 100% saturated with the flowing fluid (a fixed and immobile phase may be present and often is)
  - The flowing fluid viscosity is constant
  - Isothermal conditions prevail
  - The porous rock is homogeneous and isotropic
  - Porosity and permeability are constant
  - The flow is laminar
  - Gravitational forces are negligible.
  -

118

Permeability-Porosity Relationships

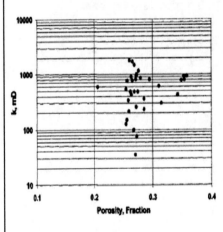

- Figure shows a semi-log plot of K vs. ϕ obtained from a large number of samples of a sandstone formation.
- Even though formation is considered very uniform and homogeneous, no specifically defined trendline between K and ϕ.
- Relationship is qualitative and not directly or indirectly quantitative.
- ϕ can be high and k is nearly zero if there is no connectivity.

119

Permeability-Porosity Correlations

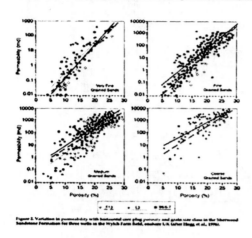

Figure 2. Variation in permeability with horizontal core plug porosity and grain size class in the Sherwood Sandstone Formation for three wells in the Wytch Farm field, onshore UK (after Hogg et al, 1996).

120

Klinkenberg Effect

- **Klinkenberg Effect** - The permeability to gas is a function of the mean free path of the gas molecules and thus depends on factors which influence the mean free path such as temperature, pressure, and the nature of the gas. At high pressure, the permeability k_g should be expected to approach that of a liquid.

$$K_g = K_\infty \left(1 + \frac{b}{P_m} \right)$$

- K_g = gas permeability
- $K\infty$ = newly defined permeability which is the extrapolated value of kg to infinite mean pressure Pm
- b = constant for a given gas in a given medium
- P_m = mean pressure

121

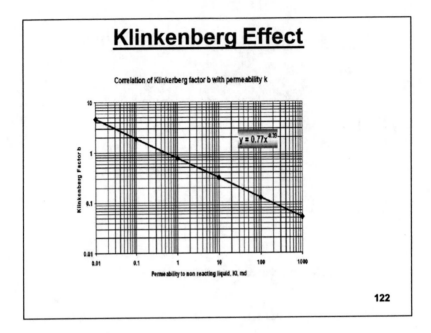

122

Fluid Saturation

- Fluids in most reservoirs are believed to have reached a state of equilibrium and separated according to their density, i.e., oil overlain by gas and underlain by water.

- Irreducible water is distributed throughout the oil and gas zones in addition to bottom (or edge) aquifer water.

- The forces retaining the water in the oil and gas zones are referred to as capillary forces because they are important only in pore spaces of capillary size.

- Connate (interstitial) water saturation S_{wc}:

- Generally not uniformly distributed throughout the reservoir but varies with permeability, lithology, and height above the free water table.

123

Fluid Saturations

- Fluid Saturations
  - Critical oil saturation Soc
  - Residual oil saturation Sor
  - Movable oil saturation So
  - Critical gas saturation Sgc
  - Critical water saturation Swc
- Average Saturations:

$$S_o = \frac{\sum_{i=1}^{n} \phi_i h_i S_{oi}}{\sum_{i=1}^{n} \phi_i h_i} \quad \& \quad S_w = \frac{\sum_{i=1}^{n} \phi_i h_i S_{wi}}{\sum_{i=1}^{n} \phi_i h_i} \quad \& \quad S_g = \frac{\sum_{i=1}^{n} \phi_i h_i S_{gi}}{\sum_{i=1}^{n} \phi_i h_i}$$

124

Absolute, Effective, and Relative Permeability

- ***Absolute permeability*** - Permeability of a rock to a fluid when the rock is 100% saturated with that fluid

- ***Effective permeability*** - measures the ability of a porous material to conduct a fluid when the saturation of that fluid in the pore space is less than 100%

- ***Relative permeability***: provides a mechanism to quantify the flow amount, for each phase, in a multiphase environment.
  - Relative Permeability is the ratio of effective permeability of a fluid at a given saturation to the effective permeability of that fluid at 100% saturation (absolute permeability). It is a form of normalization.

125

<u>Relative Permeability Ratio</u>

- Relative (or effective) permeability ratio is as follows:
  - $K_{rw} = K_w/K$ for water; $K_{ro} = K_o/K$ for oil
  - K_w = effective permeability due to water
  - K_o = effective permeability due to oil
  - K = absolute permeability
- It expresses the ability of a reservoir to permit the flow of one fluid as related to its ability to permit flow of another fluid
- Permeability ratio varies from 0 to 1.0
- Relative permeability ratio plotted against saturation on a semi log paper shows a linear trend over a wide saturation range.
- **3-Phase Relative Permeability** exists when three fluids exist in the rock simultaneously. Hence, 2-phase relative-k data must be amplified and extended for 3-phase systems.

126

<u>Capillary Pressure</u>

- The rise in height of fluid in capillary tube is due to the attractive forces (adhesion tension) between the tube and the liquid and the small weight represented by the column of liquid in the tube.
- The pressure difference existing across the interface separating two immiscible fluids, one of which wets the surface of the rock in preference to other.
- Petroleum engineering convention defines capillary pressure as the pressure in the oil phase minus the pressure in the water phase. In an oil wet system the value can be negative.

127

Capillary Pressure

- The equation for capillary pressure is as follows:

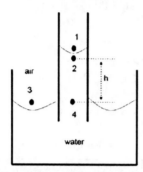

$$P_c = P_1 - P_2$$

$$P_3 = P_1 + \rho_a g h$$

$$P_4 = P_2 + \rho_w g h$$

$$P_3 = P_4$$

$$P_1 + \rho_a g h = P_2 + \rho_w g h$$

$$\boxed{P_c = P_1 - P_2 = (\rho_w - \rho_a) g h}$$

The density for air is generally taken as zero. The air can be replaced with oil and the equation will be the same. Only the subscripts will change

128

Capillary Pressure for Oil-Water System

- Oil-Water System - A reference depth is required. The original water-oil contact is often used. It may be located using core analysis or logging.

- Threshold capillary pressure, Pct exists at the initial water-oil contact and the free-water level is calculated by rearranging the previous equation for a known Pc.

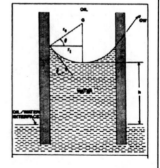

129

<u>Application of Capillary Pressure</u>

- FREE-WATER LEVEL: The imaginary level marked by an oil-water interface that has capillary pressure equal to zero.
- WATER-OIL CONTACT: Depth in the reservoir below which there is essentially no oil (Sw =100%).
- THRESHOLD CAPILLARY PRESSURE (Pct) -The pressure required to overcome the capillary pressure of the largest pores and force the first small drop of non-wetting fluid into the rock.
- WATER-OIL TRANSITION ZONE: The vertical zone of variable water saturation between the water-oil contact and the point above this contact where the water saturation equals irreducible saturation.

130

<u>Capillary Pressure and Saturation Distribution</u>

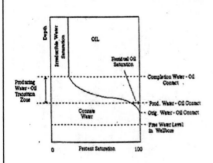

1. Gravity and capillary pressure control the initial distribution of fluids.
2. Gravity causes less dense fluids to locate higher in the reservoir.
3. Capillary pressure causes wetting fluids to locate in small pores and non-wetting fluids in large pores.

131

Wettability of Reservoir Rocks

- Wettability is the preference of one fluid to spread on or adhere to a solid surface in the presence of other immiscible fluids
- Wettability is a complex function of fluid and solid properties
- Reservoir rock wettability is due, in part, to the presence or absence of minute quantities of polar compounds
- The interaction between these polar compounds and the rock surface mineralogy determines wettability

132

Wettability of Reservoir Rocks

- Molecular forces in place
  - Cohesive forces – between similar molecules
  - Adhesive forces – between different molecules

$$\cos \theta = \frac{\sigma_{so} - \sigma_{sw}}{\sigma_{wo}}$$

If $\theta = 0$, completely water wet

If $\theta = 90$, cohesive forces = adhesive forces, intermediate wet

If $\theta = 180$, completely oil wet

- Wettability Determination:
  - Wettability contact angle measurement using uncontaminated reservoir oil
  - Wettability displacement method using preserved core samples (to maintain original relative permeability) **133**

Reservoir Rock Sampling and Analysis

- Methods of rock sample acquisition
  - Continuous sampling while drilling (cuttings)
  - Selective sampling while drilling (conventional coring)
  - Side-wall core sampling
- Advantages of Rock Sample Analysis
  - Provides direct qualification and quantification for
    - Rock type and lithology
    - Rock properties
    - Hydrocarbon presence and type

134

Reservoir Rock Sampling and Analysis

- Short comings of rock sample analysis are:
  - Bringing the samples to the surface alter in-situ conditions
  - Removing overburden stress may significantly alter porosity and permeability values
  - Representatively of samples recovered in fractured formations
  - Significance of measured oil saturation values
- Different Aspects of Rock Samples Analysis
  - Sample recovery and handling
  - Sample preparation
  - Properties measurement
  - Data interpretation and analysis

135

Core Analysis

- **Core analysis is the laboratory measurement of petrophysical properties of core samples.**
- **Core analysis allows visual examination of a portion of a formation and direct measurement of several important characteristics of the rock.**
- **A synergism exists between direct physical measurement on core samples and downhole well logs and other well-test data.**
- **Good engineering utilizes all evaluation tools to make the best possible development decisions**

136

Core Analysis

- **Three types of analysis exists:**

    > *Conventional (plug) analysis*

    > *Whole core (full diameter) analysis*

    > *Sidewall core analysis*

- **Plug analysis is the most commonly used. The size of the plug is usually 1" in diameter and 1" long.**
- **Full-diameter core analysis allow testing of complex lithology, such as heterogeneous carbonates.**
- **Sidewall samples are small cores recovered from projectiles fired into the drill hole., used extensively in softer sand formations.**

137

Reservoir Engineering:

Reservoir Fluid Properties
Well Testing
Reservoir Modeling

1

Properties of Reservoir Fluids

- The compositions of fluids in any particular reservoir are unique, and their physical properties vary from a gas substantially lighter than air, through clear volatile liquids to black-tar-like substances too viscous to flow
- Information describing the physical properties of reservoir Hc fluids and water, and their behavior in response to changes in pressure and temperature is essential to all phases of petroleum engineering and operations.
- The reservoir engineer is principally concerned with:
  - The volumes of fluids present in a reservoir an their distribution
  - Physical properties which affect fluid flow in reservoir rock, and
  - Volumes and properties change in response to changes in reservoir pressure and sometimes temperature.

2

<u>Equation of State for an Ideal Gas</u>

- This is the combination of Boyle, Charles and Avogadro Laws
- Imagine a two-step process for combining Boyle and Charles Laws

$$(V_1 \text{ at } p_1, T_1) \xrightarrow[T_1 = \text{constant}]{\text{Step 1}} (V \text{ at } p_2, T_1)$$

$$\xrightarrow[p_2 = \text{constant}]{\text{Step 2}} (V_2 \text{ at } p_2, T_2). \qquad (3-3)$$

3

<u>Equation of State for Ideal Gas</u>

$$p_1 V_1 = p_2 V \text{ or } V = \frac{p_1 V_1}{p_2},$$

$$\frac{V}{T_1} = \frac{V_2}{T_2} \text{ or } V = \frac{V_2 T_1}{T_2}.$$

$$\frac{p_1 V_1}{T_1} = \frac{p_2 V_2}{T_2}.$$

- From Eq. 3.7, we can say PV/T = constant. This constant is R for one molecular weight of gas. Is R the same for all ideal gases? We need to apply Avogadro's Law.

4

Gas Density

- Since it is more convenient to measure the specific gravity of gas than the gas density, *specific gravity* is more commonly used.

- Specific gravity is defined as the ratio of the density of a gas at a given temperature and pressure to the density of air at the same temperature and pressure, usually near 60°F and atmospheric pressure.

- Whereas the density of gases varies with temperature and pressure, the specific gravity is independent of temperature and pressure when the gas obeys the ideal gas law.

5

Mixture of Ideal Gases

- Since the PE primarily is concerned with gas mixtures, the laws governing the behavior of mixtures of ideal gases will be introduced.

- This will later lead to an understanding of the behavior of mixtures of real gases.

- The laws are

    - Dalton's law of partial Pressures

    - Amagat's law of partial volumes

6

Behavior of Real Gases

- Real gases comprise
  - Dry gases
  - Wet gases
  - Gas Condensates
- Look at the correlations describing the properties of real gases
- Use the textbook examples to illustrate the methodology

7

Behavior of Real Gases

- There are over 100 EOS for real gases. One of them is compressibility EOS.
- Compressibility is one of the most useful data in the petroleum industry. The Z-factor is a correction for converting ideal gas equation to real gas EOS
- But it has its limitations

8

Behavior of Real Gases

$$pV = znRT, \; pV_M = zRT, \; pv = \frac{zRT}{M}, \text{ and } \rho_g = \frac{pM}{zRT}, \quad (3\text{--}39)$$

- This equation is known as compressibility EOS. Also known as compressibility equation or real gas equation.
- The z is called deviation factor or super-compressibility.
- Note: sometimes 1/Z is called super-compressibility

9

Behavior of Real Gases – Z-factor

- The Z-factor = volume occupy by a gas at a given P and T to the volume the gas would occupy at the same T and P if it behaved like an ideal gas. That is,

$$z = \frac{V_{actual}}{V_{ideal}}$$

- Z-factor is not a constant
- Z = f(gas composition, T & P)
- It must be determined experimentally.

10

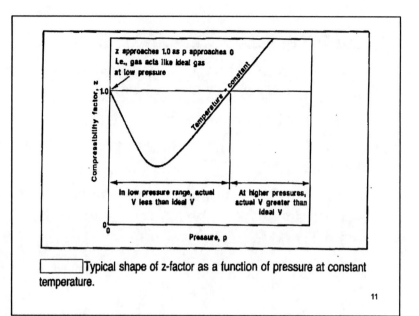

Typical shape of z-factor as a function of pressure at constant temperature.

11

Real Gas Equation of State

- PV=ZnRT (Real Gas); PV=nRT (Ideal Gas = 1.00)
- For one mole, Z represents
  - $Z=(PV/RT)_{real}/(PV/RT)_{ideal}$
  - Really, it's just the ratio of the real molar volume to the ideal molar volume
- Z-factors from the following
  - Standing-Katz chart
  - Dranchuk and Abou-Kassem correlation
  - Find the z-factor as a function of pseudo-reduced pressure and pseudo-reduced temperature

12

Compressibility EOS for Gas Mixtures

- The shapes of the isotherms of z-factors are very similar. This true for all real gases. This led to the following:

- Laws of Corresponding States, and definition of $T_{reduced}$ (T_r), and $P_{reduced}$ (P_r).

- The law of Corresponding States says that all pure gases have the same z-factor at the same T_r and P_r.

- $T_r = T/T_c$ and $P_r = P/P_c$

13

Compressibility EOS for Gas Mixtures

- Law of corresponding states is extended to cover gas mixtures which are closely related.

- Since obtaining Tc and Pc of multicomponent mixtures is difficult, pseudo critical temperature (pTc) and pseudo critical pressure (pPr) are used.

- pTc and pPr are devised simply for use in correlating physical properties.

- Pseudo critical properties are not equal to the actual critical properties of the mixture.

$$T_{pc} = \sum_j y_j T_{cj} \text{ and } p_{pc} = \sum_j y_j p_{cj} .$$

14

Compressibility EOS for Gas Mixtures

- Recall z-factors are function of the type of gas (composition) as well as of temperature and pressure. Fortunately, most of the components of natural gas are HCs of the same family. Hence, a correlation of this type is possible.
- Correlation is useful because the components of most natural gases appear in approximately the same ratio to one another.

15

Pseudo critical Properties: C_7+

- For petroleum fluid, all components heavier than hexane are generally lumped together as heptanes plus.
- The z-factor for C7+ can be correlated with the specific gravity.

16

Fluid Composition Unknown

- The z-factor for the fluid can be correlated with the specific gravity.

- This calculation is surprisingly accurate.

- Remember, the z-factor charts did not have any information about composition.

- It's the same thing here.

17

Other Equations of State

- The limitation of compressibility EOS is that z-factor is not constant

- Hence, mathematical manipulations cannot be made directly, but must accomplished by graphical or numerical methods

- Other commonly used EOS were developed so that the coefficients which correct the ideal gas for reality may be assumed constant.

- This permits these equations to be used in mathematical calculations involving differentiation or integration.

18

<u>Other Equations of State</u>

- Most common equations of state used (accepted) in the petroleum industry are:
  - – van-der Waals
  - – Redlich and Kwong and
  - – Peng-Robinson
- These three EOSs are cubic and have two empirical constants
- These equations have been used widely to calculate physical properties and vapor-liquid equilibria of HC mixtures

19

<u>van-der Waals EOS</u>

$$\left(p + \frac{a}{V_M^2} \right) (V_M - b) = RT$$

- Equation was proposed in 1873
- Differs from ideal by adding a/V_m^2 to pressure and subtracting constant b from molar volume.
- a/V_m^2 corrects for the attraction between the gas molecules
- B corrects for the volume of the gas molecules
- a and b are characteristics of the particular gas and R is universal gas constant.

20

Redlich-Kwong EOS

Redlich-Kwong Equation of State

Redlich and Kwong proposed an equation of state which takes into account the temperature dependencies of the molecular attraction term in a manner similar to Clausius.☐

$$\left[p + \frac{a}{T^{\frac{1}{2}} V_M (V_M + b)} \right] (V_M - b) = RT \quad \boxed{}$$

The advantage over the Clausius equation is that a third empirical constant is not included.

21

Peng-Robinson EOS

Peng-Robinson Equation of State

Peng and Robinson proposed a slightly different form of the molecular attraction term.☐

$$\left[p + \frac{a_T}{V_M(V_M + b) + b(V_M - b)} \right] (V_M - b) = RT \quad \boxed{}$$

The term a_T is temperature dependent as in the Soave-Redlich-Kwong equation of state; however, it does not have exactly the same values. The coefficients are calculated as follows.

$$b = 0.07780 \frac{RT_c}{p_c} \text{ and } a_c = 0.45724 \frac{R^2 T_c^2}{p_c}, \quad \boxed{}$$

$$a_T = a_c \alpha,$$

where

$$\alpha^{\frac{1}{2}} = 1 + m(1 - T_r^{\frac{1}{2}}), \quad \boxed{}$$

and

$$m = 0.37464 + 1.54226\omega - 0.26992\omega^3.$$

22

Gas Formation Volume Factor

- Units,
  - Reservoir bbl/SCF or
  - ft³/SCF

$$Bg = \frac{\dfrac{ZnRT}{P}}{\dfrac{Z_{SC}nRT_{SC}}{P_{SC}}}$$

23

Gas Formation Volume Factor

$T_{sc} = 520°R$, $p_{sc} = 14.65$ psia, and for all practical purposes $z_{sc} = 1$, then

$$B_g = \frac{zT(14.65)}{(1.0)(520)p} = 0.0282 \frac{zT}{p} \frac{cu\ ft}{scf}.$$

Also,

$$B_g = \left(0.0282 \frac{zT}{p} \frac{cu\ ft}{scf}\right)\left(\frac{bbl}{5.615\ cu\ ft}\right) = 0.00502 \frac{zT}{p} \frac{res\ bbl}{scf},$$

24

<u>Isothermal Compressibility</u>

Isothermal compressibility is defined as follows:
$$C_g(P,T) = -\frac{1}{V}\frac{\partial V}{\partial P}\bigg|_T$$

For an ideal gas

The simplest equation of state is that for ideal gases.

$$pV = nRT \text{ or } V = \frac{nRT}{p}$$

We wish to eliminate the term $\partial V/\partial p$ in Equation 6–4, so we derive this term from Equation 3–14 as

$$\left(\frac{\partial V}{\partial p}\right)_T = -\frac{nRT}{p^2}.$$

Combining Equation 6–5 with Equation 6–4 gives

$$c_g = \left(-\frac{1}{V}\right)\left(-\frac{nRT}{p^2}\right)$$

$$c_g = \left(-\frac{p}{nRT}\right)\left(-\frac{nRT}{p^2}\right) = \frac{1}{p}.$$

25

<u>Isothermal Compressibility</u>

For a real gas
$$V = nRT\frac{z}{p}$$

Thus,

$$\left(\frac{\partial V}{\partial p}\right)_T = nRT\frac{p\left(\frac{\partial z}{\partial p}\right)_T - z}{p^2}$$

$$c_g = -\frac{1}{V}\left(\frac{\partial V}{\partial p}\right)_T$$

$$c_g = \left[-\frac{p}{znRT}\right]\left\{\frac{nRT}{p^2}\left[p\left(\frac{\partial z}{\partial p}\right)_T - z\right]\right\}$$

$$c_g = \frac{1}{p} - \frac{1}{z}\left(\frac{\partial z}{\partial p}\right)_T.$$

26

Importance of Classification of Reservoir Fluids

- The PE must determine the type of fluid early in the life of the reservoir; because fluid type is the deciding factor in many of the decisions concerning the reservoir: Decision such as:
  - Method of sampling
  - Types and sizes of surface equipment
  - Calculation procedures for determining volume of oil and gas in place.
  - Techniques of predicting oil and gas reserves
  - Plan of depletion
  - Selection of enhanced oil recovery.

27

Identification of Reservoir Fluids

- Reservoir fluid type can be confirmed only by laboratory observation, but readily available production information can also indicate fluid type.

- Three properties are readily available the initial producing GOR (the most important factor); the gravity of the stock tank liquid, and the color of the stock-tank liquid.

- The stock-tank liquid (not a good indicator alone) but with the color are useful in confirming the fluid type indicated by the producing GOR.

28

Identification of Reservoir Fluids

- If all 3 indicators do not fit within the ranges given in rule of thumb, the rules fail and laboratory observation must be used.

- Comparison of fluid types as defined here does not fit those by regulatory agencies. The agencies' definitions usually do not bear any relationship with engineering definitions and at times they are contradictory.

29

Surface Facilities

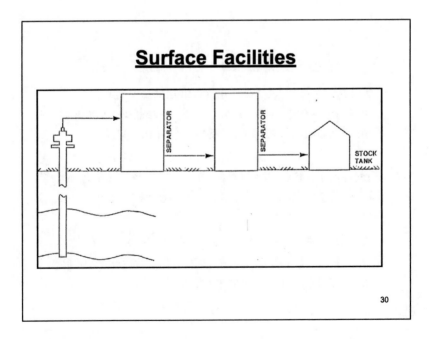

30

Black Oil Phase Diagram

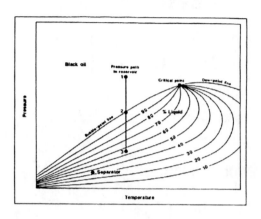

31

Black Oil Phase Diagram

- Lines within the phase envelope represent constant liquid volume measured as % total volume. The lines are called iso-vols or quality lines. The iso-vols are spaced evenly within the envelope.

- 123 indicates reduction in pressure at constant T

- 12 – indicates the oil is undersaturated.

- At point 2, oil is at its bubble point and is said to be saturated. A reduction in p will release gas to form free gas phase in the reservoir.

- 23 - additional gas is evolved and anywhere along 23, the oil is saturated.

32

Black Oils

- Not really 'black'. Hence the name is a misnomer. It is also called low shrinkage crude oil or ordinary oil.
- Color due to the heavier molecules like asphaltenes and resins
- Could be various shades of brown or reddish-brown.

33

Comments and Rule of Thumbs

- Black oils have a larger percentage of heavy molecules
- Phase diagram tends to have a large temperature range
- The iso-vol lines are spaced somewhat evenly in the phase diagram
- Field Identification of Black Oil
  - Initial GOR ≤ 2000SCF/STB
  - Stock-tank oil, gravity < 45°API
  - Stock-tank oil is very dark
- Laboratory
  - B_{oi} ≤ 2 res.bbl/STB
  - C7+ > 30 mole %, an indication of heavy HCs in black oil.

34

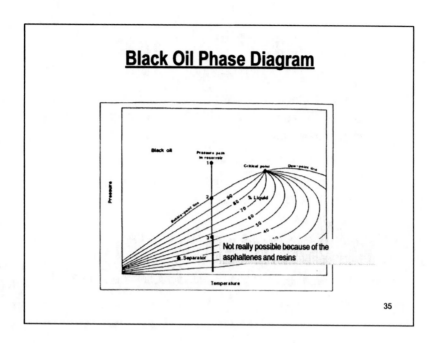

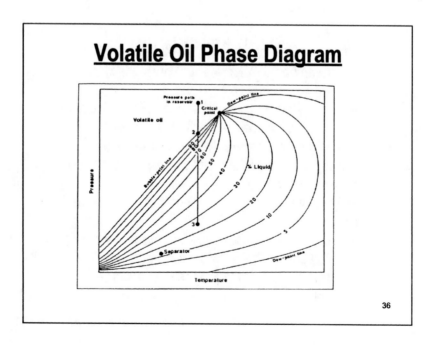

Volatile Oil Phase Diagram

- Volatile oil is defined as ethane through hexanes
- Temperature range is somewhat smaller and the critical temperature is much lower than for black oil and is close to the reservoir temperature.
- Iso-vols are not evenly spaced.
- Vertical line shows path of constant temperature. A small reduction in pressure below P_B, point 2 causes a large release of gas in the reservoir.
- A volatile oil may become as much as 50% gas in the reservoir at only few 100 psi below P_B.
- An iso-vol with a much lower % liquid crosses the separator conditions.

37

Volatile Oils

- Contain fewer heavy molecules but more of the intermediates (defined as ethane through hexanes)
- Consequently, the temperature range covered by the phase diagram is smaller (in comparison with black oils)
- The iso-vol lines are not evenly spaced (lesser heavier molecules), consequently more gas comes out of solution for the same pressure drop
- Volatile oils also called high-shrinkage crude oil or near-critical oils.
- Material balance equations for black oil does not work for volatile oils because the gas associated with black oil is assumed to be dry while the gas associated with volatile oil is reach in condensate.

38

Comments and Rule of Thumbs for Volatile Oils

- Field Identification of Volatile Oil
  - 2000 SCF/STB ≤ initial GOR ≤ 3300 SCF/STB
  - Stock-tank oil, gravity > 45°API
  - Stock-tank oil is colored (usually brown, orange and sometimes green)
- Laboratory
  - For a fluid to be volatile its critical temperature must be greater than reservoir temperature
  - B_{oi} > 2 res.bbl/STB (i.e.. high shrinkage)
  - 12.5% <C7+ < 30 mole %,
  - Volatile oil must be produced through three or more stages of separation in order to minimize shrinkage

39

Gas Condensates (Retrograde Gases)

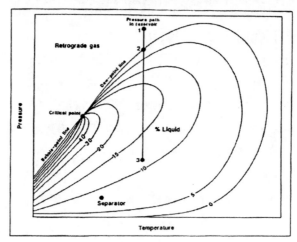

40

Retrograde Gas Phase Diagram

- Phase diagram is smaller and critical point is further down the left of the envelope.
- Contains fewer of heavy HCs than do oils.
- Has a Tc < reservoir temperature and a cricondentherm temperature > reservoir temperature.
- Initially, the retrograde gas is totally gas in the reservoir, point 1. As reservoir pressure is reduced, liquid condenses from gas to form a free liquid in the reservoir. This liquid will normally not flow and cannot be produced.
- They drop condensate in the reservoir as pressure decreases (an abnormal behavior)
- Have a much smaller percent of heavy molecules

41

Comments and Rule of Thumbs for Retrograde Gas

- Field Identification of Retrograde Gas
  - Lower limit of initial GOR is 3300 SCF/STB and upper limit of over 150,000 SCF/STB has been observed.
  - Stock-tank oil, gravity between 40°API and 60°API
  - Stock-tank oil is lightly colored, brown, orange, greenish or water-white
- Laboratory
  - 12.5% <C7+ < 12.5 mole %

42

Retrograde Gas Phase Diagram

- Retrograde gases also called retrograde gas-condensate, retrograde condensate gases, gas condensates or condensates, but the right name is RETROGRADE GAS because initially the fluid is gas in the reservoir and exhibits retrograde behavior.

- Stock tank liquid is called condensate and liquid produced in the reservoir is retrograde liquid.

- Surface gas is reach in intermediates and often processed to remove liquid propane, butanes, pentanes and heavier HCs. The liquids are called plant liquids.

43

Wet Gases

- The 'wet' in wet gases refers not to water vapor (which is present in all gases).

- It refers to the liquid dropout that occurs at the surface (at atmospheric pressure and temperature)

- Do not drop condensate in the reservoir, only in the surface separator.

44

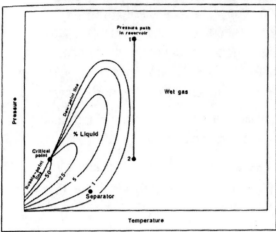

Wet Gas Phase Diagram

45

Wet Gas Phase Diagram

- The phase diagram lies below the reservoir temperature.
- Exists solely as a gas in the reservoir throughout the pressure reduction.
- The pressure path 12 does not enter the phase envelope and no liquid is formed in the reservoir.
- Separator conditions lie within the phase envelope and thus some liquid is formed at the surface.

46

Comments and Rule of Thumbs for Wet Gas

- Field Identification of Wet Gas
  - Same as for Lower limit of initial GOR is 3300 scf/STB and upper limit of over 150,000 scf/STB has been observed.
  - Stock-tank oil, gravity between 40°API and 60°API
  - Stock-tank oil is lightly colored, brown, orange, greenish or water-white
- Laboratory
  - 12.5% <C7+ < 12.5 mole %

47

Wet Gas Phase Diagram

- Surface liquid is called condensate
- Wet gas refers to the HC liquid with which condenses at surface conditions. Reservoir gas is normally saturated with gas.
- Same range of gravities as for retrograde gases. Stock-tank liquid does not change during the life of the reservoir.
- Stock-tank liquid is usually water-white
- True wet gas will have high producing GOR and will remain constant during the life of the reservoir.
- For engineering purposes, a gas that produces 50,000 scf/STB can be treated as wet gas

48

Dry Gases

- Have a large percent of methane (~90 mole %); that is primary methane with some intermediates.
- They remain 'dry' in the reservoir and at the surface because the normal surface separator conditions lies outside the envelope.
- Dry means does not contain heavy HCs that can turn to liquid; but may condense water on surface.
- A set of equations called gas material balance equations has been developed to determine Gi and predict gas reserves.

49

Dry Gases

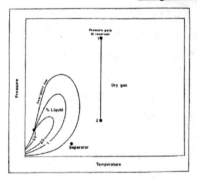

- Figure shows the phase diagram for a dry gas
- Have a large percent of methane (~90 mole %)
- They remain 'dry' in the reservoir and at the surface.

50

Properties of Dry Gas

- Concerned with several properties which commonly are normally used by PEs.
- Since volume of gas varies with T & P must state the conditions at which the volume is reported.
- Standard conditions (also known as Base conditions) are 60°F and 14.7 psia. Hence the unit SCF.

51

Properties of Dry Gas

- Dry gases are the easiest to deal with because no liquid condenses
- The composition of the surface gas = composition of the gas in the reservoir (i.e.. ρ_g) reservoir = (ρ_g) on surface.
- Hence sample of gas analyzed on surface given us a clue as to the properties in the reservoir.

52

Value of Z-Factor

- The value is obtained by many methods
- If an experimental value is available, use it
- If composition of the gas is known, the pseudo reduced properties can be calculated and z-factor get from plot of z vs. pTr and pPr.
- If only specific gravity is known, then we can use empirical graph
- Where necessary adjust for nonhydrocarbon components

53

Gas Formation Volume Factor

- Defined as ratio of volume of gas at reservoir conditions to volume of gas at standard surface conditions
- Units: res. cu.ft/SCF; res. bbl/SCF
- Reciprocal of formation volume factor is sometimes called gas expansion factor

54

Properties of Dry Gas

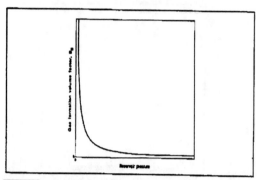

☐ Typical shape of gas formation volume factor as a function of pressure at constant reservoir temperature.

55

Coefficient of Isothermal Compressibility

The *coefficient of isothermal compressibility* is defined as the fractional change of volume as pressure is changed at constant temperature. The defining equations are

$$c_g = -\frac{1}{V}\left(\frac{\partial V}{\partial p}\right)_T \text{ or } c_g = -\frac{1}{V_M}\left(\frac{\partial V_M}{\partial p}\right)_T$$

$$\text{or } c_g = -\frac{1}{v}\left(\frac{\partial v}{\partial p}\right)_T.$$

- The partial derivative is used since only one independent variable, P is allowed to vary with T constant.
- C_g has unit of 1/psi or psi^{-1}
- For Eq. 6.4 to be useful it must be combined with an EOS.

56

Coefficient of Isothermal Compressibility

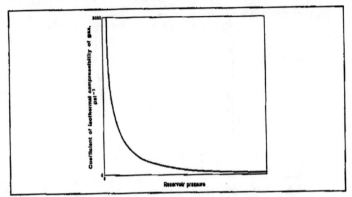

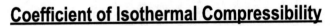 Typical shape of the coefficient of isothermal compressibility of a gas as a function of pressure at constant reservoir temperature.

57

Coefficient of Isothermal Compressibility of Ideal Gas

The simplest equation of state is that for ideal gases.

$$pV = nRT \text{ or } V = \frac{nRT}{p}$$

$$\left(\frac{\partial V}{\partial p}\right)_T = -\frac{nRT}{p^2}.$$

$$c_g = \left(-\frac{1}{V}\right)\left(-\frac{nRT}{p^2}\right)$$

$$c_g = \left(-\frac{p}{nRT}\right)\left(-\frac{nRT}{p^2}\right) = \frac{1}{p}.$$

58

Coefficient of Isothermal Compressibility of Real Gas

- At low pressure, the z-factor decreases as pressure increases. Hence the partial derivative of z-factor with respect to P is negative and c_g is larger for an ideal gas.

- At high pressure, z-factor increases with pressure and the partial derivative of z-factor with respect to P is positive and hence c_g is less for real gas (i.e.. less than in the case of an ideal gas).

59

Gas Viscosity

- The viscosity of a Newtonian fluid is defined as the ratio of the shear force per unit area to the local velocity gradient. Gas viscosity decreases as pressure decreases since the molecules are simply further apart at lower pressure and move past each other more easily. At low pressures, gas viscosity increases as temperature increases; however, at high pressure gas viscosity decreases as temperature increases.

- The viscosity of a pure gas depends upon temperature and pressure, but for gas mixtures it is also a function of the composition of the mixture. For natural gases, the correlations of Carr, Kobayashi and Burrows (1954) is widely used.

   where; $\mu_1 = \varphi\{M, T\}$; and $\mu/\mu_1 = \varphi\{p_r, T_r\}$
   μ = low pr. or dilute-gas viscosity; μ_1 = gas viscosity at high pr.

- The correlation presented above is compatible with the correlations for compressibility factors since both are based on the concept of corresponding states.

60

The Coefficient of Viscosity of Gas

- The coefficient of viscosity is a measured of resistance to flow. It is measured in centipoise (cp). This viscosity if called dynamic viscosity to differentiate it from kinetic viscosity which is (dynamic viscosity/density) and has the unit of centistokes.

- Gas viscosity decreases as reservoir pressure decreases. At low pressure, the viscosity increases with increase in temperature. However, at high pressure, viscosity increases with increase in temperature.

- The reciprocal of viscosity is called fluidity.

- Difficult to measure viscosity of gas, hence we use viscosity correlations.

61

Viscosity of Gases

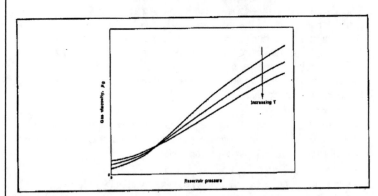

Typical shape of gas viscosity as a function of pressure at three reservoir temperatures.

62

Viscosity of Ethane

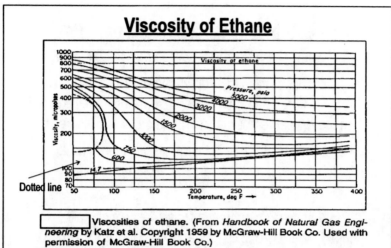

Dotted line

Viscosities of ethane. (From *Handbook of Natural Gas Engineering* by Katz et al. Copyright 1959 by McGraw-Hill Book Co. Used with permission of McGraw-Hill Book Co.)

The dotted line is the saturation line and the point of maximum temperature on the dotted line indicates critical point. At saturation pt. $\mu_l = \mu_g$. Isobars above the saturation line gives the μl and below gives μ_g

63

Single Component Phase Diagram

P_C

Solid

Liquid

Melting-point line

Pressure

Triple Point

Vapor-pressure line

Gas

C

T

Sublimation Curve

Temperature

T_C

64

Pure Substances (Single Component)

- The vapor-pressure line = dew point = bubble point
- The critical point is the upper limit of the V-P line. The temperature and pressure represented by this point are called critical temperature Tc and critical pressure Pc.
- For a pure substance Tc is the temperature above which a gas cannot be liquefied regardless of the pressure applied.
- Triple point represents the P and T at which solid, liquid, and gas coexist under equilibrium conditions. It is the lower end of the V-P line

65

Multi-Component Systems

- As the number and complexity of the molecules in a mixture increase the separation between the bubble-point and dew-point lines on the phase diagram becomes greater.
- There is wide variety of Pc and Tc and different positions of the critical points on the saturation envelope.
- There is also a large separation between the Tc and the cricondentherms in all instances and Pc for the lighter HCs mixtures.

66

Change of Hydrocarbon Liquids Volume

- Liquid properties cannot be determined with broadly applicable equations as for gases but must be either measured in the laboratory or estimated from published correlations.

- *Oil Formation Volume Factor (B.):*

  B_o is defined as the volume of reservoir oil to the volume of STB yielded by reservoir volume.

- *Solution Gas Oil Ratio (R.)*

  Solution gas-oil ratio, (R_s) is defined as the volume of gas "dissolved" in a volume of reservoir oil equivalent to a stock tank barrel.

- *Total Formation Volume Factor (B.)*

  $$B_t = B_0 + (R_s - R)B_g$$

67

Oil Formation Volume Factor

Definitions

$$Bo = \frac{\text{Volume of Oil + Dissolved gas at Reservoir Pressure \& Temp.}}{\text{Volume of Oil entering Stock tank at Tsc, Psc}}$$

Vai'ts - Ra.d"YYOr boiTcJa. (bbl) /Stock tonk bon-eta (STS)

General Shape of Bo

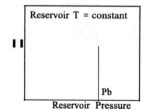

Reservoir T = constant

Pb

Reservoir Pressure

68

Solution Gas Oil Ratio (Rs)

Solution Gas Oil Ratio (Rs)

• How much gas is dissolved in the oil volume per volume basis
• Rs depends upon pressure

Units [=] SCF gas /STB oil

General Shape of Solution Gas Oil Ratio (Rs)

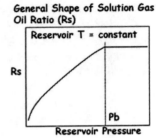

69

2-Phase Formation Volume Factor

Total Formation Volume Factor Bt

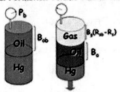

Definition of Bt

• Also called Two-phase formation volume factor

$$B_t = B_o + B_g \left(R_{sb} - R_s \right)$$

Units...

bbl/STB + bbl/SCF * (SCF/STB)

General Shape of Bt

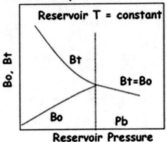

70

Viscosity of Oil

- Dynamic viscosity measures a fluid resistance to flow. Very sensitive to temperature and to a lesser degree to pressure.
- Kinematic viscosity is the ratio of dynamic viscosity over density.

Variation of Oil Viscosity

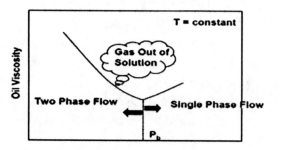

71

Introduction to Well Test Analysis

- **What is Well Test Analysis?**

  - The goal of well testing is to collect information about flow conditions in the well, around the immediate vicinity of the well, in the virgin portions of the reservoir not influenced by the drilling operations and simulation treatments, and information about the boundaries of the reservoir.

  - The well flow rate is varied and the resulting pressure transients are measured. The measurement of variation of pressure with time provides a pressure transient data which then can be analyzed to determine the formation parameters that characterize the flow conditions that exist in the system.

72

Introduction to Well Test Analysis

- Well test analysis can be considered as a systems analysis technique.

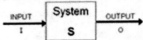

- The system "S" represents the wellbore and the formation that it is in communication with.
- The input "I" represents the constant withdrawal of the reservoir fluid and it can be considered as a forcing function applied to the system "S". The response of the system (output), "O" which represents the change in reservoir pressure is measured during the test.

73

Introduction to Well Test Analysis
Forward and Inverse Problems

- The response of the reservoir can be computed for specific initial and boundary conditions for a reservoir-well system with known properties. This is called the "forward (direct)" problem and is expressed as

    $- [I] \times [S] \rightarrow [O]$

- In an "inverse" problem we are attempting to find a well defined system whose response to the input signal is measured as output. This type of problem is known in mathematics as the inverse problem and is expressed as

    $- [O]/[I] \rightarrow [S]$

74

Why Well Test?

- The purpose of well testing is to collect pressure transient data which can be utilized after analysis, for decision making in exploration as well as reservoir engineering.
- Well Test Analysis as an Algorithmic Protocol
  - Models used for the pressure transient analysis are always built in the same manner. The basic models are typically homogeneous. Heterogeneous models have been the subject of many recent developments and solutions with various inner boundary conditions are now widely used.
  - In well test analysis, perhaps the most critical step is the identification of the model that will be used in interpreting the test data.

75

Well Testing

- Provides measurements on the dynamic behavior of the reservoir
- Provides good quality on in-situ reservoir properties
- Provides information on reservoir pressure formation deliverability, reservoir permeability and on skin effects
- Helps identifying the reservoir limits and continuity
- Helps in reconciling geologic correlations with the inferred depositional model

76

General Concepts – Porous Media as Continuum

- The microscopic scrutiny of a porous medium reveals that its local properties may vary widely depending on the volume over which the scrutiny is performed.
- Instead of a microscopic description, the usual way of approaching a description of a porous media and the fluids within it is to use the *continuum approach*. Fluid properties and porous medium properties are treated as varying "continually" in space.

77

Fundamental Equations for Well Test Analysis

1. *The Continuity Equation -,* describes how mass is accumulated and transmitted within the system as a result of fluid flow.
2. *The Equation of State* - describes how fluid density changes with pressure and temperature.
3. *The Energy Equation* - describes how energy is accumulated and transmitted within the system as a result of heat transfer.
4. *The Momentum Equation* - describes how momentum is accumulated and transmitted within the system through the action of forces.
5. *The Constitutive Equation* - describes the deformation of the fluid as a result of normal and tangential forces applied to it.
- In well test analysis, the three single most important equations are: the *continuity equation,* the *equation of state* and *Darcy's law.*

78

Introduction to Reservoir Modeling

- **What is Reservoir Modeling?**

- This is the process of developing a simulator that captures the physical processes occurring in the reservoir.

- The process combines physics, mathematics, reservoir engineering, and computer programming to develop a tool for predicting performance of the reservoirs under various operating conditions.

- Reservoir simulation consists of three basic components including: (1) simulator, (2) user (simulation engineer), and (3) reservoir description.

- There are two main environments in reservoir simulation studies: engineering environment and computer environment.

79

Introduction to Reservoir Modeling

- There are five basic steps involved in conducting a reservoir simulation study. These steps are:
  - Model Selection
  - Data Preparation
  - Computer Runs
  - Analysis and Interpretation
  - Performance Prediction

80

Why Reservoir Modeling?

- <u>**Newly Discovered Fields**</u>
  - To integrate the available pieces of information as accurately as possible in order to construct an overall picture of the reservoir.
  - To establish the most optimum field development plan, optimum operation strategies, recovery mechanisms, and take into account future needs for improved recovery.
- <u>**Mature Fields**</u>
  - To increase or maintain the recovery of the reservoir.
  - To investigate the appropriate locations and number of in-fill wells in the in-fill drilling program or the improved recovery applications.
- <u>**Project Screening**</u>
  - To provide supporting information pertaining to whether or not such project should be implemented.
- <u>**Modeling as a Dynamic Process**</u>
  - Reservoir modeling is a dynamic process in which the reservoir model is continuously refined as field development and operation progresses and more information becomes available

81

Why Reservoir Modeling?

What Questions Can a Reservoir Model Answer?

- <u>Basic Simulation Approaches</u>
  - *Analytical approach* - providing an exact solution to an approximate problem. This approach is utilized in classical well test analysis.
  - *Numerical approach* - providing the approximate solution to an exact problem. This approach attempts to solve the more realistic problem with very limited assumptions.
- <u>Reservoir characterization</u>
  - A reservoir simulator can be used to characterize the reservoir under study through a process called *history matching* in which the reservoir parameters are adjusted or tuned to match the past performance of the reservoir.

82

Why Reservoir Modeling?

What Questions Can a Reservoir Model Answer?

- Forecasting
  - After the simulation model is being adjusted and validated through the history matching process, it can be used to forecast future reservoir performance.
  - It allows an engineer to investigate reservoir performance under various production and operation strategies in order to come up with a well-designed strategy for field development, and field operation.

- Feasibility analysis
  - Results from the simulation study can then be used to perform cost and revenue calculations in order to select a feasible production and operation strategy for the field.

83
83

Mathematical Model of a Reservoir

- **Model types:**

  (a) geological; (b) Detailed simulation model; (c) Simplified simulation model; (d) Material balance model

(a) Geological model makes use of cores and logs

(b) Detailed simulation model

84

Reservoir Simulation

- Reservoir simulation offers many incentives to the reservoir management engineer
- Economics
  - Scoping studies
  - New field development plans
  - Reservoir deliverability studies
  - Alternative development plans
  - Operating strategies
  - Recovery processes
  - Decision analysis
- Other Incentives
- Integration of data and physics; credibility and reliability; unbiased calculations for third party (partners and government); performance monitoring (well test analysis & decline curve analysis; education (parametric studies); research efforts; training tool; & non-elective studies

85

Uncertainty of Reservoir Simulation

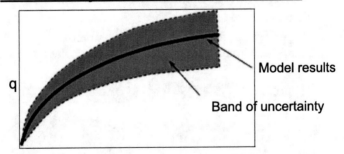

q

Model results

Band of uncertainty

Time

- Clearly, we need a good "checks and balances" to reduce the size of the uncertainty band.
- The sources of uncertainty are: data quantity & quality; geology; scale up; & mathematical model

86

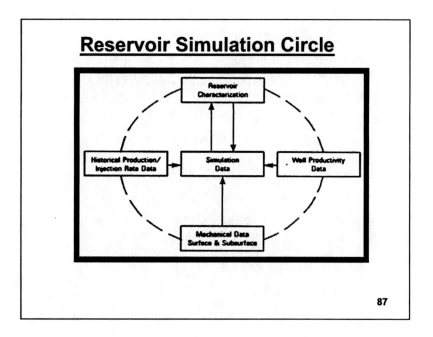

Reservoir Simulation Circle

Reservoir Characterization

Historical Production/ Injection Rate Data

Simulation Data

Well Productivity Data

Mechanical Data Surface & Subsurface

87

Reservoir Simulation

- Major steps used in the development of reservoir simulators are:
  - Formulation
  - Discretization
  - Well representation
  - Solution
  - Validation & application

88

Reservoir Simulation

- Reservoir simulation as a tool in the hands of a reservoir management engineer offers several incentives: economics and others
- Economics
  - Scoping studies
  - New field development plans
  - Reservoir deliverability studies
  - Alternative development plans
    - Operating strategies
    - Recovery processes
  - Decision analysis

89

Reservoir Simulation

- Other Incentives of reservoir simulation:
- Integration of data and physics
- Credibility and reliability
  - Unbiased calculations
  - Third party review
- Performance monitoring
  - Well test analysis
- Education
  - Parametric studies
  - Research efforts
  - Training tool
- Non-elective studies

90

Uncertainty and Evolution of a Geologic Model

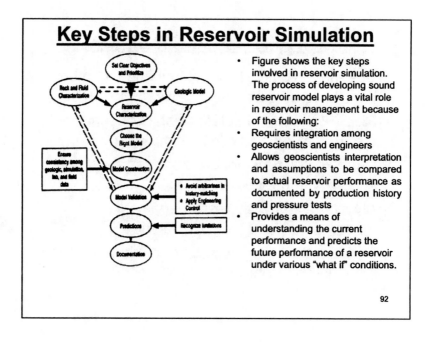

Exploration Concept
– Structural

Development Concept
– Structural
– Stratigraphic

Development Concept
– Structural
– Stratigraphic
– Lithologic
– Areal continuity
and scale

Continued Refinement With Time

91

Key Steps in Reservoir Simulation

- Figure shows the key steps involved in reservoir simulation. The process of developing sound reservoir model plays a vital role in reservoir management because of the following:
- Requires integration among geoscientists and engineers
- Allows geoscientists interpretation and assumptions to be compared to actual reservoir performance as documented by production history and pressure tests
- Provides a means of understanding the current performance and predicts the future performance of a reservoir under various "what if" conditions.

92

Integrated Geoscience Software System

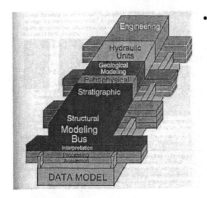

- A major breakthrough in reservoir modeling is the advent of integrated geoscience (reservoir description) and engineering (reservoir production performance) software designed to manage reservoirs more effectively and efficiently.

93

Intelligent Use of Simulators

- Use the SIMPLEST model, minimum number of cells, largest time step, fewest runs
- RELIABILITY of simulator results depends on the ACCURACY of critical input data
- Use SHORT runs until you are certain of the accuracy of the results
- Always CHECK key results against fundamentals and industry practice
- DESIGN the study according to manpower and computer availability

94

Sources of Reservoir Data and Data Implementation in Reservoir Engineering

1

Data Acquisition and Analysis

- Most of the data except for the production and injection data are collected during delineation and development of the fields.

- An effective data acquisition and analysis program requires careful planning and well coordinated team efforts of interdisciplinary geoscientists and engineers throughout the life of the reservoir.

- Justification, priority, timelines, quality, and cost-effectiveness should be the guiding factors in data acquisition and analysis . It will be more effective to justify to management data collection if the need for the data, the cost, and the benefits are clearly defined.

2

<u>Data Acquisition and Analysis</u>

- Coring, logging, and initial reservoir fluid sampling should be made at the appropriate times using the proper procedures and analyses.

- Normally, all wells are logged, however, an adequate number of wells should be cored to validate the log data.

- It is also beneficial to measure pressures in all wells at least every two to three years to aid in calibrating reservoir models.

3

<u>An Efficient Data Flow Diagram</u>

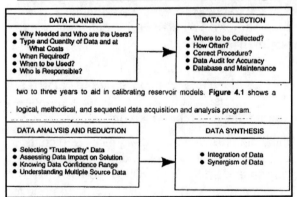

DATA PLANNING	DATA COLLECTION
● Why Needed and Who are the Users? ● Type and Quantity of Data and at What Costs ● When Required? ● When to be Used? ● Who is Responsible?	● Where to be Collected? ● How Often? ● Correct Procedure? ● Data Audit for Accuracy ● Database and Maintenance

two to three years to aid in calibrating reservoir models. **Figure 4.1** shows a

logical, methodical, and sequential data acquisition and analysis program.

DATA ANALYSIS AND REDUCTION	DATA SYNTHESIS
● Selecting "Trustworthy" Data ● Assessing Data Impact on Solution ● Knowing Data Confidence Range ● Understanding Multiple Source Data	● Integration of Data ● Synergism of Data

This figure shows a logical, methodical, and sequential data acquisition and analysis program   4

Data Validation

- Field data are subjected to may errors such as sampling, systematic, and random errors. Hence, the collected data need to be carefully reviewed and checked for accuracy as well as for consistency.

- Core and log analyses data should be carefully correlated and their frequency distributions made to identify different geologic facies.

- Log data should be carefully calibrated using core data for porosity and saturation distributions, net sand determination, and geological zonation of the reservoir.

- The reservoir fluid properties can be validated by using equation of state calculations and by empirical correlations.

5

Data Validation

- The validity of geological maps must be established by using the knowledge of depositional environment. Also. The presence of faults and flow discontinuities as evidenced in a geological study can be investigated and validated by pressure interference, pulse, and tracer well tests.

- The reservoir performance should be closely monitored while collecting routine production data including reservoir pressures. If past production and pressure data are available, classical material-balance techniques and reservoir modeling can be very useful to validate the volumetric original hydrocarbon-in-place (OHCIP) and aquifer size and strength.

- Empirical correlations can be used to generate data that are not always available. Data such as laboratory rock properties (oil-water and gas-oil relative permeabilities) and fluid properties (PVT data).

6

Data Storing and Retrieval

- The acquired, recorded, and validated data from various sources must be stored in a common computer database accessible to all interdisciplinary end users. As new geoscience and engineering data are available, the database must be updated. The stored data are used to perform multipurpose reservoir management functions including monitoring and evaluating the reservoir performance.

- A major challenge for the industry is due to the non-communicating nature of the network system. The problems are:

  1. Incompatibility of the software and data sets from different disciplines, and

  2. Database usually do not communicate with each other.

7

Data Application

A better representation of the reservoir is made from 3-D seismic information. The cross-well tomography provides interwell heterogeneity. Geological maps such as gross and net pay thickness, porosity, permeability, saturation, structure, and cross-section are prepared from seismic, core and log analyses data. These maps are used for reservoir delineation, reservoir characterization, well locations, and estimates of oil-in-place and gas-in-place.

The more commonly used logging systems are:

- Open-hole or petrophysical data logs

  - Resistivity, induction, spontaneous (SP), and gamma ray logs

  - Density, sonic compensated neutron, sidewall neutron logs

  - Porosity, dielectric, and caliper logs

- Cased-hole logs

  - Gamma ray, neutron, carbon/oxygen, chlorine, pulse neutron, and caliper logs

8

Data Application

- Production logs can also be used to identify remaining oil saturation in undeveloped zones in existing production and injection wells. Time-lapsed logs observation wells can detect saturation changes and fluid contact movement.

- Core analysis is classified into conventional, whole-core, and sidewall analyses. The most commonly used conventional or plug analysis involves the use of a plug or a relatively small sample of the core to represent an interval of the formation to be tested.

- Unlike wire-line log analysis, core analysis give direct measurement of the formation properties, and the core data are used for calibrating well log data. These data can have a major impact on the estimates of hydrocarbon-in place, production rates, and ultimate recovery.

9

Data Application

Fluid properties determined from laboratory studies or correlations are used for the following:

- Volumetric estimates of reservoir oil and gas in place, reservoir type and reservoir performance analysis.
- Estimates of reservoir performance (IPR), wellbore hydraulic (TPR), flowline pressure losses, and equipment design.

The well test data are very useful for reservoir characterization and reservoir performance evaluation. Pressure build-up or falloff tests provide the best estimate of the effective permeability-thickness (kh) of the reservoir in addition to reservoir pressure, stratification, and presence of faults and fractures. Pressure interference and pulse tests provide reservoir continuity and barrier information. Multi-well tracer test used in waterflood and enhanced oil recovery projects give the preferred flow paths between the injectors and the producers. Single-well tracer tests are used to determine residual oil saturation in waterflood reservoirs.

10

Data Acquisition and Analysis

- Knowledge of the reservoir is gained through an integrated data acquisition and analysis program. Data analyses require a great deal of effort; scrutiny; and innovation.
- The key steps in data acquisition are
  - (1) Plan justify time and prioritize
  - (2) Collect and analyze;
  - (3) Validate and store (database).
- An efficient data management program is needed for sound reservoir management because a lot of data are collected.

11

Reservoir Data Needed for Reservoir Management

Classification	Data	Acquisition Timing	Responsibility
Seismic	Structure, stratigraphy, faults, bed thickness, fluids, interwell heterogeneity	Exploration	Seismologists, Geophysicists
Geological	Depositional environment diagenesis, lithology, structure, faults, and fractures	Exploration, discovery & development	Exploration & development geologists
Logging	Depth, lithology, thickness, porosity, fluid saturation, gas/oil, water/oil and gas/water contacts, and well-to-well correlations	Drilling	Geologists, petrophysicists and engineers
Coring		Drilling	Geologists, drilling and reservoir engineers, and laboratory analysts
Basic	Depth, lithology, thickness, porosity, permeability, and residual fluid saturation		
Special	Relative permeability, capillary pressure, pore compressibility, grain size, and pore size distribution		
Fluid	Formation volume factors, compressibilities, viscosities, gas solubilities, chemical compositions, phase behavior, and specific gravities	Discovery, delineation, development, and production	Reservoir engineers and laboratory analysts

12

Reservoir Data Needed for Reservoir Management

Classification	Data	Acquisition Timing	Responsibility
Well Test	Reservoir pressure, effective permeability-thickness, stratification, reservoir continuity, presence of fractures or faults, productivity and injectivity indices, and residual oil saturation	Discovery, delineation, development, production and injection	Reservoir and production engineers
Production & Injection	Oil, water, and gas production rates, and cumulative productions, gas and water injection rates and cumulative injections, and injection and production profiles	Production & injection	Production & reservoir engineers

13

Data Required for Reservoir Performance Analysis

Data Group	Volumetric	Decline Curve	Material Balance	Mathematical Models
Geometry	Area, thickness	No	Area, thickness Homogeneous	Area, thickness Heterogeneous
Rock	Porosity, saturation	No	Porosity, saturation, relative permeability, compressibility Homogeneous	Porosity, saturation, relative permeability, compressibility, capillary pressure Heterogeneous
Fluid	Form. vol. factors	No	PVT Homogeneous	PVT Heterogeneous
Well	No	No	PI for rate vs. time	Locations Perforations PI
Production & Injection	No	Production	Yes	Yes
Pressure	No	No	Yes	Yes

14

Integrated Reservoir Model

- Geologic Model
- Geo-cellular Model
    - Fine Scale (Layering Based on the Scale of Data Acquisition: Approximately 1/2 -ft)
    - Reservoir Properties Populated Geostatistically
- Simulation Flow Model
    - Coarse Scale (Layering Based on Computational Considerations)
    - Reservoir Properties Populated from Scale-Up of the Fine Grid Model
    - Fluid Properties (PVT) and Rock Fluid Interaction Properties (Relative Permeability and Capillary Pressure)
    - History Match ⇐ Interactive
    - *Predictions / Forecasts* ⇐ *Main Objective*

15

General Categories of Reservoir Hydrocarbons

- Exploration
- Discovery
    - Discovery Well
    - Delineation Wells
- Appraisal
    - Delineation Wells
    - Appraisal Wells
- Development
    - Production Wells
    - Injection Wells
- Objective of Discovery: Delineation, and Appraisal Wells; Obtain Reservoir Data
- Objective of Development Wells: Produce Oil and Gas (Generate Revenue)

16

Stages of Field Development

- Discovery Wells
  - Always logged (open-hole)
  - If hydrocarbons are present, generally tested (DST)
  - May be temporarily suspended as future development well
- Delineation and Appraisal Wells
  - Always Logged (Open-Hole)
  - Cored as Required
  - Tested (DST) as Required
  - May be Temporarily Suspended as Future Development Well
- Development Wells
  - Always Logged (Open-Hole)
  - Generally not Cored
  - Generally not Tested (DST)
  - Flow Tested as Required
  - May Be Logged (Cased-Hole) as Required

17

Seismic Data Applications to Integrated Reservoir Models

- Provide Top Structure of Surfaces Bounded by Strong Seismic Reflectors
- Estimate Gross Reservoir Thickness
- Identify Large Scale Geologic Features
  - Faults
  - Pinch-Outs
  - Erosional Surfaces
- Identify Fluid Contacts
- Provide Seismic Attributes for Conditioning Geostatistical Models

18

Seismic Data Issues

POSITIVE

- Main Exploration Method in Petroleum Industry
- Capable of Making Measurements of Subsurface Formations without Drilling a Well

NEGATIVE

- Seismic Resolution
- Requires a Velocity Model
- Long Lead Time
  - Schedule Seismic Crew /Ship (May Need to Schedule One Year in Advance)
  - Shoot Data (May Take Several Months)
  - Process Data (May
  - Take One Year)

19

Geological Processes – Depositional Environment

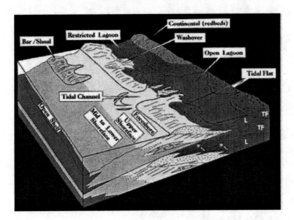

20

Core Data

- Core Description
  - Lithology
  - Net/Gross Thickness
- Routine Core Analysis (PKS Analysis)
  - Porosity
  - Permeability
  - Current Saturations
- Special Core Analysis (SCAL)
  - Relative Permeability
  - Capillary Pressure/J-Function
  - End-Point Saturations (Critical Saturations)
  - Compressibility
  - Electrical Properties: Archie Parameters (a, m, and n)
  - Mineralogy (X-Ray Diffraction)

21

Core Data

Applications to Integrated Reservoir Models

- Identification of Depositional Environment through Core Description
- Calibration of Log Derived Properties
  - Initial Saturations
  - Porosity
  - Permeability
- Mechanical Properties
  - Formation Compressibility (required input for flow simulation)
  - Rock Strength
- Development of Permeability-Porosity Transforms
- Relative Permeability Data (Core Scale: "Rock Curves")
- Capillary Pressure/J-Function Data (Core Scale: "Rock Curves")
- End-Point Saturation Data (Core Scale)

22

Core Data Issues

POSITIVE

- Data Analyzed under Controlled Laboratory Conditions
- Only Method Available to Obtain Full Relative Permeability and Capillary Pressure Curves
- Can be Used to Calibrate Log Data
- Data Coverage/Sampling: Core Description Interprets Entire Section of Interest
- Lead Time: 2-3 Months for Routine Core Analysis

NEGATIVE

- Expensive (Acquisition / Transport / Analysis / Preservation / Storage)
- Core Scale versus Reservoir Scale Issues
- Data Coverage/Sampling: A Few Plugs per Wellbore for Routine/Special Core Analysis
- Radius of Investigation: Inches (Removed from the Reservoir)
- Different Analyses (SS / USS Tests, Ambient / Reservoir Conditions)
- Lead Time: 9-12 Months for Special Core Analysis
- Life-Cycle of Field Issues
- – Changes in Service Companies with time
- – Changes in Technology with Time
- Core Recovery not always Possible: Unconsolidated or Fractured Rock

23

Core Data

Good Reservoir Management Practices

- Use core analysis to analyze some non-reservoir rock: in many instances the movement of fluids is governed by distribution of non-reservoir rock and non-reservoir rock properties
- Retain all data at the scale they were measured until the scale-up process - perform scale-up as last step in model construction
- Remove any laboratory artefacts prior to using core data

24

Open-Hole Log Data

- Definition: Well Logs Run Immediately after a Well has been Drilled and Prior to Casing and Cementing Operations
- Generally Run on all Wells to Identify Intervals to Perforate
- Modern Open-Hole Log Suite (Basic):
  - Gamma Ray Log *; Induction Log (ILD)
  - Compensated Neutron Log (CNL) *; Density Log
  - Sonic Log; Caliper Log
- Additional Modern Open-Hole Logs:
  - Spectral Gamma Ray
  - Magnetic Resonance Imaging Log (MRIL): (Total Porosity, Free Fluids, Bound Fluid)
  - Image Logs
- Wireline Formation Tester (RFT/SFT)
  May also be run in cased-hole environment

25

Open-Hole Log Data Applications of the Gamma Ray Log to Integrated Reservoir Models (When Run in Open-Hole Environment)

- Correlate zones well-to-well (identify marker surfaces)
- Identify lithologies
  - Shales: high gamma ray count (API gamma ray units)
  - Sandstones: low gamma ray count
- Estimate Volume of Shale - V_{sh}
- Aid in Interpretation of deposition environment (in conjunction with Induction Log - ILD)
  - Coarsening Upward Sequence
  - Coarsening Downward Sequence

26

Open-Hole Log Data
Induction Log (ILD Log)
- Induction Log Tool Measures Conductivity (Reciprocal of Resistivity) of Subsurface Formations
  - Deep Induction Log (ILD):True Total Resistivity of Formation
  - Medium Induction Log (ILM): Resistivity of Invaded Zone
  - Spherically Focused Induction Log (SFL): Resistivity of Flushed Zone
- The Presence of Reservoir Brine Reduces Total Resistivity, therefore, the ILD Measures Water Saturation
- Main Objective: Identify Intervals to Perforate (in Conjunction with Gamma Ray Log)
- Interpreted using Archie Equation:

$$S_w = \left(\frac{aR_o}{\phi^m R_t} \right)^{\frac{1}{n}}$$

–Rt: total resistivity measured by ILD tool
– Rw: water resistivity measured in laboratory from produced water samples (corrected to downhole conditions)
– ϕ: porosity Interpreted from CNL, Density, and Sonic Logs
– m = cementation factor; n = saturation exponent
– a, m, n: Archie parameters measured from core and depend on the type of rock

27

Open-Hole Log Data
Applications of the ILD Log to Integrated Reservoir Models
- Initial Conditions:
  - Estimate Initial Water Saturations (Archie Equation)
  - Identify Initial Oil-Water Contact: OWC
  - Estimate Height of Capillary Transition Zones
  - Estimate Reservoir Scale Value of Swir (Initial Water
- Saturation above Transition Zone)
  - Aid In interpretation of deposition environment (in conjunction with gamma ray log)
    - Coarsening Upward Sequence
    - Coarsening Downward Sequence
- Swept Conditions (Infill Wells):
  - Estimate Current Water Saturation (Archie Equation)
  - Identify Current Oil-Water Contact
  - Estimate Reservoir Scale Sorw (Fully Swept Intervals)

28

<u>Open-Hole Log Data</u>

Compensated Neutron Log (CNL)

- Compensated neutron logging tool measures total hydrogen content of subsurface formations
- All reservoir fluids contain hydrogen, therefore, the CNL measures formation porosity
- May be run in open-hole or cased-hole environments
- Main objectives:
  - Estimate formation porosity
  - Identify intervals to perforate (avoid free gas zones in oil reservoirs)
  - Identify free gas saturation (in conjunction with density log)
  - Identify gas-oil contact: GOC
- Slightly sensitive to lithology
- Slightly sensitive to saturating fluids (gas effect)
- Calibrated to core porosity

29

<u>Open-Hole Log Data</u>

Applications of the CNL Log to Integrated Reservoir Models

- Estimate formation porosity

- Estimate gas saturation (in conjunction with

- Density log: "neutron/density cross-over"

- Identify gas-oil contact: GOC

30

Open-Hole Log Data
Density Log

- Density logging tool measures electron density of subsurface formations
- Electron density is related to mass density, therefore, the density log measures formation bulk density
- Main objectives:
  - Estimate formation porosity
  - Identify free gas saturation (in conjunction with compensated neutron log - CNL)
  - Identify gas-oil contact: GOC
  - Identify intervals to perforate (avoid free gas zones in oil reservoirs)
- Slightly sensitive to lithology
- Very sensitive to saturating fluids (free gas)
- Interpreted with the relationship:

$$\phi = \left(\frac{\rho_{ma} - \rho_b}{\rho_{ma} - \rho_f} \right)$$

ρ_b = bulk density measured by density logging tool
ρ_f = density of filtrate measured in laboratory (corrected to downhole conditions)
ρ_{ma} = density of formation matrix Measured in Laboratory

31

Open-Hole Log Data

Applications of the Density Log to Integrated Reservoir Models

- Estimate Formation Porosity

- Estimate Free Gas Saturation (in Conjunction with Compensated Neutron Log - CNL: "Neutron/Density Cross-Over")

- Identify Gas-Oil Contact: GOC

32

Open-Hole Log Data

Sonic Log

- Sonic logging tool measures the sonic transit time of compressional sound waves
- Main objective: estimate formation porosity
- Interpreted with the Wyllie relationship:

$$\phi = \left(\frac{\Delta t_{ma} - \Delta t_{log}}{\Delta t_{ma} - \Delta t_f} \right)$$

- Δt_{log} - sonic transit time measured sonic logging tool
- Δt_{ma} - sonic transit time of formation matrix (measured in laboratory)
- Δt_f - sonic transit time of wellbore fluids (measured in laboratory)

33

Open-Hole Log Data

Wireline Formation Tester

- Wireline formation tester measures formation pressure versus depth
- Probably the most useful tool in integrated reservoir modeling (provides dynamic data)
- Industry versions of the wireline formation tester:
  - Single-probe tool: performs a miniature build-up test (several cc's of fluid withdrawal)
    - Schlumberger: RFT (repeat formation tester)
    - Halliburton: SFT (sequential formation tester)
  - Multi-probe tool: multiple probes allow for a miniature build-up test and vertical interference test
    - Schlumberger: MDT (modular dynamics tester)
    - Halliburton: RDT (reservoir dynamics tester)
- New innovation: unlimited volume of fluid withdrawal

34

<u>Open-Hole Log Data</u>

Applications of Wireline Formation Tester Pressures to Integrated Reservoir Models

- Estimate formation pressures and pressure gradients
- Initial conditions:
    - Estimate oil-water contact when masked by lithological effects
    - Estimate gas-oil contact when masked by lithological effects
    - Identify reservoirs which are not in hydraulic communication (cannot be used to identify reservoirs which are in hydraulic communication)
- Pressure depleted conditions:
    - Define flow units using dynamic reservoir data
    - Identify vertical communication barriers
    - Determine levels of pressure depletion in the reservoir
    - Provide vertical pressures for history matching
    - Provide estimate of pressure range for static bottom-hole pressure in history matching

35

<u>Open-Hole Log Data Issues</u>

POSITIVE

- Data acquired at reservoir scale
- Well logging performed on almost every well (identify intervals to perforate)
- Easy to justify
- Relatively inexpensive compared to core data (may still be as much as 20% of total well cost)
- Good vertical resolution: data acquired at ½-ft intervals
- Good vertical coverage (entire section of interest)
- Logging while drilling (LWD) aids in steering non-conventional wells
- Radius of investigation: 0-5-ft

NEGATIVE

- Life-cycle of field issues
    - "Legacy" logs (e.g. SP logs)
    - Changes in service companies with time
    - Changes in technology with time
    - Changes in log analysts with time
- Original/altered state issues with infill wells
- Thin bed effects

36

Open-Hole Log Data

Good Reservoir Management Practices

- For identification of flow units, run wireline formation tester on development wells and infill wells after some pressure depletion
- Above the transition zone, open-hole water saturations can provide estimates of initial water saturation, S_{wi}, and reservoir scale irreducible water saturation, S_{wir}
- In swept zones of infill wells, open-hole water saturations can provide estimates of reservoir scale residual oil saturation, S_{oir}
- To estimate OWC, wireline formation tester pressure data (discovery wells and appraisal wells) can be extrapolated from lower known oil (LKO) and higher known water (HKW) prior to pressure depletion
- To estimate GOC, wireline formation tester pressure data (discovery wells and appraisal wells) can be extrapolated from HKO and LKG prior to pressure depletion

37

Cased-Hole Log/Production Log Data

- Definition: well logs run after a well has been cased, cemented, and put on production
- Used to monitor development wells during the life of the reservoir
- Used to diagnose problem wells and interpret profile control problems (water breakthrough/gas breakthrough)
- Common cased-hole logs:
  - Gamma ray log
  - Pulsed neutron capture log (PNC)
  - Carbon / oxygen logs (C/O)
  - Cement evaluation logs
  - Production logs (PLT)

38

Cased-Hole Log Data
Gamma Ray Log

- Gamma ray tool measures natural gamma rays emitted by subsurface formations (identical to open-hole tool)
- May be run in open-hole or cased-hole environments
- Applications to an integrated reservoir model (when run in cased-hole environment):
  - Correlate depths with open-hole gamma ray log
  - Identify deposition of radioactive scale (indication of injection fluid breakthrough)

39

Cased-Hole Log Data
Pulsed Neutron Capture Log (PNC Log)

- Pulsed neutron capture tool measures decay time of pulsed, high energy neutrons in subsurface formations
- Decay time is proportional to the chlorine content, therefore, the PNC log measures salt water saturation
- Generally run in salt water environment
- Industry versions of the pulsed neutron capture log:
  - Schlumberger: thermal decay time log (TDT log)
  - Halliburton: thermal multi-gate decay log (TMD log)

40

Cased-Hole Log Data

Carbon / Oxygen Log (C/O Log)

- Carbon/oxygen tool measures spectral energy of gamma rays
- Volume of carbon and oxygen are determined by spectral energy signature:
  - High carbon: high oil volume
  - High oxygen: high water volume
- Generally run in fresh water environment
- Has the same objectives as the pulsed neutron capture (PNC) log

41

Cased-Hole Log Data Applications of the PNC and C/O Logs to Integrated Reservoir Models

- Water saturations versus time
- Layer breakthrough times for well-by- well water cut match current water saturation maps for history
- Matching flood front advance
- Reservoir scale value of residual oil saturation, S_{or}

Applications of the Cement Integrity Logs to Integrated Reservoir Models

- Identify wells with poor cement bonds
- Prevent changing reservoir data to match mechanical wellbore problem

42

Cased-Hole Log Data
Production Logging Tool (PLT)
- Production logs measure relative contribution of perforation sets to total well production / injection
- Main objectives:
  - Identify fluid entry/exit points
  - Determine relative contribution of each set of perforations to total well production/injection

Applications of PLT Data to Integrated Reservoir Models
- Identify fluid entry/exit points
- Identify relative contribution of each perforation set
- — Production wells
  - Oil entry; • Water entry (well water cut match)
  - Gas entry (well GOR match)
- — Injection wells
  - Zones taking injection fluids; • Only tool available to determine where injection fluids are going for history matching

43

Open-Hole Log Data Issues

POSITIVE	NEGATIVE
• Data acquired at reservoir scale	• Life-cycle of field issues
• Good vertical resolution: data acquired at 1/2 foot intervals	– Changes in service companies with time
• Good vertical coverage (entire section of interest)	– Changes in technology with time
• Provide dynamic data	– Changes in log analysts with time
• Can be used to analyze flow/ tubing/cement/reservoir properties	
• Can be used in model construction and history match phases of model development	
• Easy to justify to management/partners because of positive impact on production/injection	

44

Cased-Hole/Production Log Data
Good Reservoir Management Practices

- Run baseline PNC or C/O logs prior to production on early development wells to calibrate cased-hole saturations with open-hole saturations

- Use PLT data from injection wells to identify layers taking injection fluids: use this information in history match !

- Do not change *reservoir properties* during a history match based on *mechanical wellbore problems*

- PNC or C/O saturations in swept zones can provide an estimate of reservoir scale S_{or} and S_{gr}

- PNC or C/O saturations can be used to generate water saturation maps and identify by-passed oil

45

Pressure Transient (Well Test) Data

- Definition: tests run on wells in which wellbore pressures are monitored in response to planned, imposed changes to production/injection rates

- Main objectives: identify well and near-well reservoir properties:
  - Well conductivity (k*h); Well damage (skin)
  - Well productivity index; Wellbore storage
  - Static reservoir pressure

- Common types of well tests:
  - Drill stem test (DST); Build-up test
  - Drawdown test; Multi-rate test
  - Interference test

46

Pressure Transient (Well Test) Data

Applications of the Pressure Transient Data to Integrated Reservoir Models

- Estimate properties required for well model:
  - K*h (permeability from a well test is the effective permeability in the presence of water - must be corrected for comparison with absolute permeability from core data)
  - Skin (must be corrected for layering effects: Odeh,1985)
  - Productivity indices (field measured productivity indices: must be corrected for grid size effects: Peaceman, 1978)
- Develop core to build-up permeability transforms
- Estimate static pressures for history matching
  - Well pressures for individual well match (must be corrected for grid size effects: Peaceman, 1978, Odeh, 1985)
  - Develop of isobaric maps   47

Pressure Transient (Well Test) Data

POSITIVE	NEGATIVE

• Data acquired at reservoir scale	• Life-cycle of field issues
	  – Changes in technology with time:
	    • Computer assisted interpretation
	    • Pressure derivative analysis
	    • New reservoir idealizations

• Radius of investigation: on the order of 10^1 to 10^2 feet	– Changing well conditions (wax/scale deposition) – Changes in pressure transient analysis from test to test

• Used to calibrate core scale permeability to reservoir scale permeability	• Expensive:
	  – For DST: rig time plus deferred production
	  – For producing well: deferred production

• Only method available to obtain well parameters	• Many of the properties obtained from well tests require adjustments for use in an integrated reservoir model (grid/layering dependencies) • Requires idealization of reservoir

48

Data Conflicts
Potential Areas of Conflict

- Generalities may be superseded by specific knowledge of the reservoir
- Porosity
  - Core: core scale, high confidence
  - Open-hole logs: reservoir scale, high confidence
- Permeability
  - Core permeability: core scale, high confidence
  - Open-hole logs: reservoir scale, low confidence (derived)
  - Pressure transient permeability: reservoir scale, high confidence (may represent averaged value over several flow units within test interval)
- Reservoir pressure
  - Formation wireline tester: reservoir scale, high confidence (individual flow units)
  - Pressure transient analysis:reservoir scale, medium confidence (may represent averaged value over several flow units within test interval

49

Data Conflicts
Potential areas of conflict

- Initial saturations
  - Core: core scale, medium confidence (represents small sample)
  - Open-hole logs: reservoir scale, high confidence
  - Cased-hole logs: reservoir scale, medium confidence (may represent swept conditions)
- End-point saturations
  - Core: core scale, high/medium confidence
  - Open-hole logs:
    - Swir: reservoir scale, high confidence
    - Sor: reservoir scale, medium/low confidence (uncertainty over whether Sor has been reached, only found in in-fill wells)
  - Cased-log derived:
    - Sor: reservoir scale, high/medium confidence (uncertainty over whether Sor has been reached)

50

Data Conflicts
Potential areas of conflict

- Transition zone heights
  - Core: core scale, medium confidence (capillary pressure from core represents small sample)
  - Open-hole logs: reservoir scale, high confidence
- Fluid contacts
  - Scale is not an issue
  - Open-hole logs: high confidence (may be masked by lithological effects)
  - Wireline formation tester: high confidence
  - Seismic: medium/low: seismic flat events not always apparent, requires velocity model)

51

Data Conflicts
Potential Areas of Conflict

- Always give more weight to the following data
- Data that contain a high degree of confidence
  - Pressure transient permeability versus log derived permeability
  - Unsteady-state relative permeability versus steady state relative permeability
  - Bottom-hole PVT samples versus recombined separator samples
- Data that are measured at the appropriate scale for the reservoir model
  - Log versus core data
  - Pressure transient data versus core data
- Data that are representative of the processes occurring in the reservoir
  - Differential (variable composition) PVT data versus flash (constant composition) PVT data
  - Imbibition (increasing wetting phase saturation) versus drainage (decreasing wetting phase saturation) relative permeability and capillary pressure data for water wet reservoirs
- Preliminary versions of the flow model can be used to screen conflicting data in order to determine further course of action   52

Introduction to Production Analysis

- Field production data can be used as an independent means of verifying the reservoir description and volumes in place using static data.

- Static reservoir descriptions provide an initial representation of the reservoir that must be evaluated using dynamic flow information.

- Decline curves, produced flow ratios and tracer production are examples of dynamic data. A study that purports to be an integrated study must incorporate all available static and dynamic data.

53

Introduction to Production Analysis

- Production data analysis methods include empirical and analytical techniques.
  - Empirical techniques fit a curve to production data. An empirical technique is decline curve analysis. Empirical techniques are widely used to prepare reserves estimates and yield information that can help differentiate between reservoir realizations.
  - Analytical techniques include analytical aquifer models, pressure transient test models, and material balance. Semi analytic techniques combine empirical relationships and physical models. Analytic and semi analytic methods can be used to identify flow regimes and characterize reservoirs.

- Dynamic field production data provides information about reservoir heterogeneities like channeling and compartmentalization. It is useful for preparing a flow model, and building confidence in the resulting reservoir representation..

54

Objectives of Production Analysis

- Understand what production analysis (PA) does

- Understand limitations

- Encourage use

- Recognize problems

- Promote better reservoir management

- What to expect of software

55

Data Used in Production Analysis

- Production Analysis involves using routinely-gathered production information to characterize, understand, and predict reservoir performance.

- Production rates and cumulative production (information we usually have)

- Flowing wellhead pressures (information which could help a great deal)

- Flowing bottom hole pressures (even better information)

56

Production and Reservoir Forecasts

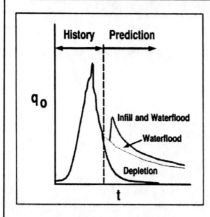

- Figure shows that the economic viability of a petroleum recovery project is greatly influenced by the reservoir production performance under the current and future operating conditions.
- Hence, the evaluation of the past and present reservoir performance and forecast of its future behavior is an essential aspect of the reservoir management process,
- Classical volumetric, material balance, and decline curve analysis, and high technology reservoir simulators are used.

57

Economic Optimization

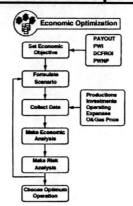

Figure shows the key steps involved in economic optimization which is the ultimate goal selected for reservoir management

58

Evaluation Reservoir Management – Surveillance/Monitoring?

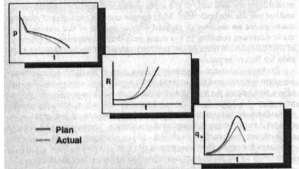

— Plan
— Actual

Figure shows the evaluation of a typical project performance. Here the actual performance of reservoir pressure, GOR, WOR, and production is compared routinely with the expected performance

59

Hydrocarbon Phase Behavior

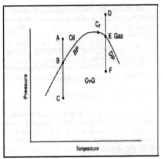

- Figure shows the natural producing mechanisms influencing the primary reservoir performance .
- Oil Reservoir
  - Liquid and rock expansion (A-B)
  - Solution gas drive (B-C)
  - Gas-Cap drive (O+G)
  - Aquifer water drive
  - Gravity segregation
  - Combination drives
- Gas Reservoirs
  - Gas depletion or expansion (D-E-F)
  - Aquifer water drive
  - Combination drives

60

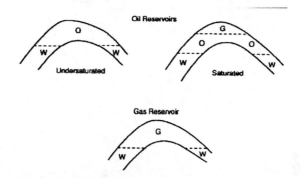

Hydrocarbon Reservoirs

Figure shows the different types of reservoirs

61

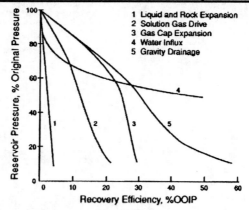

Influence of Primary Producing Mechanisms on Reservoir Pressure and Recovery Efficiency

1 Liquid and Rock Expansion
2 Solution Gas Drive
3 Gas Cap Expansion
4 Water Influx
5 Gravity Drainage

Figure shows important characteristics of natural producing mechanisms of oil reservoir

62

Characteristics of Various Driving Mechanisms

		Characteristics			
Mechanisms	Reservoir Pressure	GOR	Water Production	Efficiency	Others
1. Liquid and rock expansion	Declines rapidly and continuously $P_i > P_b$	Remains low and constant	None (except in high S_w reservoirs)	1–10% Avg. 3%	
2. Solution gas drive	Declines rapidly and continuously	First low, then rises to maximum and then drops	None (except in high S_w reservoirs)	5–35% Avg. 20%	Requires pumping at an early stage
3. Gas cap drive	Falls slowly and continuously	Rises continuously in up-dip wells	Absent or negligible	20–40% Avg. 25% or more	Gas breakthrough at a down-dip well indicates a gas cap drive
4. Water drive	Remains high. Pressure is sensitive to the rate of oil, gas, and water production	Remains low if pressure remains high	Down-dip wells produce water early and water production increases to appreciable amount	35–80% Avg. 50%	N calculated by material balance increases when water influx is neglected
5. Gravity drainage	Declines rapidly and continuously	Remains low in down-dip wells and high in up-dip wells	Absent or negligible	40–80% Avg. 60%	When k > 200 mD, formation dip > 10° and μ_o low (< 5 cP)

63

Comparison of Reservoir Performance Analysis and Reservoir Estimation Techniques

	Volumetric	Decline Curve	Material Balance	Mathematical Models
1. *Applicability/Accuracy*				
Exploration	Yes/Questionable	No	Yes/Questionable	Yes/Questionable
Discovery	Yes/Questionable	No	Yes/Questionable	Yes/Questionable
Delineation	Yes/Questionable	No	Yes/Questionable	Yes/Fair
Development	Yes/Better	No	Yes/Better	Yes/Good
Production	Yes/Fair	Yes/Fair	Yes/Good	Yes/Very Good
2. *Data Requirements*				
Geometry	Area, thickness	No	Area, thickness homogeneous	Area, thickness heterogeneous
Rock	porosity, saturation	No	porosity, saturation, rel. permeability, compressibility homogeneous	porosity, saturation, rel. permeability, compressibility capillary pressure heterogeneous
Fluid	Form. Vol. Factors	No	PVT homogeneous	PVT heterogeneous
Well	No	No	PI for rate vs. time	Locations Perforations PI
Production & Injection	No	Production	Yes	Yes
Pressure	No	No	Yes	Yes
3. *Results*				
Original Hydrocarbon in Place	Yes	No	Yes	Yes
Ultimate Recovery	Yes with rec. eff.	Yes	Yes	Yes
Rate vs. Time	No	Yes	Yes with PI	Yes
Pressure vs. Time	No	No	Yes with PI	Yes

64

Production Analysis

- **Production Analysis (PA) vs. Pressure Transient Analysis (PTA)**
  - Pressure transient data are acquired as a part of a controlled "experiment," performed as a specific event.
  - Production data are generally considered to be surveillance/monitoring data – with little control and considerable variance occurring during the acquisition of the production data.

65

Production Analysis

Production Analysis (PA) vs. Pressure Transient Analysis (PTA)

PRODUCTION ANALYSIS	PRESSURE TRANSIENT ANALYSIS
• Long term	• Short term
• Data gathering is routine	• Data gathering is costly
• "Coarse" data	• Dense data
• Little or no pressure data	• Accurate pressures
• Routine data gathering	• Purposeful data gathering
• Emphasis on rate/cum data	• Emphasis on pressures
• Total reservoir	• Near wellbore
• Macroscopic	• Microscopic
• Gross properties	• Precise properties
• Observation	• Controlled experiment
• Little historical literature coverage	• Substantial literature coverage and academic interest

Similarities: Based on diffusivity equation; Darcy's Law; & Conservation of Mass

66

Decline Curve Analysis

- Estimates of the ultimate recovery, and of the producing life of a well or a lease are commonly made by extrapolating past performance trends.

- This technique is most useful after initial development when production patterns have been established.

- The method of extrapolating a trend, for purposes of estimating future performance depends upon the assumption that the factors caused changes in the past will operate in the same way in the future.

67

Reasons (Whys) for Decline Curves

- Why decline curves?
    - Traditional
    - Data are available
    - Easily applied
    - Widely accepted
- Why not decline curves?
    - Easily mis-applied
    - Subjective
- Choices:
    - Exponential vs. hyperbolic
    - Transient vs. pseudo-steady state

68

Decline Curve Analysis

- Decline curve analysis techniques offer an alternative to volumetric and material balance methods, and history matching with reservoir simulation for estimating original gas in place and gas reserves.
- The performance trends that are most frequently used are the rate versus time and rate versus cumulative production.

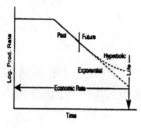 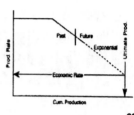

69

Arps Decline Curves

- Arps proposed an *empirical* method of characterizing historical production and using it to forecast production. His work provided a rational method to forecast and, thus, to evaluate, reserves.
- Arps in 1945 observed that future production could be predicted by fitting an exponential equation to historical decline rates.
- The exponential equation worked well for many reservoirs, but did not adequately represent the behavior of some producing wells in depletion drive reservoirs. A better fit was obtained using a hyperbolic decline equation for these wells

70

Arps Decline Curves

- Assumptions of Arp's Decline Curves
  - Well analyzed is produced at constant BHP.
  - Well analyzed is produced from an unchanging drainage area (fixed size) with no-flow boundaries.
  - Well analyzed has constant permeability and skin factor.
  - Decline curve analysis must be applied only to boundary dominated (stabilized) flow data.

71

Arps Decline Curves

- The various methods of decline-curve analysis are based on the manner in which the rate of decline varies with time, production rate, etc. The drop in production rate per unit of production rate per unit time is called the nominal decline rate D *(a constant):*

$$D = \frac{dq/dt}{q} = Kq^n$$

- q = production rate, barrels per day, month, or year
- t = time, day, month, or year
- K = constant;   n = exponent
- The decline rate in the above equation can be constant or variable with time yielding 3 basic types of production decline – **exponential**, **hyperbolic**, and **harmonic**

72

Arps Decline Curves Equations

- **Exponential** $$D = -\frac{dq/dt}{q} = K = \ln\frac{\left(\frac{q_t}{q_i}\right)}{t}$$ n = 0

- **Hyperbolic** $$D = -\frac{dq/dt}{q} = K = \ln\frac{\left(\frac{q_t}{q_i}\right)}{t}$$ 0 <n<1

- **Harmonic** $$D = -\frac{dq/dt}{q} = Kq$$ n =1

  - q_i = initial production rate
  - q_t = production rate at time t
  - Both exponential and harmonic are special cases of the hyperbolic decline   73

Decline Curve Analysis

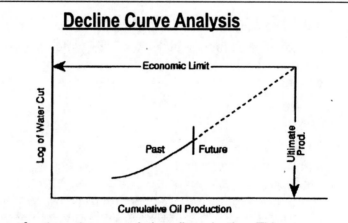

Log of water cut vs. cumulative oil production. This type of plot is use when economic production rate is dictated by the cost of water disposal.

74

Decline Curve Analysis

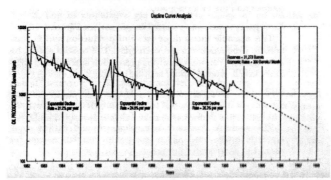

A typical past history match and reserves forecasts. This example shows
three cycles of production and fluctuations in production rates in each
cycle. The reasons should be investigated. Teamwork involving
production and reservoir engineers would produce more realistic results.

75

Another Use of Arps Decline Curves

- For wells that produce water the Arps plots can be
  made   and extrapolated using oil cut, f_o, in place of
  oil rate
  - $Log(f_o)$ vs.t
  - f_o vs. Q
  - $Log(f_o)$ vs. Q
- Many factors influence production rates and
  consequently decline curves. These are proration,
  changes in production methods, workovers, well
  treatments, pipeline disruptions, and weather and
  market conditions. Hence, care must be taken in
  extrapolating the production curves in the future

76

Fetkovich Type Curves

- In 1980, Fetkovich introduced rate-based decline curves. The curves included both transient and pseudo-steady state behavior. Rather than the P_D vs. t_D presentation of Agarwal-Ramey pressure transient analysis type curves, Fetkovich curves presented q_D vs. t_D

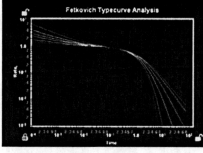

$t_{DA} = D_i t$ Decline curve Dimensional time

$$t_D = \frac{0.00634 kt}{\phi \mu c_t r_w^2}$$ Dimensional time

$$t_{Dd} = \frac{t_D}{\frac{1}{2}\left[\left(\frac{r_e}{r_w}\right)^2 - 1\right]\left[\ln\left(\frac{r_e}{r_w}\right) - \frac{1}{2}\right]}$$

Relationship between t_{Dd} and t_D

77

Fetkovich Type Curves

- A set of dimensionless cumulative production curves, Q_D vs. t_D, was added later to make curve matching easier.

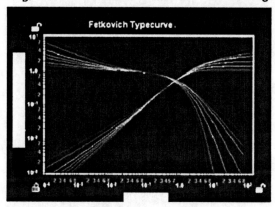

78

Decline Curve Dimensionless Rate

Decline curve Dimensional time

$$q_{Dd} = \frac{q(t)}{q_i}$$

Dimensional time for oil

$$q_D \; oil = \frac{141.2q(t)\mu B}{kh\left(p_i - p_w\right)}$$

Dimensional time for gas

$$q_D \; gas = \frac{50300Tq(t)p_{sc}}{T_{sc}kh\left(m\left(p_i\right) - m\left(p_w\right)\right)}; \; with \; m(p) = \int\limits_0^p \frac{2pdp}{\mu Z}$$

Relationship between Decline curve Dimensional time and Dimensional time

$$q_{Dd} = q_D\left[\ln\left(\frac{r_e}{rw}\right) - \frac{1}{2}\right]$$

79

Fetkovich Analysis

- The Fetkovich decline type curves are based on analytical solutions to the flow equations for production at constant BHP from a well centered in a circular drainage area with no-flow boundaries. Some characteristics of Fetkovich decline curve:
  - Includes transient
  - Improved confidence
  - Eliminates many mis-applications
  - Better reserve estimates
  - Reduces annual reserve revisions
  - Limited model selection
  - Provides opportunity for shrewd investors

80

Fetkovich Analysis

- **Implicit Assumptions for Fetkovich and Arps**
  - **– Constant flowing BHP**
  - **– Unchanging drainage area**
  - **– Unchanging wellbore conditions**
- Fetkovich suggested that the problem of constantly declining FBHP, a real occurrence, could be accommodated by using rate normalized by p (q/Δp) (when we have pressure data).

81

Limitations of Decline Type Curves

- Assumption of constant BHP production strategy
- Assumption of constant *k* and *S*
- Assumption of true boundary dominated flow conditions
- Assumption of volumetric reservoir

82

Modern Production Analysis

- The technology of Modern Production Analysis, a term appropriated shamelessly from Pressure Transient Analysis (remember "Modern Well Test Analysis"?), means the use of advanced type curves. The advanced type curves require BHP data as well as production data. If we don't have BHPs, we can't perform Modern Production Analysis.

- Installation of permanent downhole pressure gauges in critical wells is becoming more commonplace. High dollar wells both offshore and onshore use permanent instrumentation routinely. The cost of such instrumentation is small compared to total well cost and operators see advantages in permanent pressure gauges. Reservoir and production engineers should lobby for such instrumentation in order to manage their reservoirs better.

83

Modern Production Analysis (cont'd)

- A second best option involves routine gathering of flowing wellhead(tubing) pressures for conversion to FBHPs using good multiphase flow software packages. Some PA software packages have the ability to calculate FBHP from FTHP using simple correlations.

- The last and least desirable option is to simply assume values of FBHP. With a data set of production rates only, assuming values of FBHP would allow the user to apply advanced type curves. Some software packages will not allow entry into the advanced type curve analysis without a set of FBHP's.

- Typically, one might assume a constant value for FBHP. This would be a reasonable assumption for a pumping well.

84

Blasingame Type Curve

- **Blasingame Type Curve** - This plot is called a Blasingame plot, or more generally, a *normalized rate plot*. The y-axis is normalized rate or $q/\Delta p$. $q/\Delta p$ is a form of productivity index (PI). The x-axis is Material Balance Time, N_p/q. Note that there is only one depletion stem on this plot. All of the "b" depletion stems gone except for the b=1 stem. The use of Material Balance Time forced this representation.

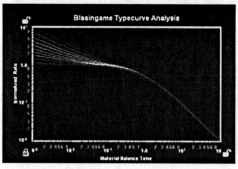

85
85

Blasingame Type Curve

- If we were to integrate each of the curves and then divide by the elapsed material balance time, we would get a *cumulative average* value of q/ p at each material balance time. We'll call this value the *integral*.

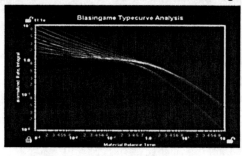

86

Blasingame Type Curve

- The last operation we'll do is take the derivative of the integral curve with respect to the natural log of material balance time and call the new curve the *derivative*

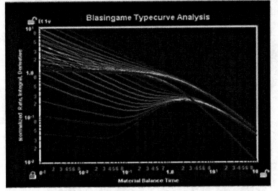

87

Inverted Productivity Index Type Curve (IPI)

- The use of the *pressure-normalized rate* (q/Δp) and material balance time converts the variable rate, variable pressure data to a constant rate solution.

- Alternatively, we can use *rate-normalized pressure* (Δp/q) and material balance time to convert the data to a constant pressure solution. This plot might be called an inverted PI (IPI) or reciprocal PI (RPI) plot.

- Inverting the long time approximation of the steady-state flow equation

88

IPI Type Curves

- On a log-log plot, the equation plots as a straight line with slope = 1

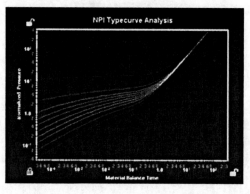

89

IPI Type Curves

- On the IPI Type curve, the transient behavior seems to be a little better defined than on the PI Type curve.

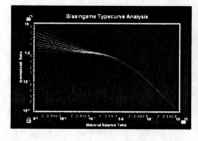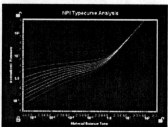

90

IPI Type Curves

- Performing an integral operation on the IPI as we did on the PI, we have

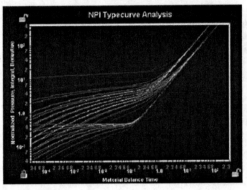

91

Type Curve Matching

- Why do we go to the effort to calculate an integral and a derivative of the integral?
  - It gives us more curves to match on.
  - The integral smoothens the data

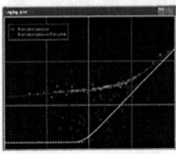 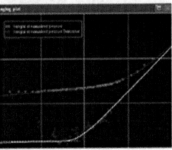

92

Diagnostics

- **Two diagnostic categories will be considered:**
  - Diagnosis of the near-wellbore geometry and condition
  - Diagnosis of production problems
- **Near-wellbore Geometry and Conditions**
  - Near-wellbore effects appear during transient flow; it is not possible to assess near-wellbore effects during steady-state flow. Software writers provide many wellbore models in their software packages.
  - As reservoir or production engineers, we should know what near-wellbore geometry exists for each well we work with – hydraulic fracture, partial penetration, highly deviated well, horizontal well, etc. and should select the appropriate type curve for analysis.
  - In a perfect world, where there is good data, looking at the shape of the transient data and comparing it to the shape of the theoretical transient stems should allow the engineer to characterize the near-wellbore geometry.
  - However, in the real world, production data gathered and reported on a weekly or, more typically monthly, basis is likely too coarse and noisy to use in fine differentiation in fracture length or precise estimation of skin and permeability.

93

Diagnostics Near-Wellbore Geometry & Condition

- Diagnosis of production data is much more art than science. That is, one must know how to the following:
  - Diagnosing production data to determine an appropriate model (pattern recognition)
  - Characterization of production mechanism from production data.
- The production engineer and the reservoir engineer must both realize that production analysis provides only a single bite at the apple. One cannot extract new information from the near-wellbore region after the transient period is past. That information gathering effort falls into the realm of pressure transient analysis. One might, however, determine from steady- state behavior that something is a miss in the near- wellbore without being able to diagnose or quantify it.

94

Production Problem Diagnosis

- We must frequently and strongly of staying constantly aware of non-ideal (<u>real</u>) well responses which are not considered in theoretical models.

- It has been said that if the only tool one has is a hammer, every problem looks like a nail. Production analysis doesn't fit every situation and many times the "nails"

- Production data sets are bent and rusted and should not be hammered. It is important to gain the knowledge of when to dive into analysis and when to confidently walk away. The following pages should help engineers in making those decisions.

95

Production Problem Diagnosis

- **Problems which can make production analysis difficult or impossible are:**
  - Changing reservoir model
  - Changing wellbore model
  - Changing fluid properties
  - Changing rock properties
  - Changing saturations
  - Changing production mode
  - Wellbore effects
  - Allocation errors
  - And, most importantly, BAD DATA!

96

Guidelines for Modern Production Analysis

- Analogy between the engineer who practices production analysis and an airplane pilot.
  - Most of the important work of an airplane pilot is done during pre-flight and post-flight checks. Flying is the fun part and is made enjoyable by well done pre- and post-flight efforts.
  - Similarly, much of the important work of production analysis is accomplished during the problem set-up and the post-mortem check.
- Based on this similarity here are some guidelines:
  - Review production data for consistency
  - Review well history (recompletions/stimulation)
  - Gather reservoir and PVT data
  - Perform diagnostic analysis
  - Review data and check data correlation
  - Perform simplified analysis of production data
  - Establish reservoir model with diagnostic plots
  - Review analysis and see if it makes sense

97

Analysis of Problem Wells

- A well problem is different from a reservoir problem, and an analyst should know which is which.

- A comprehensive problem well analysis is essential and as such careful analytical work should be completed before a work over rig is moved in. This is the least costly part of a work over operation.

- Problem well analysis may be handled on a reservoir basis, an area basis, or by study of an individual well.

98

Analysis of Problem Wells

- The conclusion of a comprehensive problem well analysis will usually result in one of the following recommendations:
  1. Work over the well. That is, move a rig in to do the job. This is an expensive recommendation.
  2. Pressure maintenance. That is gas lift or pump. This implies that the hardware to achieve this is already in place.
  3. Continue production until the economic limit is reached.
  4. Enhance recovery operation (e.g.. Gas injection or water flooding). This is an expensive operation.
  5. Shut the well in. This will result in loss of revenue.
  6. Abandon the well. May need a rig to set cement plugs.

99

Classification of Problem Wells

Problem wells may be classified as

- Limited production rate
- Excessive water production
- Excessive gas production if oil wells, and
- Mechanical failures

100

<u>Limited Production Rate</u>

This may be a result of one of the following:

(1) Low permeability

(2) Low reservoir pressure

(3) Formation damage - already discussed under stimulation.

(4) Wellbore, tubing or flowline plugging

(5) High viscosity oil

(6) Excessive back pressure on formation

101

<u>Excessive Water Production Problems</u>

Water problems may result from:

1. Natural water drive or water drive aggravated by conning or fingering.

2. Extraneous sources including casing leaks or cementing failures.

3. Fracturing or acidizing into water zones and etc.

102

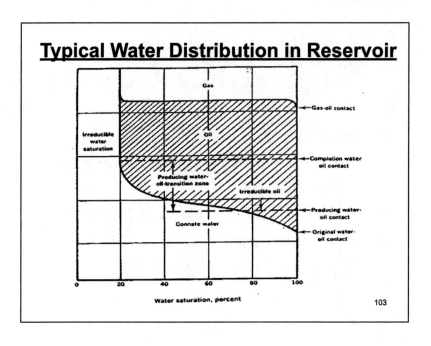

Excessive Gas Production

- The primary sources of gas in oil wells are: (a) gas dissolved in the oil, (b) primary or secondary gas caps, (c) gas flow through channels or from zones or reservoirs adjacent to the oil zone.
    - Gas can be a benefactor in oil wells if used expertly; but often times operational or design miscalculations can make gas an adversary to oil production.
- Gas problems can be listed as follows: (a) gas flowing in the reservoir, (b) gas conning, and (c) gas fingering.

104

Excessive Gas Production

- In a dissolved gas drive reservoir, gas saturation increases as oil withdrawals continue and P_R declines. When the P_R goes below the P_B then gas will start flowing as a second phase, lowering the k_o.
  - The popular solution for this problem is to maintain the reservoir pressure by recycling gas back into the reservoir.
- With high pressure drawdown, gas conning might occur in wells with continuous k_h without an appreciable decline in P_R.

105

Excessive Gas Production

- In stratified reservoirs, premature fingering of gas may occur with high pressure drawdown at the well bore, with gas usually flowing through the high permeability zones.
- Fingering is prevalent in reservoirs where permeabilities vary appreciably between zones.
- Gas flow zones above or below the oil zones may be due to: (a) casing leaks, (b) cement bond failure, (c) fractures communicating with gas zones, and (d) acidizing into gas zones.

106

Mechanical Failures

- Common mechanical failures are: (1) primary cement failure, (2) casing, tubing, and packer leaks, (3) failure of artificial lift equipment, and (4) wellbore communication in multiple completions.
- Casing-cement-formation bond failures are frequently caused by applying fracture pressure on a well during matrix acid job.
  - Temperature surveys and other production logs are beneficial in locating casing leaks.
  - Wellbore communication can usually be detected by packer leakage tests, by abrupt changes in producing characteristics, or by observing equal shut-in pressure on two concentric string of pipe.

107

Problem Well Analysis Checklist

- An apparent well problem must be studied first based on well performance. Determine also whether similar problems exist on offset wells, the same reservoir, field or geologic trend. The next step is to analyze the problem based on well history.
- Analysis of well history: This includes analysis of all the operations conducted from the start of drilling to the time the analysis is made.

108

Problem Well Operations

The operations include:

(a) Analyze the drilled rock from core samples, cuttings, logs, and etc.. This will give us the rock and fluid property of the reservoir.

(b) Analyze the extent of damage from drilling and completion fluids.

(c) Analyze and evaluate the primary cement job. (d) Analyze the data available on initial completion (i.e. date of completion, test data, completion interval, completion details, etc.)

109

Problem Well Operations

(e) Study in details well stimulation and results

(f) Analyze details on all jobs - work over, re-completion, well stimulation, well servicing and chemical treatments.

(g) Analyze production history - i.e.. current production test, old production test records, and evaluation of ultimate recovery.

(h) Analyze P_R - study annual subsurface pressure surveys and note changes in production behavior with depletion.

110

Problem Well Operations

(h) Also, compare subsurface pressure history of well with that of offset wells, the area and reservoir.

(i) Examine structure, isopachous, isobaric, water percentage, and gas-oil ratio maps if as available.

(j) Study stratigraphic cross-sections.

111

Exploration and Field Development Phases

112

Field Development Process and Maximizing Project Value

- In Maximizing Project Value our attempts towards maximizing project value are typically improving the net present value. Here are guidelines:
  - Pay utmost attention to safety and environmental rules and regulations
  - Think about how to minimize the number of wells to be drilled
  - Have robust procedures in place for estimating reserves
  - Devise and implement optimum production strategies
  - Minimize capital and operating expenditures
  - Devise a rapid construction schedule to minimize the time between discovery and development
  - Minimize risks
  - Ensure that flexibility and simplicity factors are all incorporated
  - Think about contingency and emergency situations
  - Use the previously gained knowledge and experience
  - Acknowledge the external factors and work them into your plan as imposed boundary conditions   113

Field Development Plan — Data Flow and Organization

114

Project Development Team

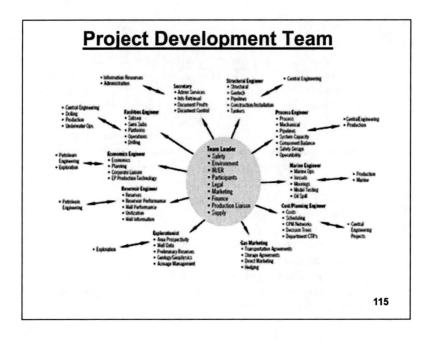

115

What Brings Success to a Field Development Plan?

- Starting with a plan of action including all functions
  - Plan should include all of the disciplines involved
- Having a flexible plan
  - The plan must be adaptable to changing economic, legal and environmental circumstances
- Having management support
  - Higher level management should get involved from "day one"
- Having commitment from field personnel
  - A direct imposition does not work

116

Reasons for Failure of a Field Development Plan

- Un-integrated Approach (Lack of Team Work)
  - Reservoir engineering, production engineering, gas and chemical engineering, environmental and legal aspects, economics and management, research and service labs production operations, drilling, design and construction engineering geology and geophysics
- Starting too late
  - Reservoir management should not come into the picture because of a crisis that occurred
  - In contrast it should be in place to avoid the occurrence of a potential crisis.
- Lack of maintenance
  - Reservoir management is analogous to health management

117

Field Development Plan

- A prudent field development plan should provide answers to the following questions:
  - How are the fluids moving in the reservoir?
  - What is the optimum production rate?
  - How can it be achieved?
  - What is the optimum well spacing?
  - Can the well completion methods be improved?
  - When will artificial lift be needed?
  - What kind of artificial lift should be used?
  - What is the current recovery mechanism?
  - How can it be improved?
  - Are enhanced recovery methods needed?
  - When should injection facilities be installed?
  - What size should they be?

118

The Development Plan: Process and Content

- A development plan must contain the following contents:
  - Executive Summary
  - Field Description
  - Development and Management Plan
  - Annual Field Report
    - Field Description
    - Development and Management
  - Cessation of Production

119

The Development Plan: Process and Content

- Executive Summary
  - Description of reserves, development strategy, facilities and pipelines
  - Outline map showing the limits of the field
  - Project schedule, total capital cost
  - Estimate of ultimate recovery with expected production profiles
  - Field management plan

120

The Development Plan: Process and Content

- Field Description
    - Seismic interpretation and structural configuration
    - Geological interpretation and reservoir description
    - Petrophysics and reservoir fluids
    - Hydrocarbons-in-place
    - Well performance
    - Reservoir units and modeling approach
    - Improved recovery techniques
    - Reservoir development and production technology

121

The Development Plan: Process and Content

- Development and Management Plan
    - Drilling program and well locations
    - Drilling and production facilities
    - Process facilities
    - Cost estimates
    - Field management plan

122

The Development Plan: Process and Content

- Annual Field Report
  - Field Description
    - Hydrocarbons initially in place and recoverable reserves
    - Well status and operations
    - Geology and geophysics
    - Field facilities and infrastructure
  - Development and Management
    - Field management
    - Studies
    - Improved oil recovery
    - Forecasting
    - Proposed changes to the development plan
    - Field operating costs

123

The Development Plan: Process and Content

- Cessation of Production
  - Definition of economic limit
  - Determination of cut-off rates and timing
  - Cash flow over the period up to this limit and approximately two years beyond
  - The form and costs of abandonment
  - Possible options for extending field life
  - Production and injection profiles together with projections through to economic limit and approximately two years beyond
  - Details of any remaining license obligations
  - Maps indicating the estimated location and distribution of remaining technically recoverable hydrocarbons

124

Environmental Evaluation Aspects of Reservoir Management

- Finding and producing oil and gas while minimizing adverse environmental impact requires an understanding of the complex issues facing the upstream petroleum industry.
- These issues concern operations that generate wastes, their potential influence on the environment, mechanisms and pathways for waste migration, effective ways to manage wastes, treatment methods to reduce their volume and/or toxicity, disposal method, remediation methods for contaminated sites and all applicable regulations.

125

Environmental Evaluation Aspects of Reservoir Management

- **SOURCES OF WASTES:**
- Produced water
- Drilling wastes
- Associated wastes (sludges, pit wastes, scrubber wastes, stimulation wastes, etc.)
- **ENVIRONMENTAL IMPACT OF WASTES:**
- The primary measure of the environmental impact of petroleum wastes is their toxicity to exposed organisms.

126

Environmental Evaluation Aspects of Reservoir Management

WASTE MIGRATION:

- The environmental impact of released wastes would be minimal if the wastes stayed at the point of release; however most wastes migrate from their release point to affect a wider area

MANAGING WASTES:

- Develop and maintain an effective waste management plan.
- Such a plan should identify the materials and wastes at a particular site and list the best way to manage, treat and dispose of those wastes

127

Environmental Evaluation Aspects of Reservoir Management

WASTE TREATMENT METHODS FOR HYDROCARBON WASTES:

- Washing by agitation in a jet of high velocity water
- Flotation
- Bioremediation
- Distillation
- Solvent extraction
- Incineration
- Critical/supercritical fluid extraction

128

Environmental Evaluation Aspects of Reservoir Management

- **WASTE TREATMENT METHODS FOR NON HC-AQUEOUS WASTES:**
- Washing by agitation in a jet of high velocity water
- Flotation
- Bioremediation
- Distillation
- Solvent extraction
- Incineration
- Critical/supercritical fluid extraction

129

Environmental Evaluation Aspects of Reservoir Management

WASTES DISPOSAL METHODS:

- These methods depend on the: type; composition and regulatory status of the waste.

DISPOSAL METHODS FOR AQUEOUS WASTES:

- The primary disposal method is to inject them via disposal wells.
- If the quality of the wastewater meets or exceeds regulatory limits, permits to discharge it onto surface waters may be obtained.

130

Environmental Evaluation Aspects of Reservoir Management

- **CLEANUP METHODS FOR CONTAMINATED SITES:**
  - Drilling contamination recovery wells
  - Removing volatile hydrocarbons by injecting air and/or pulling vacuum to vaporize those components.
  - Use of heat, surfactants, and bioremediation to remove subsurface hydrocarbons
  - Filtration or absorption with activated carbon

131

Drilling Project and Risk Management

Definition of Project Risk:

"An uncertain event or condition that, if it occurs, has a positive or a negative effect on at least one project objective, such as time, cost, scope or quality."

1

Risk is Everywhere

- Internal and External to the Project
- Partial List:
  - Budgets/Funding
  - Schedules
  - Scope or requirements changes
  - Technical Integration Issues

2

Risk of not using the right tools . . . .

Risk of improper scoping of a project . . . .

Risk Characterization

■ Risk: Is knowing the mathematical probabilities that affect the outcome but not the outcome itself, e.g. the number of rolls of a dice to obtain an outcome of 6.

■ Uncertainty: Is not knowing the mathematical probabilities, e.g. the price of oil one year from now

■ Since we do not generally know the mathematical probabilities impacting the outcomes of a particular activity in drilling a well, we are actually dealing with **Uncertainty Management.**

5

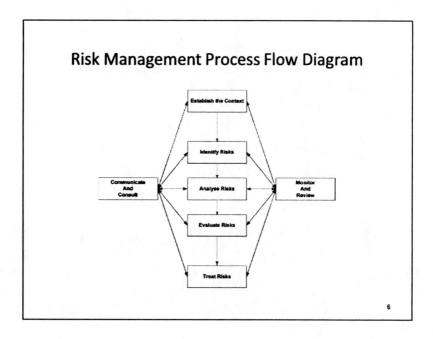

Risk Management Process Flow Diagram

6

Risk Management Planning:
Inputs

- Enterprise Environmental Factors
- Organizational Process Assets
- Project Scope Statement
- Project Management Plan

7

Risk Management Planning:
Tools and Techniques

- Planning Meeting and Analysis
  - Develop Risk Management Plan
  - Include key team and organizational members
  - Outcomes feed into Risk Management Plan

8

Risk Management Planning:
Outputs

- Risk Management Plan
  - Methodology (Example utilizing Six Sigma)
    - SIPOC – Suppliers, Inputs, Process, Outputs and Customers
      - Suppliers –The Suppliers to the process
      - Inputs – The Inputs to the Process
      - Process – The Process the team is trying to improve
      - Outputs – The Outputs of the Process
      - Customers – The Customers that get the processes output
  - Roles and responsibilities

9

Risk Management Planning:
Outputs

- Risk Management Plan, continued
  - Budgeting
  - Timing
  - Risk categories
    - Financial Resources
    - Schedule
    - Culture
    - Age of Team
    - New Technology
  - Definitions of risk and impact

10

Risk Identification:
Inputs

- Enterprise Environmental Factors
- Organizational Process Assets
- Project Scope Statement
- Risk Management Plan
- Project Management Plan

11

Risk Identification:
Tools and Techniques

- **Documentation Reviews**
- **Information Gathering Techniques**
    - Brainstorming
    - Delphi technique
    - Interviewing
    - Root Cause Analysis
    - Strengths, Weaknesses, Opportunities, and
      Threats (SWOT) analysis

12

Risk Identification:
Tools and Techniques

- **Checklist Analysis**
- **Assumption Analysis**
- **Diagramming Techniques**
    - Cause-and-effect diagram
    - System or Process flow charts
    - Influence diagrams

13

Risk Identification: Outputs

- **Risk Register**
  - List of identified risks
  - List of potential responses
  - Root causes of risk
  - Updated risk categories

14

Dynamic Risk Scale Example

- **Risk Likelihood Scale**
- **Level Descriptor Likelihood/Probability/ Frequency**
- **Negative Risk Consequence Scale**
- **Level Description Consequence**
- **Positive Risk Consequence Scale**
- **Level Description Consequence**

15

Dynamic Risk Scale, continued

- Risk Likelihood Scale
  - Risk Likelihood relates to the probability or frequency of a risk occurring.
- Level Descriptor Likelihood/Probability/ Frequency
  - R – Likely Will probably occur in most circumstances
  - Y – Possible Might occur at some time
  - G – Unlikely Could occur at some time

16

Dynamic Risk Scale, continued

- Negative Risk Consequence Scale
  - Risk Consequence relates to the outcome or impact of a risk considered in relation to the achievement of objectives. It should be noted that there can be more than one consequence and the impact can be positive or negative.

17

Dynamic Risk Scale, continued

- Level Description Consequence
  - -1 Minor - No regulator impact; low stakeholder impact; financial loss up to $10,000; no effect on operations; up to 1% impact on targets
  - -2 Moderate - Some regulatory impact; medium stakeholder impact; financial loss up to $100,000; some effect on critical operations, up to 5% impact on targets
  - -3 Major - High regulatory impact; medium stakeholder impact; major financial loss up to $ 1m; major effect on critical operations; up to 10% impact on targets

18

Dynamic Risk Scale, continued

- Positive Risk Consequence Scale
  - Risk Consequence relates to the outcome or impact of a risk considered in relation to the achievement of objectives. It should be noted that there can be more than one consequence and the impact can be positive or negative.

19

Dynamic Risk Scale, continued

- Level Description Consequence
  - +1 Minor - Minor benefit; some operational improvement; some image gain; possible financial gain
  - +2 Moderate - Some reputation enhancement; high operational improvement; high financial gain
  - +3 Major - Enhanced reputation; high operational gain; major financial gain

20

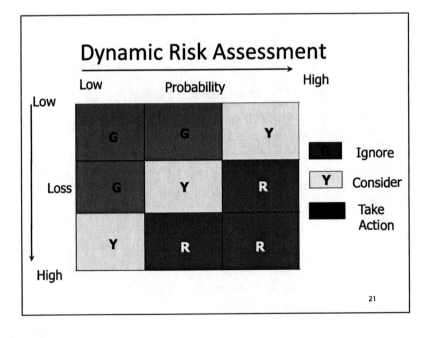

21

Risk Response Planning

- Developing options
- Determining actions related to the project's objectives
  - Enhance opportunities
  - Reduce threats
- Addresses the risks
  - Set priorities
  - Insert resources and activities into budgets, schedule, and project management plan

22

Risk Response Planning:
Inputs

- Risk Management Plan
- Risk Register

23

Risk Response Planning:
Tools and Techniques

- <u>Accept</u> – Do nothing because the cost to fix is more expensive than the expected loss
- <u>Avoid</u> – Elect not to do part of the project associated with the risk
- <u>Contingency Planning</u> – Frame plans to deal with risk consequence and monitor risk regularly (identify trigger points)
- <u>Mitigate</u> – Reduce either the probability of occurrence, the loss, or both
- <u>Transfer</u> – Outsource

24

Risk Response Planning:
Outputs

- The Risk Register (Updates)
- Project Management Plan (Updates)
- Risk-Related Contractual Agreements

25

Risk Monitoring and Control

- Identifying, analyzing, and planning for newly arising risks
- Tracking identified risks
- Reanalyzing existing risks
- Monitoring trigger conditions
- Monitoring residual risks
- Reviewing the execution of risk responses while evaluating their effectiveness

26

Definition of Drilling Project Management

■ Drilling Project Management is the provision of an inter-disciplinary team to manage a project on behalf of a Client by utilizing external resources to deliver value to the Client by achieving the Project Mission.

■ *For a drilling project, this entails delivering a completed well or wells to an agreed specification and within an agreed budget.*

■ Fundamentally, drilling a well requires a holistic approach that integrates all the key functions together

■ Achieving the project mission requires understanding of certain key principles

27

The Criteria for Drilling Success

— Whether building a house or drilling a well, the same principles of project management apply. It is just the scale that differs (drilling a well is bigger), but the objective remains unchanged.

— The essence of project management is to define the IDEA and achieve the REALITY. The criteria for success of a project are as follows: **Cost; Time; Quality; and Safety**

— These 4 criteria are mutually interdependent – cost is a function of time which is a function, at least partially of quality and safety.

— In the Oil & Gas business drilling project, there is a further complication. We can drill a well that is technically successful in terms of the 4 criteria, but the well can turn out to be dry.

28

The Essential Principles for Project Management

■ Project Mission – must be defined; clear and unambiguous

■ Creation of new value – The value is obvious and quantitative; oil and gas production or even data from a dry well

■ Discrete function – project is a discrete function because it requires drilling a well or series of wells and platform construction

■ Definable life cycle and budget – Estimate number of days & AFE

29

The Essential Principles for Project Management

- Establishment of a Project Organization & Mobilization of External Resources

  Organization defined based on project requirements; it is temporary; has no resources of its own; and hence must mobilize such resources to design, plan, and execute the project

- Organization disbanded at completion of project mission – Once the project mission has been achieved the project organization is either disbanded or moves on to the next project.

30

The Application

- Project Organization
  - Can be employed by the client as in-house drilling
  - Can also be an external resources contracted (out-sourced)
- In either case, the Client is typically an oil and gas industry E&P (such as VANKORNEFT) who defines the project mission in terms of the well objectives – surface location, total depth, target tolerance etc.
- The project organization is then responsible for achieving the project mission and fully accountable to the client for the results

31

The Application (cont'd)

– Client (e.g. VANKORNEFT) is the entity with corporate responsibility for the project mission.

– Within an E&P, the client is typically the Asset

– The Drilling Project Management organization is a Service Organization

– Where a production or development project comprises a number of wells each well could be viewed as a sub-project of the overall project comprising all the wells

– Should the Drilling Project Organization resides within the operating company? Let us look at the PROs and CONS

32

The Application (cont'd)

– PROS (Advantages)
  • Retention of knowledge
  • Uniformity of standards across the Operator between Clients
  • Loyalty and commitment to the Operator
  • Preservation of the reservoir – the core asset of the Operator

– CONS
  • Knowledge can only be retained if there is a means to capture it.
  • Uniform standards can exist without the others
  • Nowadays, loyalty and commitment are lost
  • Modern well construction technology without damage is dependent on technology and processes applied not who applied them

– Whether an in-house or external resource depends largely on the Operator

33

The Project Mission

– First order requirements (Mission Goals):
  - Number of wells
  - Spatial co-ordinates: surface and target
  - Total depth
  - Reservoir objectives
  - Evaluation objectives
  - Minimum hole size through reservoir
  - Completion design
  - HSEQ standards and procedures

34

The Project Mission

– The Client sets the mission goals and then requires that the Project achieves these goals for an agreed budget and schedule – i.e. 2nd order requirements
– Second order requirements:
  - Project Cost
  - Project Schedule
  - These two are highly interdependent with Cost being directly impacted by schedule and vice versa
– Defining the schedule and the budget is a complex time-consuming matter. It can be an iterative process requiring the project mission to be re-defined to meet a particular schedule or budget . That is, it is a continuous improvement.

35

Cost vs. Value

— The illusion of "Cheaper, Deeper"
  — "no matter what the cost"

— The necessity to maximize value NOT minimize cost

— How is Value quantified?

— How do DPM principles impact Value?

36

Selecting A Drilling Rig

— Rig A = $40,000 per day

— Rig B = $60,000 per day
  — (i) Which is the lower cost?
  — (ii) Which is the higher value?
  — (iii) Which results in the lower risk?

37

Selecting A Drilling Rig

— Rig A takes 40 days at $40,000/day
  – $1,600,000

— Rig B takes 20 days at $60,000/day
  – $1,200,000

38

Selecting A Drilling Rig

— Which is Cheaper? Better?

— Which involves the Least Risk?

— How is this determined in advance?
  • Is the rig capable and reliable?
  • Are the various rates well defined?
  • Are the personnel and mission statement (MS) adequate?
  • Are liabilities and indemnities covered?

39

Selecting A Drilling Rig

—Compliance

- Technical
- Commercial
- HSEQ
- Legal

40

Well Construction Foundation

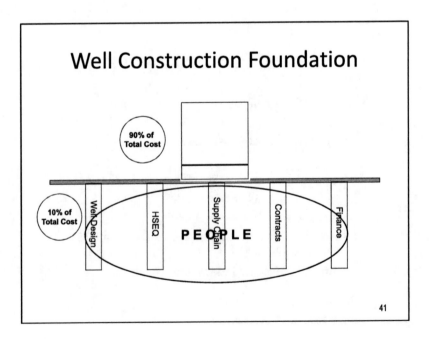

41

Contract Strategy

— There are several drilling contract strategies with the following terminologies: day rate, incentive lump sums, Malus, turnkey, footage/meterage, integrated project management

— These contract strategies are not mutually exclusive but mutual interdependent as thought by operator and much more likely by the contractor who has a vested interest in a particular strategy.

— Any and all of the above contracting strategies require drilling project management because the latter is a methodology.

42

Contract Strategy

— A drilling contractor may market turnkey drilling as a competitor for integrated services or project management without recognizing that a turnkey project could be most effectively managed by using project management principles and integrated services

— It is not a contract strategy that defines the way the well is to be drilled but the recognition that project management is the fundamental key to the success of the well in achieving its objectives and then deciding which contracting strategy can best achieve these objectives.

43

Contract Strategy

- The key to contracting strategies is RISK and REWARD.
- How much risk is either the Contractor or the Operator willing to take? This willingness to accept risk is a function of the degree of perceived risk itself – high or low – and the reward available for taking on risk.
- The higher the perceived risk, the greater the required reward.

44

Contract Strategy

0%	**Contractor**	100%

Day Rate	**RISK**	**Turnkey**

100%	**Operator**	0%

- This figure shows the limits of the Risk/Reward spectrum for a conventional, simple day rate contract and a turnkey contract. For the day rate the Contractor takes no contract risk, instead the operator accepts all the risk. In the turnkey, the contractor absorbs 100% risk.

45

Contract Strategy

Note the following:

- A good contract should require the Contractor to take on some specific operational risk related to his own performance
- In the turnkey, even though the contractor absorbs 100% risk, the Operator is still responsible for certain obligations such as pollution, liability etc, because he has the license holder or the lease. Cannot be passed to the contractor.
- Hence, in between the two contract strategies, there are various types that involve shared risk between the Operator and the Contractor

46

Interdependent Contracts

– Operator

- Drilling Contractor

  - Drilling Project Management

    - Integrated Drilling Services

Turnkey

47

Day Rate Contract

Advantages	Simple to apply
	Transparent
	Bids easy to analyze
	Readily understood

48

Day Rate Contract

Disadvantages	Favors complacency and inefficiency
	Encourages low cost bids
	Rewards poor investment, maintenance, personnel and health safety and env (HSE)
	Project value subtracted and cost increased

49

Footage / Meterage Contract

Advantages	Demands efficiency from the Contractor Simple to apply Readily applicable to specific contracts *e.g.* drill bits

50

Footage / Meterage Contract

Disadvantages	Potential for confrontation Contractor may not have full control of the operation, and hence the risk Operation/HSEQ standards can be compromised

51

Variable Incentive

Advantages	Versatile
	Relatively simple and transparent
	Incorporated into bid process
	Shared risk/reward

52

Variable Incentive

Disadvantages	Difficulty in agreeing on planned days
	Considers only time-dependent costs
	Operator can resent paying a bonus
	Contractor can be penalized for others' failure

53

Malus Contract

- A Malus contract is one in which a penalty is imposed for failure to perform to some agreed standard. Commonly this type of contract is a combination bonus/malus contract. In this contract good performance is rewarded and poor performance is penalized.

Advantages	Can apply to any service or part of service within an overall project contract
	Readily definable during bid process
	Contractor can determine the level of acceptable risk
	Flexible and quantifiable

54

Malus Contract

Disadvantages	Applies to only a specific service or part of service
	Operator "benefit" from malus can be small relative to overall loss from service failure
	Contractor can suffer consequential penalty from failure by other contractor(s)

55

Lump Sum Contract

Advantages	
	Can apply to any service or part of service within an overall project contract
	Versatile and flexible
	Readily definable during the bid process
	Contractor can determine the level of acceptable risk

56

Lump Sum Contract

Advantages (cont.)	
	Contractor accepts risk over which he has full control
	Contractor "focused" on lump sum task
	Operator's knowledge and experience can be much less than Contractor's

57

Lump Sum Contract

Disadvantages	Operator gives up control of lump sum task
	Operator can consider lump sum too expensive relative to day rate option
	Potential for windfall gains
	Expensive lump sum component can detract from an otherwise competitive bid

58

Combined Incentive Contract

Advantages	Versatility
	Can be incorporated into the bid process
	Operator and Contractor(s) can decide how much of their respective contracts are incentivized and how

59

Turnkey Contract

•*Definition:*

A turnkey contract is one where the turnkey contractor, normally the drilling contractor, agrees to drill a well to an agreed specification for a fixed price. If the turnkey contractor fails to deliver the well to the agreed specification, no payment is due to the contractor.

60

Turnkey Contract

Advantages - **Operator**	Fixed, guaranteed well cost
	Minimal supervisory resources required
	Turnkey contractor highly focused on the well, maximizing efficiency

61

Turnkey Contract	
Disadvantages - Operator	Operator abrogates control
	Contractor could compromise Operator's HSEQ standards
	Contractor could erroneously apply "cheaper, deeper" operating practices
	Contractor can "walk away," Contractor catastrophe may exceed financial resources placing liability back on Operator

62

Turnkey Contract	
Disadvantages – (cont.) Operator	Economic cost if well production is delayed due to contractor inefficiency
	Contract cost could far exceed equivalent day rate cost
	The turnkey contract can be complex and time-consuming to negotiate

63

Turnkey Contract

Advantages - Contractor	■ Contractor has full control of his own risk ■ Added value from local knowledge and experience ■ Some of the risk can be insurable ■ Potential for substantial profits

64

Turnkey Contract

Disadvantages – Contractor	■ Risk – Geographical – Geological – Operational ■ Operator interference ■ Different skills required from managing a drilling rig

65

Turnkey Contract

– Reasons for failure
- **Inability to quantify risk**
- **Greed, desperation, ignorance, incompetence**
- **Lack of drilling project management principles**
- **"Winner's Curse"**

66

Integrated Services Contract

Advantages	◾ Increased efficiency
	◾ Reduced contracts and invoices
	◾ Reduced contractor personnel
	◾ Enhanced planning, coordination and execution
	◾ Economies of scale

67

Integrated Services Contract

Advantages (cont.)	<ul><li>Consignment / buy-back more attractive</li><li>Superior technical back-up and support</li><li>Reduced logistical requirements</li><li>Incentive contracting more attractive</li><li>Enhanced teamwork</li></ul>

88

Integrated Services Contract

Advantages (cont.)	<ul><li>Greater focus by Contractor</li><li>Greater contract value – better teams</li></ul>

69

Integrated Services Contract

Disadvantages	▪ Unable to offer every service ▪ Exclusion of specialist non-IS companies ▪ Perceived loss of control by Operator ▪ Restricted to right place and right time

70

Integrated Services Contract

Disadvantages (cont.)	▪ Non-IS companies can resent contracting to the IS company rather than the Operator

71

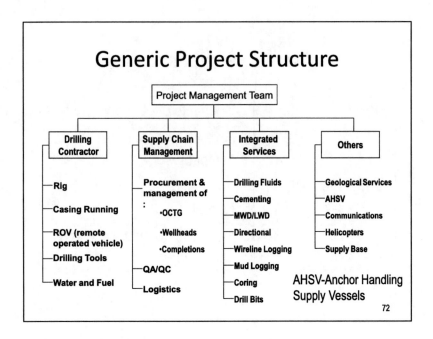

Drilling Project Management

—Where does it fit?

- It is NOT a contracting strategy
- These contracting strategies are NOT mutually exclusive
- Principles that should be applied to ANY contracting strategy
- An Operator can award a drilling contractor a turnkey contract that applies DPM principles using integrated services to construct the well

73

Cost / Value Matrix

| | Contractor | | Operator | | Partnership |
	Ideal	Reality	Ideal	Reality	Alignment
Rates	High	Low	Low	Low	Low
Value Added	High	Low	High	Low	High
Reward	High	Low	Low	Low	High
Project Cost	Low	High	Low	High	Low

74

Alliances

Asset

Project

Well

75

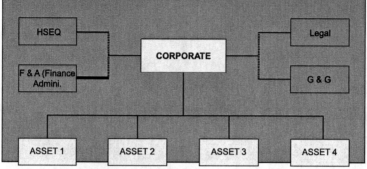

Generic Corporate Management Structure

Generic project structure to project structure – i.e. generic project team structure (# of persons in each function and in each team (and origin) will vary from project to project. The team constituents will also vary. **However, the first and foremost in project planning is selection of Project Manager.**

76

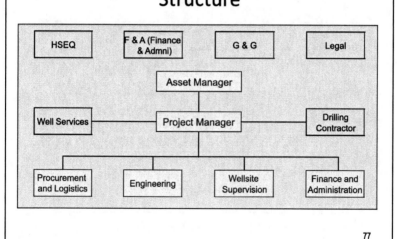

Generic Project Management Structure

77

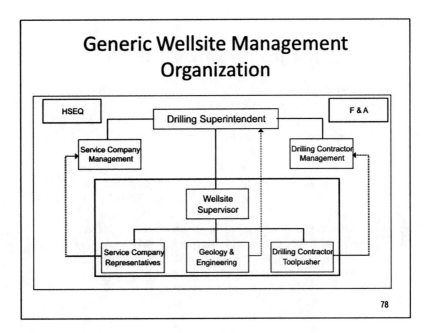

Principal Operator "Models"

1. Independents Operators
   - No in-house resource (has no operating knowledge or experience).
   - These are increasingly important component of both exploration and drilling often specializing in "stranded pools", mature marginal reservoirs that are not economical for the big operators due to high overheads and operating costs.
2. Horizontally-Disintegrated Operators
   - These are medium to large operators who have re-structured but retained only their core businesses and intend contracting in all non-core activities of which drilling is one.
   - May retain a small core of drilling professionals to act as "Informed-buyer" but their intention is that all such work is "brought-in" as an external project source.

79

Principal Operator "Models"

3.   Conventional Operators
   - In-house resource inadequate.
   - Applies to one-off projects or drilling in areas with limited experience, hence may out-source the work
4.   Third-Party Contractors
   - Combination – These are large construction and fabrication companies specializing in platform design and construction. They are contracted by the license owner to develop an oil/gas field

Note: Within the above groups, no Two Operators will be the same in how they want a project to be executed with variance even existing between clients within an operator. **The key to success is the application of drilling project management principles by the project team from wherever it is sourced.**

80

The First Step

— Select a Project Manager (Not a Drilling Manager)

   • Project manager must be selected at as early stage as possible as the key individual is critical to the development and application of the drilling project management principles to the particular project mission

— Define and Agree to Project Mission

   • This is the first task of the project manager. He or she meets with the client to review and understand the project mission in terms of the first order requirements. This then becomes the basis for the project plan.

81

Project Plan Priorities

– Team definition and selection
  • Depends on the resource requirements defined by the project schedule
– Schedule - defines the initial planning budget
– Budget
  • As time progresses, the project schedule and budget become increasingly more detailed. A minimal team must be established at the beginning to carry out this work.

Again, all the three priorities are to varying degree, mutually interdependent and iterative

82

Project Team Structure

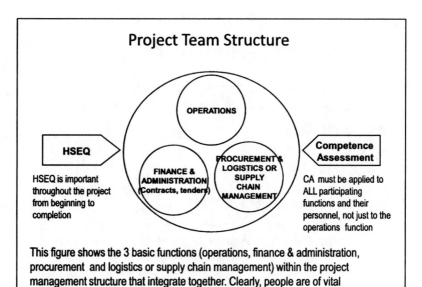

HSEQ

HSEQ is important throughout the project from beginning to completion

OPERATIONS

FINANCE & ADMINISTRATION (Contracts, tenders)

PROCUREMENT & LOGISTICS OR SUPPLY CHAIN MANAGEMENT

Competence Assessment

CA must be applied to ALL participating functions and their personnel, not just to the operations function

This figure shows the 3 basic functions (operations, finance & administration, procurement and logistics or supply chain management) within the project management structure that integrate together. Clearly, people are of vital importance to the success of project management

83

The Project Plan - Inputs

- The first act of the project manager after reviewing the project mission is the preparation of a basic and very preliminary Project Plan. This is the development of a Gantt Chart, which involves the following:
- Identify Individual Tasks (what has to be done and when)
  - Time parameters such as start date and duration
  - Task dependencies
  - Task constraints
- Identify Resources
  - Cost including what is needed and when

84

The Project Plan - Outputs

- The above data are input into a software (spreadsheet) leading to the following out puts
- Gantt Chart
- PERT Chart

  Critical path

  Resource tables for each task (tasks can be done in sequence or in parallel)

  Tracking-this follows the setting up of the project plan. The progress of each task in real time is tracked against the plan. This is reported to the client on a regular basis.

85

The Project Plan - Outputs

- While tracking the project, tasks can be added, deleted or modified as well as their duration and dependence.

- The project plan also has extensive budget and cost summary which allow cost tracking against the project planning budget.

- It is often a very worthwhile investment to have a young engineer whose principal job is maintaining the project plan and updating progress against it.

86

The Project Plan - Elements

– Defining Contracting Strategy: In drilling, the contracting strategy is always assumed as daily rate by the drilling team for the drilling contractor and tendering for service company. **This does not add value to the project.**

– Hence contracting strategy involves the following:
  - Type of Operator: conventional (major) or independent
  - Project location: onshore, offshore, deep water, etc
  - Project duration: one year or five years
  - Project mission: single wildcat or multiple wells drilling
  - Contracting market: supply (seller's market) or demand (buyer's market)

87

The Project Plan - Elements

— Importance of identifying contracting strategy at the onset of the project are as follows:

- Meeting with the contractors and service companies leads to establishment of contact and personal relationship between project team and the prospective contractors
- Assessment of equipment availability
- More formal pre-enquiry document to determine the availability and budget costs of critical equipment such as rig, supply boats, etc.

88

The Project Plan - Tangibles

— Tangible items are assets carried on the well accounts as capital expenses, and everything else being operating expenses (or intangibles)

— Long Lead Items – These are tangible items that have impact on the critical path and duration. Examples are Oil Country Tubular Goods (OCTG – casings and tubing) and Wellhead. These are important items especially when the market is active (demand is more than supply).

> Hence, maintain stock in inventory (e.g. casing)
>
> However, has the following disadvantages
>
> - Capital expense and investment
> - Maintenance/storage costs
> - Deterioration of casing and wastage
> - Common specification may lead to wastage if casing is not used later

89

The Project Plan - Tangibles

– Long Lead Items – OCTG and Wellhead
  – Casing Leasing Options to eliminate some of the disadvantages
    » Mill-sourced consignment
    » Common-stock consignment
  – Benefits
    » Operating expense
    » No inventory or surplus

90

The Project Plan – Tangibles – Casing Leasing Option (Mill-Sourced Consignment)

– Mill-sourced consignment involves casing manufacturers (about dozens in US, Japan, Germany, Italy, France, Mexico, Argentina, Brazil – Vallerec; etc.)

– This method is client specific whereby the leasing arrangement exists between an operator and a steel mill via an agent and it applies to purchasers with large requirements. That is, it is a volume business based on the casing grade, size, weight/ft, and connection.

– Advantages include uniformity in connections

– Disadvantages include high operating expenses and surplus

91

The Project Plan – Tangibles – Casing Leasing Option (Common-Stock Consignment)

- This is a method where a third party agent sets up a stock of the usual casing sizes, grades, weight, and couplings to a standard specification and lease to several individual purchasers from the common stock.
- Method is becoming increasingly popular.
- Purchasers invite these third party stockists to bid for the material supply including the consignment terms.
- Advantages include less operating expense because the stockist buy-backs the unused casing and hence no surplus
- Disadvantages include shipment of casing back from a remote location and business is restricted to specialist companies

92

The Project Plan – Tangibles - Summary

- Benefits (contd.)
  No maintenance and storage costs
  No deterioration or wastage
- But....
  May be more expensive in cash terms
  May not be optimum specification
  Not always an option

93

The Project Plan – Rig Selection

– Rig selection is the most important single task in planning a drilling project; and the rig is one that usually most constrains the critical path of the project. Why?

- Without a rig, no drilling project
- The single largest cost item (account typically for 40% or more)
- Rig performance is vital to the efficient execution of the drilling project

94

The Project Plan – Rig Selection

– Rig is selected based on technical, operational, and process. The process is covered here as it impacts the critical path of the drilling project. The process involves the following:

- **Identify general rig characteristics**: minimum water depth and type (jack-up, semi-submersible; drill ship); minimum hook load, load capacity, BOP pressure rating etc.
- **Identify potential rig sources:** this information is freely available from various rig locator database on the internet.
- **Prepare and issue pre-enquiry**

    Names and specs. of available rig(s)

    Location(s) of available rigs

    Recent and current contract details: commitments, clients, approximate date of availability; estimated mobilization and demobilization costs, daily operating rate, other future contract commitments

95

The Project Plan – Rig Availability

– Where availability is good, i.e. in a quiet market, the optimum project schedule will determine when the rig would be required. This is an ideal situation and it is then only a matter of preparing the tender documents, issuing them to drilling contractors with suitable rigs available and analyzing the bids.
– Where availability is poor in a very active market, the project schedule can be constrained entirely by when a suitable rig may be available. Here the schedule is largely outwit the project manager's control. The project manager must fully inform the client at all times of critical issues like this.
– What is the solution for this case? Operators may approach other operators directly and enquire about the possibility of sub-contracting the rig. If this is acceptable there are several methods by which this sub-contracting process can be handled.

96

The Project Plan – Rig Selection -Other Factors

– The following are other factors that have an impact on drilling project depending on the nature of the project and location
– **Supply Boats** – specification & availability; similar to rig; lump sum contract with the rig or rental (chartered); for rental boat there must be pre-enquiry and inspection.
– **Logistics** – this deals with location & weather; remote location needs critical planning for equipment and materials delivery.
  • Objective is to reduce waiting-on-material time (NPT). Critical elements of logistics are transportation (helicopter, supply boat and trucking) and storage (yard and warehouse); very complex and hence use holistic approach.

97

The Project Plan – Rig Selection - Other Factors

– Helicopters – essential for projects in remote location (jungle, offshore & deepwater); availability and crew are very important.

– Government Regulatory Authorities and Legislative Compliance – These are permit to drill, environmental, and safety issues. The project manager must assess what regulations apply, what is required to comply with and how long will it take. Very important in new countries and needed to plan approval schedule (needs substantial lag or lead time)

98

Project Design

– Project design are not the same nor perform on sequential order, but they are contemporaneous since the elements of project plans depends on the elements of the project design and vice versa.
– The key components of project design are as follows:
  • Well objectives – a derivative of the project mission
  • Well design
  • Involvement of principal contractors, i.e. short-listed ones (must be invited individually to the principal project office)
  • Key documentation – must be prepared because it has impact on the project critical path especially for approval

99

A Typical Principal Project Planning and Design Documentation for a Vankorneft Well

Document	Regulatory	Partner	Comment(s)
Prospect summary	No	No	
Well data sheets	No	No	
Tenders	No	No	Requires approval
Contracts	No	Yes	Require approval
Budget	No	Yes	
AFE	No	Yes	
Drilling program	Sometimes	Sometimes	
Testing program	Sometimes	Sometimes	
Petroleum Operations Notice	Yes	No	
Notification of Well Operations	Yes	No	
Consent to locate	Yes	No	
Consent to drill	Yes	No	
EIA (Env. Ins. Ass	Yes	No	
Rig safety case	Yes	No	

100

Project Planning - Exercise 1

– Prepare a detailed lists of tasks for a national onshore VANKORNEFT single exploration well drilling project divided as follows:

  • Preliminary Planning
  • Strategy
  • Tendering and Contracting
  • Drilling Programs and AFE
  • Government Regulations

101

Preliminary Planning – Tasks (To-Do-List)

- Objectives of the well
- Safety, Safety, safety, then environment
- Well type – purpose and profile
- Data gathering to develop well plan
- Well plan format
- Preliminary completion plan
- Casing setting depth
- Mud plan, Casing Design, Tubing Design
- Cementing plan
- Directional plan
- Well logging plan
- Well testing plan
- Bit program & Hole geometry selection
- Rig sizing & selection; AFE preparation; Drilling Time projection (SEE SLIDE 21)

102

Strategy - Tasks

- Contracts – day rate, turnkey, footage or Malus
- Selection of service companies
- Cost
- Schedule
- Logistics

103

Government Regulations - Tasks

— Government Regulations: Acquire the following:

— Permit to drill

— Environmental, and safety issues

— Assess what regulations apply, what is required to comply with and how long will it take.

— Plan approval schedule (needs substantial lag or lead time)

104

Appendix B: Useful Units and Conversion Factors

Conventional Notation	Scientific Notation	Engineering Notation
12,345.7	1.23457×10^4	12.3457×10^3
123.456	1.23456×10^2	123.456×10^0
20	2.0×10^1	20×10^0
0.675	6.75×10^{-1}	675×10^{-3}
0.0001	1×10^{-4}	100×10^{-6}

12 inches = 1 foot

3 feet = 1 yard

5½ yards = 1 rod

6 feet = 1 fathom

40 rods = 1 furlong

8 furlongs = 1 mile

1760 yards = 1 mile

5280 feet = 1 mile

60 sea miles = 1 degree

0.8684 miles = 1 sea mile

1 radian = 57.3°

1 inch = 2.54 cm

1 gallon = 231 in³

1 kilogram = 2.205 lb

1 newton = 1 kg·m/s²

1 joule = 1 N·m

1 watt = 1 J/s

1 pascal = 1 N/m²

1 BTU = 778 ft-lb

      = 252 cal

      = 1054.8 J

1 horsepower = 745.7 W

1 atmosphere = 14.7 lb/in²

           = $1.01 \cdot 10^5$ N/m²

Liquid Measure	Dry Measure
4 gills = 1 pint	2 pints = 1 quart
2 pints = 1 quart	8 quarts = 1 peck
4 quart = 1 gallon	4 pecks = 1 bushel
31½ gallons = 1 barrel	36 bushels = 1 chaldron
231 cu. in = 1 gallon	2150.42 cu. in = 1 standard bushel
	1 cubic foot = approx. 4/5 bushel

Notation	Expansion
yotta (10^{24}):	1, 000, 000, 000, 000, 000, 000, 000, 000
zetta (10^{21}):	1, 000, 000, 000, 00,0 000, 000, 000
exa (10^{18}):	1, 000, 000, 000, 000, 000, 000
peta (10^{15}):	1, 000, 000, 000, 000, 000
tera (10^{12}):	1, 000, 000, 000, 000
giga (10^{9}):	1, 000, 000, 000
mega (10^{6}):	1, 000, 000
kilo (10^{3}):	1, 000
hecto (10^{2}):	100
deca (10^{1}):	10
deci (10^{-1}):	0.1
centi (10^{-2}):	0.01
milli (10^{-3}):	0.001
micro (10^{-6}):	0.000 001
nano (10^{-9}):	0.000 000 001
pico (10^{-12}):	0.000 000 000 001
femto (10^{-15}):	0.000 000 000 000 001
atto (10^{-18}):	0.000 000 000 000 000 001
zepto (10^{-21}):	0.000 000 000 000 000 000 001
yocto (10^{-24}):	0.000 000 000 000 000 000 000 001
stringo (10^{-35}):	0.000 000 000 000 000 000 000 000 000 000 01

1/16 = .0625	9/16 = .5625
1/8 = .125	5/8 = .625
3/16 = .1875	11/16 = .6875
1/4 = .25	3/4 = .75
5/16 = .3125	13/16 = .8125
3/8 = .375	7/8 = .875
7/16 = .4375	15/16 = .9375
1/2 = .5	1 = 1.0

English System	
1 foot (ft) = 12 inches (in) 1' = 12"	in
1 yard (yd) = 3 feet	ft
1 mile (mi) = 1760 yards	yd
1 sq. foot = 144 sq. inches	sq. in
1 sq. yard = 9 sq. feet	sq. ft
1 acre = 4840 sq. yards = 43560 ft^2	sq. yd
1 sq. mile = 640 acres	acres

Metric System		
mm	Millimeter	.001 m
cm	Centimeter	.01 m
dm	Decimeter	.1 m
m	Meter	1 m
dam	Decameter	10 m
hm	Hectometer	100 m
km	Kilometer	1000 m

Measurement	Symbol	Description
Meter	m	Length
Hectare	ha	Area
Tonne	t	Mass
Kilogram	kg	Mass
Nautical mile	M	Distance (navigation)
Knot	Kn	Speed (navigation)
Liter	L	Volume or capacity
Second	s	Time
Hertz	Hz	Frequency
Candela	cd	Luminous intensity
Degree celsius	°C	Temperature
Kelvin	K	Thermodynamic temperature
Pascal	Pa	Pressure, stress
Joule	J	Energy, work
Newton	N	Force
Watt	W	Power, radiant flux
Ampere	A	Electric current
Volt	V	Electric potential
Ohm	Ω	Electric resistance
Coulomb	C	Electric charge

Multiply	By	To Obtain
Angstrom	10^{-10}	Meters
Feet	0.30480	Meters
	12	Inches
Inches	25.40	Millimeters
	0.02540	Meters
	0.08333	Feet
Kilometers	3280.8	Feet
	0.6214	Miles
	1094	Yards
Meters	39.370	Inches
	3.2808	Feet
	1.094	Yards
Miles	5280	Feet
	1.6093	Kilometers
	0.8694	Nautical miles
Millimeters	0.03937	Inches
Nautical miles	6076	Feet
	1.852	Kilometers
Yards	0.9144	Meters
	3	Feet
	36	Inches

Multiply	By	To Obtain
Feet/minute	5.080	mm/second
Feet/second	0.3048	Meters/second
Inches/second	0.0254	Meters/second
km/hour	0.6214	Miles/hour
Meters/second	3.2808	Feet/second
	2.237	Miles/hour
Miles/hour	88.0	Feet/minute
	0.44704	Meters/second
	1.6093	km/hour
	0.8684	Knots
Knot	1.151	Miles/hour

Multiply	By	To Obtain
Carat	0.200	Cubic grams
Grams	0.03527	Ounces
Kilograms	2.2046	Pounds
Ounces	28.350	Grams
Pound	16	Ounces
	453.6	Grams
Stone (U.K.)	6.35	Kilograms
	14	Pounds
Ton (net)	907.2	Kilograms
	2000	Pounds
	0.893	Gross ton
	0.907	Metric ton
Ton (gross)	2240	Pounds
	1.12	Net tons
	1.016	Metric tons
Tonne (metric)	2204.623	Pounds
	0.984	Gross pound
	1000	Kilograms

Multiply	By	To Obtain
Acres	43,560	sq. feet
	4047	sq. meters
	4840	sq. yards
	0.405	Hectare
sq. cm	0.155	sq. inches
sq. feet	144	sq. inches
	0.09290	sq. meters
	0.1111	sq. yards
sq. inches	645.16	sq. millimeters
sq. kilometers	0.3861	sq. miles
sq. meters	10.764	sq. feet
	1.196	sq. yards
sq. miles	640	Acres
	2.590	sq. kilometers

144 square inches = 1 square foot

9 square feet = 1 square yard

43,560 square feet = 1 acre

640 acres = 1 square mile

30¼ square yards = 1 square rod

40 square rods = 1 square rood

4 square roods = 1 acre

272¼ square feet = 1 square rod

Multiply	By	To Obtain
Acre-foot	1233.5	Cubic meters
Cubic cm	0.06102	Cubic inches
Cubic feet	1728	Cubic inches
	7.480	Gallons (U.S.)
	0.02832	Cubic meters
	0.03704	Cubic yards
Liter	1.057	Liquid quarts
	0.908	Dry quarts
	61.024	Cubic inches
Gallons (U.S.)	231	Cubic inches
	3.7854	Liters
	4	Quarts
	0.833	British gallons
	128	U.S. fluid ounces
Quarts (U.S.)	0.9463	Liters

Multiply	By	To Obtain
BTU	1055.9	Joules
	0.2520	kg-calories
Watt-hour	3600	Joules
	3.409	BTU
HP (electric)	746	Watts
BTU/second	1055.9	Watts
Watt-second	1.00	Joules

Conversion Formulas	
Celsius to kelvin	$K = C + 273.15$
Celsius to fahrenheit	$F = (9/5)C + 32$
Fahrenheit to celsius	$C = (5/9)(F - 32)$
Fahrenheit to kelvin	$K = (5/9)(F + 459.67)$
Fahrenheit to rankin	$R = F + 459.67$
Rankin to kelvin	$K = (5/9)R$

Multiply	By	To Obtain
Atmospheres	1.01325	Bars
	33.90	Feet of water
	29.92	Inches of mercury
	760.0	mm of mercury
Bar	75.01	cm of mercury
	14.50	Pounds/sq. inch
Dyne/sq. cm	0.1	N/sq. meter
Newtons/sq. cm	1.450	Pounds/sq. inch
Pounds/sq. inch	0.06805	Atmospheres
	2.036	Inches of mercury
	27.708	Inches of water
	68.948	Millibars
	51.72	mm of mercury

Speed of light	299,792,458 m/s
	983.6×10^6 ft/s
	186,284 miles/s
Velocity of sound	340.3 meters/s
	1116 ft/s
Gravity	9.80665 m/s square
(acceleration)	32.174 ft/s square
	386.089 inches/s square

Kitchen and Household Measurements

1 pinch = 1/8 teaspoon or less

3 teaspoons = 1 tablespoon

2 tablespoons = 1/8 cup

4 tablespoons = 1/4 cup

8 tablespoons = 1/2 cup

12 tablespoons = 3/4 cup

16 tablespoons = 1 cup

5 tablespoons + 1 teaspoon = 1/3 cup

4 oz = 1/2 cup

8 oz = 1 cup

16 oz = 1 lb

1 oz = 2 tablespoons fat or liquid

1 cup of liquid = 1/2 pint

2 cups = 1 pint

2 pints = 1 quart

4 cup of liquid = 1 quart

continued

continued

Kitchen and Household Measurements

4 quarts = 1 gallon

8 quarts = 1 peck (such as apples, pears, etc.)

1 jigger = 1½ fluid oz

1 jigger = 3 tablespoons

$$D = \frac{m}{V} \left(\frac{g}{cm^3} = \frac{kg}{m^3} \right)$$

D: density

M: mass

V: volume

$$P = \frac{W}{t}$$

P: power (W = watts)

W: work (J)

t: time (s)

$$d = v \cdot t$$

d: distance (m)

v: velocity (m/s)

t: time (s)

$$K.E. = \frac{1}{2} mv^2$$

K.E.: kinetic energy

m: mass (kg)

v: velocity (m/s)

$$a = \frac{vf - vi}{t}$$

a: acceleration (m/s²)

vf: final velocity (m/s)

vi: initial velocity (m/s)

t: time (s)

$$d = vit + \frac{1}{2}at^2$$

d: distance (m)

vi: initial velocity (m/s)

t: time (s)

a: acceleration (m/s²)

$$F = m \cdot a$$

F: net force (N = newtons)

m: mass (kg)

a: acceleration (m/s^2)

$$Fg = \frac{G \cdot m_1 \cdot m_2}{d^2}, \left(G = 6.67 \times 10^{-11} \frac{N - m^2}{kg^2} \right)$$

Fg: force of gravity (N)

G: universal gravitational constant

m_1, m_2: masses of the two objects (kg)

d: separation distance (m)

$$I = \frac{Q}{t}$$

I: electric current ampères

Q: electric charge flowing (C)

t: time (s)

$$W = V \cdot I \cdot t$$

W: electrical energy (J)

V: voltage (V)

I: current (A)

t: time (s)

$$P = V \cdot I$$

P: power (W)

V: voltage (V)

I: current (A)

$$p = m \cdot v$$

p: momentum (kg · m/s)

m: mass

v: velocity

$$W = F \cdot d$$

W: work (J = joules)

F: force (N)

d: distance (m)

$$H = c \cdot m \cdot \Delta T$$

H: heat energy (J)

m: mass (kg)

ΔT: change in temperature (°C)

c: specific heat (J/kg · °C)

Kilometer–Mile Conversion			
Kilometers	**Miles**	**Miles**	**Kilometers**
1	0.6	1	1.6
5	3.1	5	8.05
10	6.2	10	16.0
20	12.4	20	32.1
30	18.6	30	48.2
40	24.8	40	64.3
50	31.1	50	80.5
60	37.3	60	96.6
70	43.5	70	112.7
80	49.7	80	128.7
90	55.9	90	144.8
100	62.1	100	160.9
500	310.7	500	804.7
1000	621.4	1000	1609.3

Metric Tables

Capacity	**Area**
10 milliliters = 1 centiliter	100 sq. millimeters = 1 sq. centimeter
10 centiliters = 1 deciliter	100 sq. centimeters = 1 sq. decimeter
10 deciliters = 1 liter	100 sq. decimeters = 1 sq. meter (centare)
10 liters = 1 dekaliter	100 sq. meters = 1 are
10 dekaliters = 1 hectoliter	10,000 sq. meters = 1 hectare
1000 liters = 1 kiloliter (stere)	100 hectares = 1 sq. kilometer

Length	**Weight**
10 millimeters = 1 centimeter (cm)	10 milligrams = 1 centigram
10 centimeters = 1 decimeter	10 centigrams = 1 decigram
10 decimeters = 1 meter (m)	10 decigrams = 1 gram
10 meters = 1 dekameter	1000 grams = 1 kilogram (kilo)
100 meters = 1 hectometer	100 kilograms = 1 quintal
1000 meters = 1 kilometer	1000 kilograms = 1 metric ton

Metric: U.S. Weights and Measures

Dry Measure	Long Measure
1 pint = .550599 liter	1 inch = 2.54 centimeters
1 quart = 1.101197 liters	1 yard = .914401 meter
1 peck = 8.80958 liters	1 mile = 1.609347 kilometers
1 bushel = .35238 hectoliter	

Liquid Measure	Square Measure
1 pint = .473167 liter	1 sq. inch = 6.4516 sq. centimeters
1 quart = .946332 liter	1 sq. foot = 9.29034 sq. decimeters
1 gallon = 3.785329 liters	1 sq. yard = .836131 sq. meter
	1 acre = .40469 hectares
	1 sq. mile = 2.59 sq. kilometers
	1 sq. mile = 259 hectares

Avoirdupois Measure	Cubic Measure
1 ounce = 28.349527 grams	1 cu. inch = 16.3872 cu. centimeters
1 pound = .453592 kilograms	1 cu. foot = .028317 cu. meter
1 short ton = .90718486 metric ton	1 cu. yard = .76456 cu. meter
1 long ton = 1.01604704 metric tons	

Index